Geoethics

Geoethics
Ethical Challenges
and Case Studies in Earth Sciences

Max Wyss
International Centre for Earth Simulation, Geneva, Switzerland

Silvia Peppoloni
INGV – Istituto Nazionale di Geofisica e Vulcanologia, Rome, Italy;
IAPG – International Association for Promoting Geoethics, Rome, Italy

ELSEVIER

AMSTERDAM BOSTON HEIDELBERG LONDON NEW YORK OXFORD
PARIS SAN DIEGO SAN FRANCISCO SINGAPORE SYDNEY TOKYO

Elsevier
Radarweg 29, PO Box 211, 1000 AE Amsterdam, The Netherlands
The Boulevard, Langford Lane, Kidlington, Oxford OX5 1GB, UK
225 Wyman Street, Waltham, MA 02451, USA

Notices
Knowledge and best practice in this field are constantly changing. As new research and
experience broaden our understanding, changes in research methods, professional practices,
or medical treatment may become necessary.

Practitioners and researchers must always rely on their own experience and knowledge
in evaluating and using any information, methods, compounds, or experiments described
herein. In using such information or methods they should be mindful of their own safety and
the safety of others, including parties for whom they have a professional responsibility.

To the fullest extent of the law, neither the Publisher nor the authors, contributors, or editors,
assume any liability for any injury and/or damage to persons or property as a matter of
products liability, negligence or otherwise, or from any use or operation of any methods,
products, instructions, or ideas contained in the material herein.

ISBN: 978-0-12-799935-7

Library of Congress Cataloging-in-Publication Data
Geoethics : ethical challenges and case studies in earth sciences / [edited by] Max Wyss,
Silvia Peppoloni.
      pages cm
      Includes bibliographical references and index.
   ISBN 978-0-12-799935-7 (hardback)
1. Geology–Research–Moral and ethical aspects. 2. Geophysics–Research–Moral and
ethical aspects. 3. Research–Moral and ethical aspects–Case studies. I. Wyss, Max,
1939- editor. II. Peppoloni, Silvia, editor.
   QE40.G375 2015
   174′.955--dc23
                              2014035845

British Library Cataloguing in Publication Data
A catalogue record for this book is available from the British Library

For information on all Elsevier publications
visit our web site at http://store.elsevier.com/

Working together
to grow libraries in
developing countries

www.elsevier.com • www.bookaid.org

Contents

Section I
Philosophical Reflections

4.   **Ethics of Disaster Research**

Ilan Kelman

5.   **Toward an Inclusive Geoethics—Commonalities of
     Ethics in Technology, Science, Business, and Environment**

Thomas Potthast

6.   **Humans' Place in Geophysics: Understanding the
     Vertigo of Deep Time**

Telmo Pievani

Section II
Geoscience Community

7. Research Integrity: The Bedrock of the Geosciences

Tony Mayer

8. Formulating the American Geophysical Union's Scientific Integrity and Professional Ethics Policy: Challenges and Lessons Learned

Linda C. Gundersen and Randy Townsend

9. Ethical Behavior in Relation to the Scholarly Community: A Discussion on Plagiarism

Stefano Solarino

Section III
The Ethics Of Practice

Section IV
Communication with The Public, Officials and The Media

Section V
Natural and Anthropogenic Hazards

26. Geoethics and Risk-Communication Issues in Japan's Disaster Management System Revealed by the 2011 Tohoku Earthquake and Tsunami

Megumi Sugimoto

27. "When Will Vesuvius Erupt?" Why Research Institutes Must Maintain a Dialogue with the Public in a High-Risk Volcanic Area: The Vesuvius Museum Observatory

Maddalena De Lucia

28. Voluntarism, Public Engagement, and the Role of Geosciences in Radioactive Waste Management Policy-Making

Nic Bilham

Section VI
Low Income and Indigenous Communities

Contributors

Alberto Armigliato Settore di Geofisica, DIFA, Università di Bologna, Bologna, Italy

Nic Bilham The Geological Society of London, Burlington House, Piccadilly, London, UK

Roger Bilham CIRES and Department of Geological Sciences, University of Colorado, Boulder, CO, USA

Juliane C. Carneiro Graduate Program in Coastal and Oceanic Systems, Federal University of Paraná, Brazil

Denis Cardoso Graduate Program in Coastal and Oceanic Systems, Federal University of Paraná, Brazil

Marcio M. Cintra Graduate Program in Coastal and Oceanic Systems, Federal University of Paraná, Brazil

Madhumita Das Department of Geology, Utkal University, Bhubaneswar, Odisha, India

Maddalena De Lucia Istituto Nazionale di Geofisica e Vulcanologia, Osservatorio Vesuviano, Naples, Italy

Daniela Di Bucci National Department of Civil Protection, Via Vitorchiano, Roma, Italy

Giuseppe Di Capua INGV – Istituto Nazionale di Geofisica e Vulcanologia, Rome, Italy; IAPG – International Association for Promoting Geoethics, Rome, Italy

Mauro Dolce National Department of Civil Protection, Via Ulpiano, Roma, Italy

Vinod K. Gaur CSIR, Centre for Mathematical Modelling and Computer Simulation (C-MMACS), Bangalore, India; Indian Institute of Astrophysics, Bangalore, India

William Gawthrop (Retired) President of Green Mountain Geophysics, Boulder, Colorado, USA

Robert J. Geller Department of Earth and Planetary Science, Graduate School of Science, University of Tokyo, Tokyo, Japan

Armin Grunwald Karlsruhe Institute of Technology, Institute for Technology Assessment and Systems Analysis

Linda C. Gundersen U.S. Geological Survey, Reston Virginia

Jessica Heesen International Centre for Ethics in the Sciences and Humanities (IZEW), Universität Tübingen, Germany

Peter Hocke Institute for Technology Assessment and Systems Analysis, Karlsruhe Institute of Technology, Karlsruhe, Germany

Daniel Hostettler Fastenopfer, Lucerne, Switzerland

Ilan Kelman Institute for Risk & Disaster Reduction and Institute for Global Health, University College London, London, United Kingdom; Norwegian Institute of International Affairs (NUPI), Oslo, Norway

Shrikant Daji Limaye Ground Water Institute (NGO), Pune, India

Daniel F. Lorenz Disaster Research Unit (DRU), Freie Universität Berlin, Germany

Eduardo Marone Graduate Program in Coastal and Oceanic Systems, Federal University of Paraná, Brazil; Graduate Program in Geology, Federal University of Paraná, Brazil; International Ocean Institute – OC, Federal University of Paraná, Brazil

Tony Mayer President's Office, Nanyang Technological University, Singapore

Janette Merlo Department of Environmental Sciences and Hazard Management, School of Sciences, University of Colima, Colima, Mexico

Marco Mucciarelli Seismological Research Centre, OGS, Trieste, Italy

Jurgen Neuberg School of Earth & Environment, The University of Leeds, UK

Gianluca Pagnoni Settore di Geofisica, DIFA, Università di Bologna, Bologna, Italy

Gerassimos A. Papadopoulos Institute of Geodynamics, National Observatory of Athens, Athens, Greece

Silvia Peppoloni INGV – Istituto Nazionale di Geofisica e Vulcanologia, Rome, Italy; IAPG – International Association for Promoting Geoethics, Rome, Italy

Telmo Pievani Department of Biology, University of Padua, Padua, Italy

Thomas Potthast International Centre for Ethics in the Sciences and Humanities, University of Tübingen, Tübingen, Germany

Andréa Ribeiro Graduate Program in Coastal and Oceanic Systems, Federal University of Paraná, Brazil

Paul G. Richards Lamont-Doherty Earth Observatory of Columbia University, Palisades, NY, USA

Mohamad amin Skandari Disaster Management Expert, Social Security Organization, Tehran, Iran

Stefano Solarino Istituto Nazionale di Geofisica e Vulcanologia, Genova, Italy

Ingrid Leman Stefanovic Dean, Faculty of Environment, Simon Fraser University; Professor Emeritus, Department of Philosophy, University of Toronto, Canada

Carol Stellfeld Graduate Program in Geology, Federal University of Paraná, Brazil

Megumi Sugimoto Institute of Decision Sciences for sustainable Society, Kyushu University, Fukuoka, Japan

Stefano Tinti Settore di Geofisica, DIFA, Università di Bologna, Bologna, Italy

Randy Townsend American Geophysical Union, Washington, DC, USA

Martin Voss Disaster Research Unit (DRU), Freie Universität Berlin, Germany

Bettina Wenzel Disaster Research Unit (DRU), Freie Universität Berlin, Germany

Zhongliang Wu Institute of Geophysics, China Earthquake Administration, Beijing, People's Republic of China; Key Lab for Computational Geodynamics, Chinese Academy of Sciences, Beijing, People's Republic of China

Max Wyss International Centre for Earth Simulation, Geneva, Switzerland

Filippo Zaniboni Settore di Geofisica, DIFA, Università di Bologna, Bologna, Italy

F. Ramón Zúñiga Centro de Geociencias, National Autonomous University of Mexico, Mexico D.F., Mexico

Preface

GEOSCIENTISTS' PARTICIPATION IN THE STEWARDSHIP OF OUR PLANET

The time for a first book on Geoethics has come. The faster, greedier pace of society and globalization demands it. The comfortable life of scholars in the ivory tower is coming to a rude awakening. People demand understandable information on geohazards, judges condemn scientist and engineers for lack of communication, indigenous people rise in anger accusing experts of misleading them, attempts to avoid transparency in developments still exist, the helplessness of technology to deal with nuclear waste becomes more evident everyday and nature exposes shortcuts in constructing critical facilities with her own awesome force. Less dramatic, but damaging the credibility of the scientific community is the rising number of cases of plagiarism, made easy by the electronic world of communication. In addition, the community of geoscientists needs to become aware of subtle ways in which our work can be useless for or even misleading the public.

Far from complete, this book takes a first step in pulling together and offering for discussion some obvious and hidden aspects of georesearch in which ethical principles may not be observed at a respectable level. As editors, our intent is not to shake a finger at malefactors and moralize; we simply give the floor to geopractitioners for voicing ethical concerns and present case histories, which demonstrate the perceived problems. Rather than pretending to recognize and understand all the ethical problems in geosciences, we view the collection of chapters in this book as a start for discussions, which hopefully may involve a growing section of our community interested in advancing toward a better understanding and formulation of ethical obligations geopractitioners should live up to.

This book includes a section on *Philosophical Reflections*, which is in part intended to help scientists sharpen their perceptions of what is understood by terms like *ethics* and *morals*. In six chapters, this section includes contributions by philosophers as well as views of geoscientists.

A section with three chapters on the *Geoscience Community* is offering a summary of institutional efforts to define the problem of ethics in geosciences and to introduce tools to combat transgressions.

A section on *Ethics of Practice* with seven chapters reveals cases in which government is not interested in learning scientific facts and may obscure them in collusion with scientific experts in order to follow an agenda. The safety of nuclear power plants requires expertise from geoscientists, but the interest in reducing construction costs sometimes mutes high-quality hazard assessments. One chapter gives advice on how first responders in disasters must conduct themselves in Muslim countries.

Communicating with the public concerning geohazards is a major sore point where much improvement seems possible. In seven chapters, a court case is discussed, the responsibilities in communicating are analyzed, the power of maps and how they can mislead are reviewed, and the methods to simplify technical language are considered.

Although many earth scientists feel they are doing much to protect the population from *natural and anthropogenic hazards*, we present a section in which shortcomings of these efforts are detailed in six chapters. Problems in dealing with volcanoes, earthquakes, tsunamis, and nuclear waste are illuminated from various points of view.

Last, but not least, a four chapter section on *Low Income and Indigenous Communities* presents some case histories that demonstrate that geoscientists could take a more active role in assisting these less-fortunate segments of society. It deals with problems due to natural hazards as well as due to possibly ruthless development attempts.

We suspect that the collection of problems and cases presented in this book represents only the tip of an iceberg. We hope that the voices raised here will generate an echo that will lead to sounding out the depth of the hidden part of unethical behavior in matters involving geosciences. By laying this corner stone for a continued and more complete discussion of the ethical level to which we geoscientists should rise in contributing to a responsible stewardship of our planet, we expect to begin to grasp more clearly what exactly our responsibility is and how we can meet it.

Max Wyss[1] and Silvia Peppoloni[2,3]

[1]*International Centre for Earth Simulation, Geneva, Switzerland,*
[2]*INGV – Istituto Nazionale di Geofisica e Vulcanologia, Rome, Italy,*
[3]*IAPG – International Association for Promoting Geoethics, Rome, Italy*

Acknowledgments

We thank the legal and the anonymous reviewers for their comments and Louisa Hutchings for managing the assembly of the book.

Section I

PHILOSOPHICAL REFLECTIONS

Chapter 1

The Meaning of Geoethics

Silvia Peppoloni[1,2] and Giuseppe Di Capua[1,2]
[1]INGV – Istituto Nazionale di Geofisica e Vulcanologia, Rome, Italy; [2]IAPG – International
Association for Promoting Geoethics, Rome, Italy

Chapter Outline

Abstract

This chapter outlines a framework of the issues addressed by geoethics. Starting from an etymological analysis of the word "geoethics," we identify the cultural basis on which to expand the debate on geoethics, while also proposing for consideration by the scientific community some questions that may guide the development of future research and practice in geosciences.

We attempt to define some fundamental points that, in our opinion, will strengthen geoethics and help its development. The goal of geoethics is to suggest practical solutions and provide useful techniques, and also to promote cultural renewal in how humans perceive and relate to the planet, through greater attention to the protection of life and the richness of the Earth, in all its forms.

As each science does, geoethics should also be able to present an image of the world, pointing out the manner in which it can be understood, investigated, designed, and experienced.

Keywords: Etymological analysis; Geoethics; Geoscientists oath; Responsibility; Society.

THE BIRTH OF GEOETHICS

For many years, the growing impact of people on natural processes has been both recognized and documented. In 1873, the Italian geologist Antonio Stoppani

Geoethics. http://dx.doi.org/10.1016/B978-0-12-799935-7.00001-0

(1824–1891) hypothesized and defined the "Anthropozoic era," a geological time in which humans appear as a new "geological force." This is an era dominated by human activities, in many ways similar to the concept of the Anthropocene as defined by Nobel Prize recipient Paul J. Crutzen (Crutzen, 2002). The Anthropocene refers to the time when human activities became capable of modifying the Earth's ecosystem and dynamics (Ellis and Haff, 2009). The proposal to formally enter the Anthropocene into the geological time scale is still the subject of lively debate among scientists (Zalasiewicz et al., 2008) and an increasing number of disciplines, such as geoengineering (Borgomeo, 2012), are orienting their research toward the development of technologies that take into account the impact that interventions by humans (though often necessary) may have on the natural environment (Bohle, 2013).

The historic great technological and industrial development, the exponential growth of population in the last century, and the consequently huge urban expansion have increased the effects of human interference with the geosphere. In addition, the expanding use of land and the increasing demand for natural resources have highlighted the need, not only for the scientific community but for all of the society, to consider issues such as environmental sustainability and energy, protection from natural hazards, and reduction of pollution and their inevitable repercussions on health and climate change.

It becomes evident that geoscientists, as scholars and experts on these issues affecting our planet, can play a fundamental role in society, thanks to their specific and unique skills, by addressing environmental problems at the local and global scale and helping to find optimal solutions.

Thus, discussing ethics in relation to geosciences (i.e., geoethics) and considering the social implications of geological research and practice has by default become an indispensable requirement for geoscientists (Peppoloni and Di Capua, 2012b). The debate is increasingly broad inside the scientific community and includes specific technical and methodological aspects, as well as theoretical considerations and reflections on the ethical value of geological activities.

The problems that arise from the interaction between humans and nature are complex and may require various approaches and solutions. The possibility of multiple solutions and varied results creates a need for an open and widespread scientific discussion, on how to live on the planet, while respecting natural dynamics and human life. This discussion inevitably raises ethical issues that force us to consider our responsibility in this interaction with nature.

WHAT IS GEOETHICS?

Geoethics consists of research and reflection on those values upon which to base appropriate behavior and practice where human activities intersect the geosphere (IAPG, 2012; Peppoloni and Di Capua, 2012b). It deals with the ethical, social, and cultural implications of geological research and practice, providing a point of intersection for geosciences, sociology, and philosophy (Moores, 1997; Bosi et al., 2008; Peppoloni and Di Capua, 2012b; Peppoloni, 2012a,b).

The field of geosciences is wide. The Earth system is the subject of study, which includes the solid Earth, the hydrosphere, the atmosphere, and the biosphere (GSL, 2014). This implies that the themes of geoethics are numerous, often interconnected, and include theoretical and practical aspects as well as cultural and operational perspectives.

Geoethics focuses on some of the most important environmental emergencies: it encourages a critical analysis of the use of natural resources, promotes careful management of natural risks, and fosters the proper dissemination of the results of scientific studies, including the development of environmentally friendly technologies (Peppoloni and Di Capua, 2012b), while also extending its principles to planetary protection (Martínez-Frías et al., 2011).

Moreover, geoethics promotes geoeducation, aiming at organizing effective teaching tools (Bezzi, 1999) for developing awareness, values, and responsibility, especially among young people. It fosters the development of geoparks (Eder, 2004; Zouros, 2004; McKeever, 2013) and geotourism (Newsome and Dowling, 2010; Dowling, 2011) in order to create awareness of the value of a region's geological heritage (Brocx and Semeniuk, 2007; Gray, 2008) and geodiversity (Osborne, 2000). Geoethics also aims at improving the relationships between the scientific community, decision makers, mass media, and the public (Höppner et al., 2012), highlighting the social role played by geoscientists and their responsibilities, as their choices may have ethical, cultural, and economic repercussions on society (Peppoloni and Di Capua, 2012b; Bickford, 2013).

So, on the one hand, geoethics has practical objectives such as creating references and guidelines that look at providing socio-economic solutions, compatible with the respect for the environment and the protection of humans, nature, and land. On the other hand, it aims at giving a cultural, ethical, and social frame of reference on how to conduct geological research and practice in favor of public welfare (Bickford, 2013); and to give value to geosciences in a cultural sense (Moores, 1997; Peppoloni and Di Capua, 2012b; McKeever, 2013), as part of a group of disciplines capable of suggesting new ways to understand and investigate our planet, and upon which we can base a new relationship between humans and nature.

ISSUES WITH THE DEVELOPMENT OF GEOETHICS

Despite the importance of its themes, the attention given to geoethics is still limited. Only a few geoscientists make explicit reference to geoethics. Outside the scientific community, among the general public, no one even knows of geoethics. Why?

Although the concept of geoethics is increasingly present in many scientific conferences, analyses and debates on the subject have not yet command significant attention. Geoethics does not seem to be accompanied by an adequate research base, nor a satisfactory number of scientific publications. Generally, publications on geoethics are still not considered important in the scientific curricula. So, most geoscientists are hesitant to spend time writing on ethical themes.

In short, the issues of geoethics do not easily find space in most respected scientific journals, and this has severely restricted the dissemination of the concept and the development of a critical stance in the scientific community. Consequently, there are only few research projects focused on geoethics, and funds to develop activities and tools on ethical and social aspects in geosciences are scarce.

A wide scientific debate could help geoethics not to have an ideological drift (Peppoloni, 2012a), as in some cases has happened to bioethics. Bioethics has contributed greatly to the preservation and respect of human dignity. Indeed, its themes are discussed in universities, hospitals, journals, professional organizations, and are firmly established in the public consciousness. At the same time, it has raised moral and ideological controversies (Tallacchini, 2003).

The recent publication of a special issue on geoethics (Peppoloni and Di Capua, 2012a), published on a JCR Journal (i.e., included in the Journal Citation Reports by Thomson Reuters: http://thomsonreuters.com/journal-citation-reports/), has opened the way for new perspectives in protecting the environment. In recent years, successful sessions on geoethics at national and international meetings have been held (34th International Geological Congress in 2012; European Geosciences Union – General Assembly in 2012, 2013, and 2014; Geoitalia Congress in 2009, 2011, and 2013; International Association of Engineering Geology and the Environment Congress in 2014) and ongoing editorial projects, like this book, confirm that the discussion on ethics in Earth sciences is becoming an urgent matter within the scientific community.

AN ETYMOLOGICAL STARTING POINT: INDIVIDUAL AND SOCIAL RESPONSIBILITY

Many geoscientists work with a spirit of service toward society, aware that they have knowledge and skills that are essential and useful for the common good. Nevertheless, a clear reference framework of shared values in geosciences seems to be lacking. These values would consider different cultural and economic contexts and would be a call to responsibility for geoscientists. Moreover, the ethical foundations upon which discussions and actions are based are not sufficiently clear, with the result that geoethics lacks theoretical and practical strength.

An etymological analysis of the word "geoethics" may help to explain the deeper, hidden meaning of the word. Where does the term "geoethics" come from? What are its origins and connotations? What is the history of its development? What is its deeper meaning?

"Geoethics" is the union of the suffix "geo" and the word "ethics." The suffix "geo" contains a deep and ancient meaning. It refers to "*gaia*," which means

"Earth" in Greek, but its ancient Sumerian base "*ga*" refers more specifically to "home, the dwelling place." So the Earth is the place where we dwell, where our ancestors dwelt, and where our children will dwell.

The term "ethics" was defined by Aristotle (384/383–322 BC) as the investigation of and reflection on the operational behavior of humans. It identifies that part of philosophy that deals with the problem of human action (Dizionario di filosofia, 1988).

The etymological analysis of the word "ethics" is complex: ethics is derived from the Greek word "ἔθος" (*ĕthos*), which means "habit, custom." This noun has the same origin as "εἴωθα" (*eiotha*), a Greek perfect form meaning "I am accustomed to, I have the habit of, I am familiar with" (Liddell and Scott, 1996).

Words such as "accustomed" and "familiar" imply a sense of belonging to a community, either a family or a larger social group. But what determines familiarity and therefore a habit of behavior? This can be traced back to the Semitic root "*edum*" meaning "experience, to be experienced in."

In other words, I experience something (an event, a circumstance), I acquire knowledge, and I familiarize myself with this event. From now on, I am expert enough to be able to choose the behavior or custom most suitable to that particular circumstance or event.

But the meaning of the word "ethics" is more than this: "ethics" is traced also to the Greek word "ἦθος" (*ēthos*), which refers more specifically to the characteristics or habits of the individual, one's personal characteristics (Liddell and Scott, 1996). Both these nouns (*ĕthos* and *ēthos*) derive from the same root "*sweth-*" (compare with Latin word "*suesco*," meaning "I use," Ernout and Meillet, 1994), but the second term, probably more recent than the first, could be an evidence of that moment in human history when a person, who belonged to a community, became capable of perceiving himself/herself as an individual. So, the word "ethics" could have a dual meaning: one related to the social sphere and one to the individual sphere.

The same origin can be also observed going back from Greek to the Accadian language: starting from the Accadian base "*esdu*," ethics means "social foundation, social discipline," and in a wider sense "assurance of continuity." Again we meet the social dimension, the reference of the word "ethics" to the community (Semerano, 2007).

However, from the Accadian base "*betu*" comes the meaning of "home, dwelling, shelter." As such it can refer to something more personal, intimate, and deep in every human being. Moreover, from the Accadian base "*ettu*," the word ethics assumes the meaning of "character, distinguishing marks of an individual, characteristic of a person": again the individual sphere (Semerano, 2007).

In summary, it seems that a double meaning can be given to the word "ethics": on the one hand it relates to what is common, it contains a sense of belonging to a social dimension; on the other it refers to the self in relation to oneself, it expresses the personal, the individual.

It follows that ethics concerns both the common sphere, the interactions between individuals belonging to a social organization, and the personal sphere, what distinguishes an individual. Ethics means "to be part of," and at the same time "to belong to oneself." These two existential conditions (social and individual) unexpectedly coexist in the word "ethics."

By analogy, these considerations can be extended to geoethics, coming to define it on the one hand as an investigation of and reflection on the behavior of humans toward the geosphere, and on the other hand as the analysis of the relationship between the geologist who acts and his/her own action or own work. The etymological analysis of the word "geoethics" calls geoscientists to an intrinsic responsibility for their own activities.

RESPONSIBILITY OF GEOSCIENTISTS

Based on the above considerations, to see if one's behavior as a geologist is ethical, it is not sufficient to refer to the social sphere. One also has to refer to his/her own person, and clarify to himself/herself the ethical value of his/her activities.

Thus, the in-depth meaning of the word "geoethics" calls upon geoscientists to face the responsibility of ethical behavior at both a social and individual level. These two dimensions are strictly linked. In fact, an ethical attitude within an individual is reflected in his/her social behavior.

But of what does the responsibility of the geologist consist? And what motivations are needed to encourage geoscientists to practice Earth sciences in an ethical manner? The ethical commitment of geoscientists is the research and defense of truth, and a commitment to advancement of knowledge and lifelong learning (individual dimension). Ethical obligations arise from the possession of specific knowledge that has practical consequences for the public. Geoscientists are an active and responsible part of society. They should be at the service of the common good, taking care of the Earth system and promoting the awareness of every citizen's responsibility (social dimension).

A key question we must tackle is: are geoscientists well-equipped to undertake the responsibility of discovering the best solutions through the application of scientific methods, while promoting professionalism and research integrity, in a world in which our increasing interconnectedness promises increasingly complex problems and consequently the adoption of complex solutions?

AN ETHICAL COMMITMENT: THE GEOSCIENTIST'S PROMISE

The well-being and survival of humankind depend on the Earth's habitability and resources. Thus, the health of the Earth system (solid Earth, biosphere, hydrosphere, and atmosphere) must be of the utmost importance to society. It is evident that some modifications of Earth system processes, that are potentially damaging for humankind, can be and are being induced by humans, through actions that do not respect our planet's natural dynamics and equilibrium.

Problems such as the over-imprinting on the Earth's surface, the overexploitation of natural resources (some of which are not renewable), and the excessive growth of the population require an urgent transition to a sustainable world to ensure the survival of future generations. The social role of geoscientists is therefore important and their activities must be ethical. A correct and scientifically sound approach can mitigate, or at least help to avoid, many of the serious consequences that continuously arise through the irresponsible use of land. The ethical obligation of geoscientists includes bringing their knowledge to the public, to raise social awareness on environmental problems, to propose prudent choices to decision makers, and to report misguided actions. In this way, they can contribute to the conservation of the geosphere and its habitability for future generations.

The ethical obligations of modern geoscientists involve the following:

- Proper land/environment use and management. The ethics of sustainable land management must prevail, regardless of the type of commissioned project and the short-term economic interests of the client.
- Respect of truth and science. Respect for scientific truth based on updated evidence (including the recognition of data uncertainty, or of limited knowledge) and rigorous adherence to the scientific method is a must in the professional activities of geoscientists.
- Promotion of awareness of citizens' responsibilities, through effective communication and adequate education, especially of young people.
- Commitment to advancing knowledge and lifelong learning. Appreciation of the expertise of geoscientists by the public is a necessary condition in the promotion of efficient action in both the scientific and professional community. For this reason, as for physicians, the education of geologists has to be lifelong. Ethical training should be introduced into the university curriculum.

How can geoscientists, especially those with little experience, be best assisted in their acquisition of a clear and binding awareness of their ethical responsibility? In the context of a publicly growing demand for ethical behavior by those who have the ability to intervene in the public domain and act for the public benefit, the explicit acceptance of ethical responsibility by geoscientists can have the following effects:

- Increasing awareness of the social role of geoscientists, including their expertise and contributions, to strengthen their sense of belonging to professional and scientific communities.
- Fostering awareness of geoscientists within society in general and especially recognition of their social mission, essential for the public benefit, and, consequently, of their specific and unique role.
- Stimulation of cultural growth at the individual and community levels, exploitation of research, and implementation of scientific and professional skills.

For geoscientists, cultural and practical preparation has to take on an ethical dimension, starting at the university level. Through their individual commitment, young geoscientists can assume the need for continued cultural education as an ethical duty.

The explicit and conscious assumption of ethical obligations by geoscientists appears opportune and useful (Ellis and Haff, 2009; Matteucci et al., 2014). Some have suggested that the acceptance of ethical obligations may be facilitated by following the model of the Hippocratic Oath of physicians (the formula in the sidebar was proposed by the Italian Commission of Geoethics, in 2012, during the 34th International Geological Congress in Brisbane). The oath is based on the three fundamental pillars of geoethics proposed by Peppoloni and Di Capua (2012b). These three pillars form the basis of geoscientists' ethical obligations: awareness of one's responsibility, appreciation of the value of geological culture, and intellectual honesty.

OUTLOOK AND STRATEGIES FOR THE FUTURE

Geoscientists who are more aware of their role in society and of their ethical duties toward the geosphere will be able to provide a new perspective to geosciences and to build a new vision of how to understand the planet, encouraging the pursuit of the following aims:

- Raising the level of integrity of geoscientists, with the implementation of principles such as intellectual honesty, responsibility, self-criticism, cooperation, and fair comparison.
- Dissemination of scientific culture in society and involvement of the public in the idea of a geological heritage to be shared and protected.
- Creation of a new model of social and economic development, based on economic principles of sustainability and eco-friendliness.
- Protection of life and Earth.

The pursuit of these objectives can be done concretely through the following tools and actions:

- Codes of ethical conduct for geological research and practice (TGGGP, 2013).
- Guidelines for best practice, and eco-friendly and sustainable technologies in different fields (georesources, geohazards, communication, etc.).
- Action protocols for the proper management of the relationship between geoscientists and decision makers (Hays and Shearer, 1981; IAVCEI, 1999; UNDRO, 1991; Jordan, 2013; Dolce and Di Bucci, 2015).
- Regulatory framework of legal liability for geoscientists engaged in activities that have an impact on society.
- Easy access to and friendly use of the data and results of public research, and quality control of information and results.

- Constant and authoritative support of geoscientists for politicians and decision makers.
- Qualified presence in media spaces (both traditional and new), and strengthening of cooperation with mass media operators (Di Capua and Peppoloni, 2009; Marone et al., 2015).
- Exchange of knowledge and experience between the world of professionals and researchers, transfer of advanced knowledge to industry and authorities, and collaboration in the training of technicians.
- Support for both theoretical and practical innovations, which attempt to renew the way we approach and manage environmental problems and natural resources, from the perspective of sustainability for future generations (Lambert et al., 2013).
- Attention to the uniqueness of each region and to its historical, cultural, and environmental characteristics, respecting the bio/geodiversity of each environment, and using cross-disciplinary methods including consulting with experts in related fields.
- Innovative and diversified tools for geoeducation, based on exchange of experience among educators and users, to stimulate an active approach to scientific learning and a possible direct involvement in activities of social interest.
- Educational campaigns to teach people appropriate behavior in the management of energy, and water, and in the area of protection from natural hazards.
- Working groups and networks for international collaboration that actively work to develop the debate on ethical issues in Earth science, to introduce geoethics in the teaching of university courses and to include the principles of ethics and research integrity in the management and implementation of national and international research projects, that have large environmental and social impact.
- Development of research activities, without being overconfident in the results, by checking the information sources, by verifying the compatibility of results with the observations, and by following defined integrity criteria (Mayer and Steneck, 2011), as expressed in the Singapore (2010) and Montreal (2013) Statements.

CONCLUSIONS: QUESTIONS FOR A PUBLIC DEBATE ON GEOETHICS

Science does not claim to solve all problems of society: it has value especially when it is aware of its limits (Peppoloni, 2012b). In any case, science, and in particular geosciences, can play a key role in supporting society by providing useful tools to mitigate the impact of human activities on the geosphere and to deal with the environmental challenges that face humanity. Geoscientists can help society to define a framework of values on which to base the study and implementation of procedures and tools for the benefit of the public.

The following basic questions may be useful to clarify some fundamental steps to be taken in the development of geoethics to render our actions effective:

- How do we articulate an ethical criterion for geoscientists?
- How can the freedom of research be combined with the principle of sustainability?
- Where should the line be drawn between preservation and economic development of the geosphere, especially in low-income countries?
- How can the relationships between geoscientists, media, politicians, and citizens be made more fruitful, particularly in relation to protection from natural hazards?
- What communication and educational strategies should be adopted to transfer the value of the geosciences to society?

The development of geoethics, not only as a critical attitude toward the relationship between humans and the geosphere, but also as a genuine scientific discipline, will be possible only if geoscientists are able to give convincing answers to these questions. Geoscientists have an important historical responsibility in the third millennium: to demonstrate that geological knowledge is really a benefit to mankind and in particular for future generations.

The Geoscientist's Promise

- *I promise I will practice geosciences being fully aware of the involved social implications and I will do my best for the protection of the geosphere for the benefit of mankind.*
- *I know my responsibilities towards society, future generations, and the environment for a sustainable development.*
- *In my job I will put the interest of society at large in the first place.*
- *I will never misuse my geological knowledge, not even under constraint.*
- *I will always be ready to provide my professional expertise in case of urgent need.*
- *I will continue to improve my geological knowledge lifelong and I will always maintain my intellectual honesty at work, being aware of the limits of my capabilities and possibilities.*
- *I will act to foster progress in geosciences, the dissemination of geological knowledge and the spreading of a geoethical approach to the management of land and geological resources.*
- *I will honor my promise that my work as a geoscientist or certified geologist, will be fully respectful of Earth processes.*
  I promise

Source: Formula proposed by the Italian Commission of Geoethics (Matteucci et al. 2014).

REFERENCES

Bezzi, A., 1999. What is this thing called geosciences? Epistemological dimensions elicited with the repertory grid and their implications for scientific literacy. Sci. Educ. 83 (6), 675–700.

Bickford, M.E. (Ed.), 2013. The Impact of the Geological Science on Society. The Geological Society of America, Special Paper 501, p. 201. ISBN:978-0-8137-2501-7.

Bohle, M., 2013. To Play the Geoengineering Puzzle? IAPG – International Association for Promoting Geoethics Blog. http://iapgeoethics.blogspot.it/2013/12/to-play-geoengineering-puzzle-by-martin.html.

Borgomeo, E., 2012. Why I study Earth sciences. Ann. Geophys. 55, 361–363. http://dx.doi.org/10.4401/ag-5635. ISSN:2037-416X.

Bosi, C., Peppoloni, S., Piacente, S., 2008. Philosophical and epistemological debate in Italy within an ethical perspective of Earth Sciences. In: Proceedings of the 33rd International Geological Congress, Oslo, August 2008.

Brocx, M., Semeniuk, V., 2007. Geoheritage and geoconservation – history, definition, scope and scale. J. R. Soc. West. Aust. 90, 53–87.

Crutzen, P.J., 2002. Geology of mankind. Nature 415, 23.

Di Capua, G., Peppoloni, S., 2009. Scientific information: problems and responsibilities. In: Proceedings of the Mining Pribram Symposium, International Session on Geoethics. Pribram (Czech Republic). Available at: www.earth-prints.org/handle/2122/5455.

Dizionario di filosofia (Dictionary of Philosophy), 1988. Biblioteca Universale Rizzoli, Milano (in Italian).

Dolce, M., Di Bucci, D., 2015. Risk management: roles and responsibilities in the decision-making process. In: Wyss, M., Peppoloni, S., (Eds.), Geoethics: Ethical Challenges and Case Studies in Earth Science. Elsevier, Waltham, Massachusetts, pp. 211–221.

Dowling, R.K., 2011. Geotourism's global growth. Geoheritage 3, 1–13. http://dx.doi.org/10.1007/s12371-010-0024-7.

Eder, W., 2004. The global UNESCO network of geoparks. In: Proceedings of the First International Conference of Geoparks. Geological Publishing House, Beijing.

Ellis, E.C., Haff, P.K., 2009. Earth science in the Anthropocene: new epoch, new paradigm, new responsibilities. EOS Trans. 90 (49), 473.

Ernout, A., Meillet, A., 1994. Dictionnaire étymologique de la langue Latine (retirage de la quatrième édition). Éditions Klincksieck, Paris.

Gray, M., 2008. Geoheritage 1. Geodiversity: a new paradigm for valuing and conserving geoheritage. Geosci. Can. 35 (2), 51–59.

GSL, January 2014. Geology for Society. A report by the Geological Society of London . http://www.geolsoc.org.uk/geology-for-society.

Hays, W.W., Shearer, C.F., 1981. Suggestions for improving decision making to face geologic and hydrologic hazards. In: Hays, W.W. (Ed.), Facing Geologic and Hydrologic Hazards: Earth Science Considerations. Geological Survey, Professional paper 1240-B, pp. B103–B108.

Höppner, C., Whittle, R., Bründle, M., Buchecker, M., 2012. Linking social capacities and risk communication in Europe: a gap between theory and practice? Nat. Hazards 64, 1753–1778. http://dx.doi.org/10.1007/s11069-012-0356-5.

IAPG, 2012. Constitution of the International Association for Promoting Geoethics. http://www.iapg.geoethics.org/img/IAPG_Constitution.pdf.

IAVCEI Subcommittee for Crisis Protocols, 1999. Professional conduct of scientists during volcanic crises. Bull. Volcanol. 60, 323–334.

Jordan, T.H., 2013. The value, protocols and scientific ethics of earthquake forecasting. Geophys. Res. Abstr. 15. EGU2013–12789. EGU General Assembly 2013.

Lambert, I., Durrheim, R., Godoy, M., Kota, M., Leahy, P., Ludden, J., Nickless, E., Oberhaensli, R., Anjian, W., Williams, N., 2013. Resourcing Future Generations: a proposed new IUGS initiative. Episodes 36 (2), 82–86.

Liddell, H.G., Scott, R., 1996. A Greek-English Lexicon (with a Revised Supplement). Clarendon Press, Oxford.

Marone, E., Castro Carneiro, J., Cintra, M., Ribeiro, A., Cardoso, D., Stellfeld, C., 2015. Extreme sea level events, coastal risks, and climate changes: informing the players. In: Wyss, M., Peppoloni, S. (Eds.), Geoethics: Ethical Challenges and Case Studies in Earth Science. Elsevier, Waltham, Massachusetts, pp. 273–302.

Martínez-Frías, J., González, J.L., Rull Perez, F., 2011. Geoethics and deontology: from fundamentals to applications in planetary protection. Episodes 34 (4), 257–262.

Matteucci, R., Gosso, G., Peppoloni, S., Piacente, S., Wasowski, J., 2014. The geoethical promise: a proposal. Episodes 37 (3), 190–191.

Mayer, T., Steneck, N. (Eds.), 2011. Promoting Research Integrity in a Global Environment. Imperial College Press/World Scientific Publishing, Singapore, pp. 317–328. ISBN:978-981-4340-97-7.

McKeever, P., 2013. Memory of the Earth. World Sci. 11 (3), 4–13.

Montreal Statement on Research Integrity, 2013. 3rd World Conference on Research Integrity, Montreal, 5–8 May 2013. http://www.cehd.umn.edu/olpd/MontrealStatement.pdf.

Moores, E.M., January 1997. Geology and culture: a call for action. GSA Today, 7–11.

Newsome, D., Dowling, R.K., 2010. Geotourism: The Tourism of Geology and Landscape. Goodfellow Publishers, Oxford.

Osborne, R.A.L., 2000. Geodiversity: "green" geology in action. Proc. Linn. Soc. N. S. W. 122, 149–176.

Peppoloni, S., 2012a. Ethical and cultural value of the Earth sciences. Interview with Prof. Giulio Giorello. Ann. Geophys. 55, 343–346. http://dx.doi.org/10.4401/ag-5755. ISSN:2037-416X.

Peppoloni, S., 2012b. Social aspects of the Earth sciences. Interview with Prof. Franco Ferrarotti. Ann. Geophys. 55, 347–348. http://dx.doi.org/10.4401/ag-5632. ISSN:2037-416X.

Peppoloni, S., Di Capua, G. (Eds.), 2012a. Geoethics and Geological Culture. Reflections from the Geoitalia Conference 2011. Ann. Geophys. 55 (3) (special issue).

Peppoloni, S., Di Capua, G., 2012b. Geoethics and geological culture: awareness, responsibility and challenges. Ann. Geophys. 55, 335–341. http://dx.doi.org/10.4401/ag-6099. ISSN:2037-416X.

Semerano, G.M., 2007. Le Origini della Cultura Europea: Dizionari Etimologici (Origins of the European Culture: Ethimological Dictionaries). Olschki Editore, Firenze (in Italian).

Singapore Statement on Research Integrity, 2010. 2nd World Conference on Research Integrity, Singapore, 21–24 July 2010. http://www.singaporestatement.org/statement.html.

Tallacchini, M., 2003. Fuga dalla Bioetica. In: Boella, L. (Ed.), Bioetica Dal Vivo. Aut Aut, Novembre-Dicembre, pp. 107–118(in Italian).

TGGGP, 2013. Codes of Ethics. Task Group on Global Geoscience Professionalism. http://tg-ggp.org/professionalism-in-geoscience/codes-of-ethics.

UNDRO, 1991. Mitigating Natural Disasters: Phenomena and Options. A Manual for Policy Makers and Planners. United Nations Organization, New York.

Zalasiewicz, J., Williams, M., Smith, A., Barry, T.L., Bown, P.R., Rawson, P., Brenchley, P., Cantrill, D., Coe, A.E., Gale, A., Gibbard, P.L., Gregory, F.J., Hounslow, M., Kerr, A., Pearson, P., Knox, R., Powell, P., Waters, C., Marshall, J., Oates, M., Rawson, P., Stone, P., 2008. Are we now living in the Anthropocene? GSA Today 18 (2), 4–8.

Zouros, N., 2004. The European Geoparks Network – geological heritage protection and local development. Episodes 27 (3), 165–171.

Chapter 2

Geoethics: Reenvisioning Applied Philosophy

Ingrid Leman Stefanovic

Dean, Faculty of Environment, Simon Fraser University; Professor Emeritus, Department of Philosophy, University of Toronto, Canada

Chapter Outline

Abstract

This chapter argues that values—often taken for granted—affect decision making in significant ways within the field of Earth sciences. At the same time, it also suggests that the relationship between ethics and the geosciences is reciprocal: a new role for ethicists emerges from this interdisciplinary conversation, just as new questions arise for geoscientists. I begin by introducing some key philosophical concepts, particularly as they are informed by the growing field of environmental ethics. Part I clarifies some basic philosophical terminologies and explores the role of philosophy as an "applied" discipline. Part II illustrates how different value systems affect scientific judgment calls and decision making, using examples from the fields of geology and risk assessment. Part III argues for a new role for ethics and ethicists that emerges from this interdisciplinary conversation between philosophy and the Earth sciences.

Keywords: Applied ethics; Environmental ethics; Philosophy.

Our relationship to the Earth cannot be encompassed by science alone.

Robert Frodeman, Geo-Logic

Geoethics. http://dx.doi.org/10.1016/B978-0-12-799935-7.00002-2

PART I: THE PLACE OF ETHICS

Canada, my home, is internationally recognized as a resource-rich country. From forestry to mining industries, Earth scientists are playing a central role in advancing our understanding of the nature and scope of these environmental resources. Whether exploring water quality and quantity or mapping baseline scientific profiles of our landscapes, the fact is that the geosciences have an important role worldwide to play in the building of information databases and the development of environmental standards. From defining geographical boundaries in discussions of sovereign land rights to identifying new energy sources to assessing risks of environmental hazards, the Earth sciences are also increasingly involved in policy development, environmental planning, and even planetary climate change research and analysis.

In many respects, the Earth sciences emerge, therefore, as a preeminent example of interdisciplinary knowledge. Beyond the parameters of standard lab-based science, researchers are immediately drawn into the complexities of field work, as well as discussions of public policy, economics, and sociocultural matters. In the words of geologist and philosopher, Robert Frodeman, "defying categories, geologic insights today often function simultaneously as scientific statements, political truths, and poetic and metaphysical incantations" (Frodeman, 2003, p. 2).

To be sure, science seeks to empirically measure the availability of water or mineral resources; but what impact will shortages of such resources, here or abroad, have on foreign policy? How long will depleting aquifers sustain local communities and how does one plan longer term for the viability or reorganization of such communities? What is the likelihood of earthquakes or floods in parts of the country, and how does one minimize the risks to society through more informed adaptation and mitigation strategies?

And finally, as issues like these move us from science to the social sciences, the question emerges about what *ought* to be done in specific cases of environmental decision making. That prescriptive, rather than descriptive, moment moves us squarely into the humanities and indeed, into the field of ethics.

There are a number of ways in which one can discuss the role of ethics when it comes to the Earth sciences. To provide some context, let me begin with a *metaethical* discussion, addressing two major concerns. First, I consider how (and why) one might distinguish ethics from morality and why the distinction may or may not be important to the field of geoethics.

Second, I consider a common philosophical distinction between top–down and bottom–up approaches to questions of applied ethics. To take a "top–down" approach means that a general rule or moral principle is applied to a specific case to which the rule applies (Beauchamp, 2005, p. 7). On the other hand, "bottom–up" approaches focus primarily on the specific challenges of discrete, practical decisions. By listening to diverse narratives, "bottom–up" approaches avoid applying prescribed, theoretical principles in advance of a careful listening to the unique circumstances of each case (Beauchamp, 2005, p. 8).

Following that discussion, I want to argue for a more iterative approach between ethics and *praxis*, showing how values, ethical assumptions, and paradigms infuse decision making in the case of the geosciences—and how the geosciences can also influence unique philosophical reflections.

So then, let us begin with the question: how does "ethics" relate to questions of morality? Some believe that the distinction is important, particularly if one seeks to distinguish between "ethics" as a societal set of principles, on the one hand, and, on the other hand, one's own, personal sense of morality. In some cases, for instance, one's own *morals* may not coincide with a broader, explicitly articulated professional or *corporate ethic*. Sometimes, there is a sense that "morality" reflects one's subjective values, whereas "ethics" points to a shared understanding of socially accepted objective rules of conduct.

Nevertheless, for others, the relation between ethics and morality signifies a distinction without a difference. For practical purposes, the terms are seen to overlap, since the field of ethics raises questions of moral principles, and morality is guided by ethical deliberation.

My own sense is that, while there is an important role for professional ethics for Earth scientists, the growing field of "geoethics" itself signifies something larger (Peppoloni and DiCapua, 2012). To be sure, the field certainly invites the possibility of explicitly articulated codes and principles. But the Greek origins of the term *ethos* point to the much broader question of how we dwell virtuously in our social and environmental relationships. In the words of Hans-Georg Gadamer (1900–2002), to raise ethical questions is to reflect on "right living" (Gadamer, 2013, p. 286).

Beyond developing explicit, universally recognized principles or codes of behavior, the field of "geoethics" should also acknowledge the importance of deciphering utilitarian arguments and balancing moral costs and benefits in each discrete case of decision making. And frankly, the description of a new area of study called "geoethics" similarly invites the possibility of exploring and justifying implicit, often hidden moral and ethical—understood synonymously—judgment calls that affect environmental decision making.

In short, while "ethics" and "morality" might be distinguished in certain cases, I would argue that the term "geoethics" indicates, in the broadest possible way, an emerging field of interest that encompasses formalized codes of behavior; diverse processes of moral reflection; and even the possibility of interpreting, analyzing, and justifying taken-for-granted values and assumptions that implicitly underlie and affect decision making in the geosciences as a whole (Frodeman, 2000).

This conversation about the relation between explicitly articulated codes of ethics and broader, sometimes taken-for-granted moral interpretations of our actions and ways of thinking, reflects a parallel differentiation between "top–down" and "bottom–up" philosophical approaches to understanding (Beauchamp, 2005).

On the one hand, some ethicists believe that the task for philosophers is to develop a clear, universally accepted, rational set of theoretical guidelines that can subsequently be "applied" top–down to specific cases of moral controversy. So, for instance, recognizing that the Greek "deon" means "duty," one may choose to build a *deontological* moral theory, as did Immanuel Kant (1724–1804), whereby rational principles are seen to logically imply specific rights and responsible ways of acting (Kant, 1997, 2009). Or one may develop a *utilitarian* theoretical basis for decision making, explicitly arguing that right actions emerge by seeking "the greatest good for the greatest number" through an assessment of the consequences of a specific decision (Mill, 2007). In both cases, the "top–down" approach suggests that the way in which to "apply" ethics to particular cases of decision making is to clarify a rational set of theoretical guidelines and then put those guidelines into practice.

On the other hand, some point out that ethics can never have the precision of mathematics and that no universal consensus exists about which set of theoretical principles to in fact apply in all cases and unilaterally in a "top–down" fashion. Life is messy, critics argue, and the task for ethicists is to acknowledge this reality and learn from it, informing theory instead by way of a "bottom–up" approach, and on a case-by-case basis. The result is that rather than developing a single, monistic ethical theory that is universally accepted, philosophers should accept that plural theoretical frameworks can inform, and be informed by, a diversity of lived, practical challenges (Stone, 1988; Norton, 2005).

Many arguments in favor of such a "bottom–up" approach to ethical deliberation have emerged from applied fields, such as bioethics or environmental ethics. For example, no matter what kind of ethical principles or guidelines in favor of euthanasia one may argue for in the abstract, the fact is that, sitting at the bedside of a terminally ill patient and deliberately disengaging life support, invites a complexity of emotions, intellectual reasons, and human experiences that are often seen to exceed and be inadequately captured by simplistic theoretical arguments. Similarly in the environmental field, a utilitarian calculus of costs and benefits of building a shopping mall may never capture the full breadth of meaning of the land, held deeply through long tradition by local, indigenous communities. In that respect, ethicists suggest that "top–down" imposition of theoretical guidelines may miss the intricacies of the particular case at hand.

My own sense is that "top–down" application of a single set of theoretical principles is rarely sufficiently sensitive to the vagaries and complexities of decision making. Ethicists simply do not possess a single "how-to" manual because moral deliberation is simply more than a technical matter of unilaterally "applying" conclusively established and universally applicable theories to diverse, complex problems.

At the same time, to say that every practical situation uniquely informs theory "bottom–up" seems to deny the possibility of meaningful core values that may, implicitly or explicitly, guide decision making, despite changing circumstances. In other words, I see the field of ethics both informing and being informed by the vagaries of unique experiences *as well as* the commonalities of informed, theoretical deliberation. Sometimes, specific cases stretch our theories but, equally, without some form of theoretical reflection, moral decision making remains rudderless and arbitrary.

How does this iterative relation between theory and practice play out in the real world? And how might this conversation impact the growing interest in the field of geoethics? I address these questions in the following section.

PART II: GEOETHICS, RISK ASSESSMENT, AND ENVIRONMENTAL DECISION MAKING

My own approach to the field of ethics emerges from a lifelong interest in phenomenology and hermeneutics (Stefanovic, 2000, 1994). Without detracting from the purposes of this chapter with long-winded philosophical explanations, let me simply say that both the phenomenological and hermeneutic approaches recognize the importance of what Aristotle (384–322 BC) called in Book VI of his *Nicomachean Ethics*, "phronesis"—practical wisdom. The aim of the phenomenological method is, in the words of philosopher Martin Heidegger (1889–1976) to "lay bare" elements of lived experience that are so close, and thereby so taken for granted that we forget that they constitute the condition of the possibility of explicit theoretical calculation and deliberative thought (Heidegger, 1962; Stefanovic, 1994, 2000). Hermeneutics, as the art and theory of interpretation of linguistic and nonlinguistic texts, recognizes that understanding is rarely if ever value-free: rather, it emerges from within the background of larger, implicit, prethematic interpretive horizons, and contextual relations.

Phenomenologist and hermeneutic philosopher, Hans-Georg Gadamer explains (2013, p. 563) how:

> In the natural sciences, what are called facts are not arbitrary measurements but measurements that represent an answer to a question, a confirmation or refutation of a hypothesis. So also an experiment to measure certain quantities is not legitimated by the fact that these measurements are made with utmost exactitude, according to all the rules. It achieves legitimacy only through the context of research. Thus all science involves a hermeneutic component.

All scientific questions, in other words, emerge from within a taken-for-granted interpretive context of lived experience, paradigms, perceptions, and values. The answers that we arrive at in our scientific experiments very much depend upon what kind of questions are asked in the first place, and those questions are determined by larger research priorities and assumptions that exceed the narrow boundaries of the specific scientific experiment. In this connection, the aim of phenomenological ethics, then, is to "lay bare," to bring to attention and better understand those deeply embedded values that influence the decisions and interpretations that we make about a variety of factors, from how to orient ourselves within our daily lives to how we interpret and assign meaning to specific, scientific findings.

Consider, for instance, how the recent interest in the very term "geoethics" signifies a paradigm shift: while "ethics" has for centuries been interpreted in the Western world exclusively in terms of human and social concerns, "Geoethics" implicitly expands those parameters beyond narrow anthropocentric parameters to include a broader "ecocentric" awareness of the Earth and its nonhuman inhabitants. The value of a geological resource, for instance, in traditional interpretations, would be understood merely in terms of its mined use to human societies; and yet, a nonanthropocentric interpretive context often frames the discussion quite differently in terms of larger ecosystem impacts of mining, or in terms of the sacredness of the landscape for indigenous communities or even its "intrinsic" rather than merely "instrumental" value. The point is that the very language we use betrays hidden value systems and paradigms that frame the meanings that we assign to the external world.

Values and attitudes are taken for granted in other ways as well. Often, Earth scientists participate in risk assessments: burying nuclear waste (Hocke, 2015), for instance, requires understanding of geological conditions and those conditions are often open to interpretation. As philosopher and biologist, Kristin Shrader-Frechette, points out, the "period of interest" in assessing the risk of migration of radioactive waste (often called "radwaste") can be tens of thousands of years "four orders of magnitude longer than any period of observation" (Shrader-Frechette, 2000, p. 16). Simulation models operate under certain assumptions, based upon incomplete scientific knowledge—which means that value judgments are made in order to both evaluate risks as well as to frame the models' parameters. Because of these unknowns, a "serious difficulty with hydrogeological models is that scientists often do not agree on what would confirm them" (Shrader-Frechette, 2000, p. 18).

Of course, to recognize that models are not value-free does not imply that we should reject them. On the contrary, it is precisely because judgments are incorporated within the models that we are obliged to (1) explicate values and attitudes that may influence research parameters, (2) model from as large a variety of perspectives as possible, and (3) ensure that public discussion and transparency guide the decision-making process when it comes to controversial issues such as burying radwaste (Shrader-Frechette, 2000).

Similarly, ethical issues infuse a variety of other geological problems, such as those that arise through cases of acid mine drainage—contamination arising from abandoned mines. *Ought* those mines be restored and if so, to what level of safety? And who should rightly bear the costs of such restoration? (Frodeman, 2003, p. 20). Robert Frodeman (2003, p. 20). points out how philosophy's role in these conversations is broad and varied, and includes "ethical, aesthetic, epistemological, metaphysical and theological dimensions that…are more central to our concerns with the environment than we often acknowledge".

A final example: consider the problem of identifying an appropriate site for landfill disposal. It is clear that here, geological factors are clearly crucial but a number of ethical issues are imbedded in the siting decision as well. On the one hand, a utilitarian calculus may well favor a particular location, if estimates show that benefits outweigh costs. On the other hand, a cost–benefit analysis that apparently maximizes the overall good may not be the end of the story. For instance, questions of environmental justice often arise, since it has been shown that toxic waste sites are disproportionately located in neighborhoods where low-income and/or nonwhite populations reside (Wenz, 1988). In that case, deontological priorities such as basic human rights, duties, and principles of fairness may be seen to legitimately override cost–benefit calculations.

In each of these cases and many more that are documented in this book, ethical issues arise within cases of environmental decision making and deserve attention by geoscientists. If we are drawn to such a conclusion, the next question is: what is the role of geoethicists in this conversation? I address this issue in the following section.

PART III: A NEW ROLE FOR ETHICISTS

In his groundbreaking (no pun intended!) book, entitled *Geo-Logic*, Robert Frodeman suggests at least two roles for philosophers.

The first is to "provide an account of the specifically philosophical aspects of our environmental problems," encompassing questions that range from the moral to the aesthetic (Frodeman, 2003, p. 20). My sense is that providing such an account means more than simply applying theory to practical problems; instead, just as much as philosophers can contribute to understanding moral problems that arise within the geosciences, those very problems invite philosophers to ask different kinds of questions themselves. In other words, the interdisciplinary conversation is two-way, requiring that scientists incorporate philosophical reflection into their work and that philosophers inform their reflections through genuine engagement with the geosciences.

The second role for philosophers described by Frodeman (2003, p. 20) is to "offer a synopsis of how the various disciplines relate within a given problem". Encouraging interdisciplinary dialogue, philosophers have a responsibility to do more than simply engage in conversations among themselves but instead, to step into a different kind of collaboration that investigates relationships between the disciplines, engages in *praxis*, and redefines the meaning of "applied ethics."

To use the words of German philosopher, Jürgen Habermas (1929–present), the role for philosophers and ethicists is less to stand back and fashion speculative theory than to serve as "stand-in interpreters," assisting communities to identify, critically analyze, and justify value claims and norms as they relate to environmental decision making (Habermas, 1990). What are a community's core values? Which values are negotiable? Which are peripheral to the conversation? Those kinds of questions are best addressed by ethicists who seek to "apply" their discipline in a responsible manner, in a diversity of ways (Morito, 2010). Complex problems require both the interpretation of different standpoints, as well as negotiation among competing interests. Deciding what is the "right" or "fair" thing to do in such cases of competing interests, only benefits from the input of ethicists.

In that sense, philosophy and ethics are much more than merely of academic concern. If the field of geoethics is to do justice to both the findings of the Earth sciences as well as to a new vision of what constitutes genuine philosophical reflection, then ethicists themselves have to engage in different kinds of conversations. They need to speak less among themselves and more to the broader community as a whole. They need to learn to translate their discipline-based terminology to different audiences, in order to effectively impact upon the way in which our world is evolving. They need to do what Aristotle advocated, that is, insert themselves firmly into the lived center of the "agora"—the meeting place where practical decisions are actually made, rather than simply contemplated. In the process, philosophy will be transformed into a more meaningful discipline, just as it transforms the geosciences by inviting them to join in ethical reflection.

The field of geoethics is new and growing. As we move forward, the challenge is to find ways in which to engage in this interdisciplinary conversation in such a way that ethics informs science, as much as science informs philosophical practice. Only then are both the fields of geology and philosophy enhanced and strengthened.

REFERENCES

Beauchamp, T., 2005. The nature of applied ethics. In: Frey, R.G., Wellman, C.H. (Eds.), A Companion to Applied Ethics. Blackwell Publishing Ltd, Malden, MA, pp. 1–16.

Frodeman, R. (Ed.), 2000. Earth Matters: The Earth Sciences, Philosophy and the Claims of Community. Prentice Hall, Upper Saddle River, NJ.

Frodeman, R., 2003. Geo-logic: Breaking Ground between Philosophy and the Earth Sciences. State University of New York Press, Albany.

Gadamer, H.G., 2013. Truth and Method. Bloomsbury Academic, London.

Habermas, J., 1990. Moral Consciousness and Communicative Action: Reason and the Rationalization of Society (T. McCarthy, Trans.). Beacon Press, Boston.

Heidegger, M., 1962. Being and Time (J. Macquarrie and E. Robinson, Trans.). Harper and Row, New York.

Hocke, P., 2015. Nuclear waste repositories and ethical challenges. In: Wyss, M., Peppoloni, S.(Eds.), Geoethics: Ethical Challenges and Case Studies in Earth Science. Elsevier, Waltham, Massachusetts, pp. 359–367.

Kant, I., 1997. Critique of Practical Reason (M.J. Gregor, Trans.). Cambridge University Press, New York.

Kant, I., 2009. Groundwork of the Metaphysics of Morals (H.J. Paton, Trans.). Harper Collins, New York. First published in 1948.

Mill, J.S., 2007. Utilitarianism. Dover Publications Inc., London. First published in 1871.

Morito, B., 2010. Ethics of climate change: adopting an empirical approach to moral concern. Hum. Ecol. Rev. 17 (20), 106–116.

Norton, B.G., 2005. Values in nature: a pluralistic approach. In: Cohen, A.I., Wellman, C.H. (Eds.), Contemporary Debates in Applied Ethics. Blackwell Publishing Ltd, Malden, MA.

Peppoloni, S., DiCapua, G. (Eds.), 2012. Geoethics and Geological Culture. Reflections from the Geoitalia Conference 2011. Annals of Geophysics, vol. 55, No. 3 at: http://www.annalsofgeophysics.eu/index.php/annals/issue/view/482. accessed 02.04.14.

Shrader-Frechette, K., 2000. Reading the riddle of nuclear waste. In: Frodeman, R. (Ed.), Earth Matters: The Earth Sciences, Philosophy and the Claims of Community. Prentice Hall, Upper Saddle River, NJ, pp. 11–24.

Stefanovic, I.L., 2000. Safeguarding Our Common Future: Rethinking Sustainable Development. State University of New York Press, Albany, NJ.

Stefanovic, I.L., Spring 1994. What is phenomenology? In: Brock Review, vol. 3, No. 1, pp. 57–77.

Stone, C., 1988. Earth and Other Ethics: The Case for Moral Pluralism. Harper Collins, New York.

Wenz, P., 1988. Environmental Justice. State University of New York Press, Albany.

Chapter 3

The Imperative of Sustainable Development: Elements of an Ethics of Using Georesources Responsibly

Armin Grunwald

Karlsruhe Institute of Technology, Institute for Technology Assessment and Systems Analysis

Chapter Outline

Abstract

The pressure on natural resources is still increasing. Modern industrialized societies are increasingly confronted with scarcities: energy carriers, land surface, water, metals, biodiversity, and so on. This also holds for the space under our feet, the "underground." This space is subject of usage and exploitation to a quickly increasing extent: traditional fossil energy carriers but also shale gases are extracted, raw materials, e.g., metals, are mined, geothermal energy is looked for, waste is disposed of, and the underground is even seen as a solution to climate change by storing greenhouse gases (carbon capture and storage technology). This situation has led to competing uses and raises questions about our responsibility for future generations.

Geoethics. http://dx.doi.org/10.1016/B978-0-12-799935-7.00003-4

These issues are discussed internationally under the vision of sustainable development. This chapter will explain this vision, present criteria, and procedures for its practical application, and create references to geosciences and georesources.

Keywords: Future ethics; Responsibility; Sustainability principles.

CHALLENGE

The increasing pressure on the underground, in particular on georesources, leads to competition between different ways of making use of the underground and to social conflicts in two directions. On the one hand, conflicts between different groups of actors arise today. Distributional justice is a relevant issue because benefits and costs/risks of geotechnological intervention will be distributed among several parties. In particular, environmental justice has become a major issue over the last years (Anand, 2004). On the other, making use of the underground according to today's needs may endanger the opportunities of coming generations to fulfill their needs because georesources will already have been consumed by the generation living today. This observation results in the imperative of taking responsibility for future generations (Jonas, 1984; Brown-Weiss, 1989). The vision of sustainable development integrates the ethical issues of responsibility for the future, taking stewardship for our planet and taking care for distributional justice today (Grunwald and Kopfmüller, 2012). According to the most widely accepted definition today, sustainable development is "development that meets the needs of the present without compromising the ability of future generations to meet their own needs" (WCED, 1987).

Sustainable development is not a term that can only be determined scientifically but a sociopolitical and thus *normative* vision, based on two ethical principles: actively taking responsibility for future generations and considerations of equity among those living today. Responsibility for the future is about *preservation* of natural and cultural resources, including georesources, in the interest of future generations. Distributional justice mainly relates to the problem of development, i.e., the relationship between the rich countries in the North and the poor countries in the South, also with regard to the use of natural resources. Thus, the vision of sustainable development must by definition include both long-term considerations and the global dimension. Pursuing this vision implies that societal processes and structures should be reoriented so as to ensure that the needs of future generations are taken into account and to enable current generations in the Southern and Northern Hemispheres to develop in a manner that observes the issues of equity and participation.

Sustainable development therefore concerns the relationship between human economy, the social foundations of societies, and the natural resources at the global level. Governance activities at local, regional, national, and global levels are required in order to better meet the demands of sustainable development than hitherto. In this context, actors are confronted with the uncertainty and incompleteness of knowledge about the complex natural and societal systems and their interactions (Grunwald, 2007), with existing assessments that are partly incompatible and dominated by different interests, with the limited controllability of the systems, and with the variety and conflict potential of the proposed measures to improve sustainability.

Major challenges for sustainable management of georesources are related to the finiteness of nonrenewable resources such as fossil energies, to present and future risks arising from the use of the earth's crust as a sink for (e.g., radioactive) wastes, to justice in the use of available resources and distribution of risks, as well as to the processes of decision-making and opportunities for those affected to participate in them. Massive interventions in geosystems, such as dams (e.g., the Three Gorges Dam in China) or large-scale surface mining (e.g., for coal), raise not only ecological, but also serious social questions. In the fields of mining, tunneling, fracking, and geothermal energy different kinds of risks have to be considered, ranging from induced earthquakes to potential contamination of drinking water and accidents (Potthast, 2013).

To meet these challenges, the vision of sustainability must be made operational. There is considerable need for orientation knowledge on how to fill the vision of sustainable development with substance. Certain essential criteria have to be met for the vision to gain practical relevance: (1) a clear *object relation*, i.e., it must be clearly defined what the term applies to and what subjects should be covered by the assessments, (2) the *power of differentiation*, i.e., clear and comprehensible differentiations between "sustainable" and "non- or less sustainable" must be possible, and concrete ascriptions of these judgments to societal circumstances or developments have to be made possible beyond arbitrariness, (3) the possibility to *operationalize*, i.e., the definition has to be substantial enough to define sustainability indicators, to determine target values for them, and to allow for empirical "measurements" of sustainability. The following section will present an integrative approach that has already been applied in many fields (Kopfmüller et al., 2001).[1]

MAKING THE VISION OF SUSTAINABLE DEVELOPMENT WORK

The integrative concept of sustainable development (see Kopfmüller et al., 2001) identified three constitutive elements of sustainable development out of the famous Brundtland Commission's definition (WCED, 1987; see above), taking into account also the results of the 1992 Earth Summit at Rio de Janeiro and from ongoing research and debate: (1) the global perspective, (2) the justice postulate, and (3) the anthropocentric point of departure.

---

1. The presentation of the integrative approach to sustainable development closely follows Grunwald and Rösch (2011).

1. An essential aspect of sustainable development is its global orientation. The assignment given to the Commission by the UN General Assembly was that "a global agenda for change" was to be formulated to help define "the aspirational goals for the world community," and how they could be realized through better cooperation "between countries in different stages of economic and social development" (WCED, 1987). The chances for overcoming the global ecological and development crisis depend, in their view, on the extent to which we succeed in making sustainability a model for a "global ethic" (WCED, 1987).

2. The most important criterion for sustainability is, from the viewpoint of the Brundtland Commission, that of justice. Fundamental for their interpretation of justice is, first of all, the mutual interdependence of intra- and intergenerational justice: a (more) just present is the prerequisite for a just future. In this case, "justice" is primarily defined as distributive justice. The present inequalities of distribution as regards access to natural resources, income, goods, and social status is regarded to be the cause of global problems and conflicts, and a more just distribution or, rather, a redistribution of rights, responsibilities, opportunities, and burdens is required.

3. The third element, which is characteristic for the Brundtland Report's understanding of sustainability, is the anthropocentric orientation. The satisfaction of human needs is, in this concept, the primary goal of sustainable development—today, and in the future. The conservation of nature is not taken as an objective in its own right, but as a prerequisite for lasting societal progress—that is to say, nature is seen as a means to mankind's ends. Humanity is responsible for nature because humans, as natural beings, are dependent on certain ecosystem services, the functioning of natural cycles, and growth processes.

These constitutive elements are made more specific by deriving sustainability principles related to three general goals of sustainable development: securing human existence, maintaining society's productive potential, and preserving society's options for development and action. The principles claim to identify minimum conditions for sustainable development that ought to be assured for all people living in present and future generations (see Table 1). They unfold, on the one hand, the normative aspects of sustainability as goal orientation for future development and as guidelines for action; on the other, they provide criteria to assess the sustainability performance of particular societal sectors, spatial entities, technologies, policies, and so on.

In applying this general approach to the field of geotechnologies, georesources, and geosciences, we have to determine which principles of sustainability are, in particular, relevant to geoethics. While the answer might depend on the specific case under consideration, the following principles can *prima facie* be considered relevant to geoethics in a general sense: protection of human health, securing the satisfaction of basic needs, sustainable use of renewable resources, sustainable use of nonrenewable resources, sustainable use of the

TABLE 1 The Three Superordinate Goals and Their Respective Minimum Requirements (Principles of sustainability)

Goals	Securing Mankind's Existence	Upholding Society's Productive Potential	Keeping Options for Development and Action Open
Rules	1. Protection of human health	1. Sustainable use of renewable resources	1. Equal access to education, information, and occupation
	2. Securing the satisfaction of basic needs	2. Sustainable use of nonrenewable resources	2. Participation in societal decision-making processes
	3. Autonomous self-support	3. Sustainable use of the environment as a sink	3. Conservation of the cultural heritage and of cultural diversity
	4. Just distribution of chances for using natural resources	4. Avoidance of unacceptable technical risks	4. Conservation of nature's cultural functions
	5. Compensation of extreme differences in income and wealth	5. Sustainable development of real, human, and knowledge capital	5. Conservation of "social resources"

Kopfmüller et al. (2001) in the reception of Grunwald and Rösch (2011).

environment as a sink, avoidance of unacceptable technical risks, participation in societal decision-making processes, and equal opportunities.[2]

SUSTAINABILITY PRINCIPLES RELEVANT TO GEOETHICS

Protection of Human Health

Dangers and intolerable risks for human health have to be avoided. Production, use, and disposal of technology often have impacts, which might negatively affect human health both in the short and long term. On the one hand, this includes accident hazards in mining and other geotechnical activities (work

---

2. The presentation again follows Grunwald and Rösch (2011) based on Kopfmüller et al. (2001), complemented by giving examples specific to geoethics.

accidents). On the other, there are also "creeping" technology impacts, which can cause harmful medium- or long-term effects by emissions into environmental media and the underground. Among such creeping hazards are, e.g., gradual contamination of groundwater and underground storage of CO_2 without any guarantee of the closeness and tightness of the storage media. This principle of sustainability also concerns the question of the degree to which it is ethically tolerable to establish human settlements or industrial plants in regions potentially affected by geological hazards. Risks may, in particular, arise from earthquakes, tsunamis, volcanic eruptions, landslides, floods, and avalanches. However, the elements causing the recurrent disasters are not only the natural phenomena—as often communicated by politicians and the media—but also preceding decisions of licensing authorities or community leaders have had a place in the chain of causes and consequences. (For example, in the case of the Fukushima nuclear plant, a wall only 6 m high was built to protect the plant, although there are 22 Japanese tsunamis on record which topped 10 m (ITDB/WRL, 2005).) Geoethics also raises the question of what geological conditions would be required in order to responsibly allow human presence in the form of settlements and industrial plants. In the light of the often inevitable uncertainty of knowledge about geological hazards, the precautionary principle also needs to be considered in this context (Harremoes et al., 2002).

Securing the Satisfaction of Basic Needs

A minimum of basic services (accommodation, nutrition, clothing, health) and the protection against central risks of life (illness, disability) have to be secured for all members of society. Technology plays an outstanding role in securing the satisfaction of basic human needs through the economic system; resources supply is essential for this. This applies directly to the production, distribution, and operation of goods to satisfy the needs (e.g., technical infrastructure for the supply of water, energy, mobility, and information; waste and sewage disposal; building a house; household appliances). Geotechnical interventions such as mining have an ambivalent effect here: On the one hand, they serve the local population as workplace and opportunity to earn a living; on the other, many developing countries have extremely poor working conditions and low pay. There is often a problem of environmental justice, when the benefits from the mining activities mainly flow into the value-added and living standards in industrialized countries, while the local population receives only a fraction, but has to take the risk.

Sustainable Use of Renewable Resources

The usage rate of renewable resources must neither exceed their replenishment rate nor endanger the efficiency and reliability of the respective ecosystem. Renewable natural resources are renewable energies (wind, water, biomass, geothermal energy, solar energy), groundwater, biomaterials for industrial use (e.g., wood for building houses), and wildlife or fish stock. From the beginning, the concept of sustainability has been closely associated with sustainable use of renewable resources in forestry and fishery. It contains two statements. On the one hand, it is essential that resources are extracted in a gentle way to protect the inventory. Human usage shall not consume more than can be replenished. On the other hand, it has to be ensured that the respective ecosystems are not over-strained, e.g., by emissions or serious imbalances. Here technology plays an important role in using the extracted resources as efficiently as possible and in minimizing problematic emissions. In geoethics, the use of geothermal energy is particularly relevant in this context. On the one hand, the local usage rate must not be too high. On the other hand, no ecologically problematic technologies may be applied for the usage.

Sustainable Use of Nonrenewable Resources

The reserves of proven nonrenewable resources have to be preserved over time. The consumption of nonrenewable resources like fossil energy carriers or certain materials such as scarce metals calls for a particularly close link to technology and technological progress. The consumption of nonrenewable resources may only be called sustainable if the temporal supply of the resource does not decline in the future. This is only possible if technological progress allows for such a significant increase in efficiency of the consumption in the future that the reduction of the reserves imminent in the consumption does not have negative effects on the temporal supply of the remaining resources. So a *minimum speed* of technological progress is supposed. The principle of reserves directly ties in with efficiency strategies of sustainability; it can be really seen as a *commitment* to increase efficiency by technological progress and respective societal concepts of use for the consumption of nonrenewable resources. Regarding the material resources, the ideal of recycling management includes the idea of recycling the used materials to the largest extent and in the best quality possible; thus, the available resources would hardly decline in amount and quality. However, this ideal reaches its limits, since recycling normally includes high energy consumption and material degeneration.

Sustainable Use of the Environment as a Sink

The release of substances must not exceed the absorption capacity of the environmental media and ecosystems. Extraction of natural resources, processing of materials, energy consumption, transports, production processes, manifold forms of use of technology, operation of technical plants, and disposal processes produce an enormous amount of material emissions, which are then released into the environmental media water, air, and soil. These processes often cause serious local and regional problems, especially concerning the quality of air, ecosystems, biodiversity, and freshwater. Technology plays a major role in all strategies for solving these problems. On the one hand, as an "end-of-pipe" technology, it can reduce the emissions at the end of technical processes, e.g., in form of carbon capture and storage. On the other hand, this is the more innovative approach; technical processes can be designed in a way that unwanted emissions do not occur at all. The underground, in addition, is subject to contamination by wastes on a permanent basis. This includes radioactive wastes (Hocke, 2015) on the one hand, and other hazardous wastes, mainly toxic chemicals and heavy metals, on the other. This "final disposal" poses particular challenges in terms of geological and geotechnical care, but also requires adequate monitoring and control strategies in order to prevent deficits and negligence, with potentially serious consequences of the release of hazardous material into the groundwater.

Avoidance of Unacceptable Technical Risks

Technical risks with potentially disastrous impacts for human beings and the environment have to be avoided. This principle is necessary since the way to handle disastrous technical risks is insufficiently described in the three "ecological management principles" mentioned above. These management principles refer to "failure-free normal operation" and disregard possible incidents and accidents as well as unintended "spontaneous side effects." They are rather intended for long-term and "creeping" processes such as the gradual depletion of natural resources or the gradual "poisoning" of environmental compartments. The risk principle refers to three different categories of technical risks: (1) risks with comparatively high-occurrence probability where the extent of the potential damage is locally or regionally limited, (2) risks with a low probability of occurrence but a high-risk potential for human beings and the environment, (3) risks that are fraught with high uncertainty since neither the possibility of occurrence nor the extent of the damage can currently be sufficiently and adequately estimated. This principle is closely linked to the precautionary principle (Harremoes et al., 2002). It could be applied to the problems discussed in the context of a severe nuclear reactor accident (worst-case scenario such as happened in 2011 at Fukushima, Japan, due to a tsunami), for securing the long-term safety of final repositories for highly toxic waste, or possible risks of massive interventions into the surface of planet Earth such as deviating rivers or building huge dams (e.g., the Three Gorges Dam in China).

Conservation of Nature's Cultural Functions

Cultural and natural landscapes or parts of landscapes of particular characteristic and beauty have to be conserved. A concept of sustainability only geared toward the significance of natural resource economics would ignore additional aspects of a "life-enriching significance" of nature. The normative postulate to guarantee similar possibilities of need satisfaction to future generations like the ones we enjoy today can therefore not only be restricted to the direct use of nature as supplier of raw materials and sink for harmful substances, but has to include nature as subject of sensual, contemplative, spiritual, religious, and aesthetic experience. Insofar as geotechnical activities affect landscapes of cultural, aesthetic, or religious significance, this is also an issue of sustainability. The example of Yucca Mountain in the United States, a sacred site for the indigenous population, which was temporarily intended for the final disposal of radioactive waste, points to the heart of the conflict. But also the opposition in many alpine regions to new hydropower stations and the associated flooding of entire alpine valleys has cultural and aesthetic backgrounds beyond economic concerns about tourism.

Participation in Societal Decision-making Processes

All members of society must have the opportunity to participate in societally relevant decision-making processes. This principle aims at the conservation, extension, and improvement of democratic forms of decision-making and conflict resolution, especially regarding those decisions that are of key importance for the future development and shaping of the (global) society. Future energy supply and questions of risk acceptance and acceptability in case of geotechnical interventions into landscapes are examples of technological developments with considerable relevance for sustainability, which should be—according to this principle—dealt with through participative methods. However, in many countries, this principle has not been observed yet.

Equal Opportunities

All members of society must have equal opportunities regarding access to education, information, occupation, office, as well as social, political, and economic positions. The free access to these goods is seen as prerequisite for all members of society to have the same opportunities to realize their own talents and plans for life. This principle primarily relates to questions of societal organization, while geotechnologies and the use of georesources only play a minor role. However, according to the imperative of environmental justice (Anand, 2004), the principle of equal opportunities is violated in many fields of mining in developing countries (see the principle on "securing the satisfaction of basic needs" above). The opportunities of using georesources are distributed very unequally, both among developed and developing countries and within countries. Those who are privileged in the use of environmental goods usually contribute much more to negative environmental impacts—which in turn affect geographically and socially different people. The call for more environmental justice is related to reducing this inequality both in the opportunities to access and use natural resources and with regard to environmental impacts.

These sustainability principles cannot be directly transferred into guidelines or prescriptions for the design of geotechnologies or the use of georesources, because the respective *context* has to be taken into consideration: Which are the problems relevant for sustainability in the respective field, which technological and societal conditions apply, how are they connected, and how does the whole (and often quite complex) structure relate to the approach of the whole system of sustainability principles? So the sustainability principles have by no means a simply prescriptive character for the use of georesources. A number of steps of interpretation and transfer to the respective context have to be done on the way from ethical orientation to concrete design of geotechnologies and interventions into the geosphere.

The limits of the carrying capacity of the natural environment play a key role here. The aim is to maintain the characteristics of an ecosystem—to tolerate a certain extent of (anthropogenic) pollution without endangering the fulfillment of the ecological functions or the functions for human use. Relevant key concepts in this context are, e.g., ecological stability, ecological footprint, and carrying capacity of the environment. The term "vulnerability" describes the susceptibility of ecosystems to external disturbances, and "resilience" refers to their self-healing powers and regenerative capacity. With regard to human interventions in the natural environment, further principles of fault tolerance and reversibility were shaped that are included in the principles of sustainability mentioned above.

SUSTAINABILITY ASSESSMENTS FOR GEOTECHNOLOGIES AND GEOSCIENCES

Technology is of major importance for sustainable development. On the one hand, technology determines to a large extent the demand for raw materials and energy, needs for transport and infrastructure, material flows, emissions, as well as amount and composition of waste. Technology is, on the other hand, also a key factor of the innovation system and influences prosperity, consumption patterns, lifestyles, social relations, and cultural developments. The development, production, use, and disposal of technical products and systems have impacts on many principles of sustainable development. Therefore, a sustainability assessment of the development, use, and disposal of technologies is required as an element of comprehensive sustainability strategies.

In order to be able to contribute to a (more) sustainable usage of the underground, it is thus required to perform sustainability assessments in advance to geotechnical interventions. Regarding the uncertainty of knowledge about future consequences of today's interventions and the unavoidable conflicts of objectives between the different criteria of sustainability, the need for a methodologically secured approach of sustainability assessment is obvious. Technology assessment has developed a number of concepts and methods in this context (Grunwald, 2009). Crucial for *prospective sustainability assessment* of technologies is a complete coverage of the sustainability effects in the phases of *development* and *production, diffusion, use,* and *disposal.* Technical products and systems accumulate positive and negative effects on sustainability over their entire life cycle, ranging from the primary raw material deposits, transport, processing, and use up to the disposal. Sustainability assessment of technologies therefore must take into account of the entire *life cycle* of a technology.

Such sustainability assessment covers not only environmental effects (which are considered in the classical life cycle assessment (LCA) and ecobalancing) but also social problems (child labor, inhumane working conditions, etc.) and economic aspects of raw material extraction, manufacture, and disposal which can be accounted for in the life cycle costing. New methods such as social LCA are developed (Schepelmann et al., 2009) so as to be better able to balance also social aspects in a life cycle approach. In all these procedures, close cooperation of geosciences, technology assessment, social sciences, and ethics is required.

Sustainability considerations and assessments are a central part of any geoethics that includes use conflicts, responsibility issues, risk assessment, and risk management, as well as criteria and procedures for dealing with these challenges.

REFERENCES

Anand, R., 2004. International Environmental Justice: A North–South Dimension. Ashgate, Hampshire, UK.

Brown-Weiss, E., 1989. In Fairness to Future Generations: International Law, Common Patrimony and Intergenerational Equity. New York.

Grunwald, A., 2007. Working towards sustainable development in the face of uncertainty and incomplete knowledge. J. Environ. Policy Plann. 9 (3), 245–262.

Grunwald, A., 2009. Technology assessment: concepts and methods. In: Meijers, A. (Ed.), Philosophy of Technology and Engineering Sciences, vol. 9. Amsterdam, pp. 1103–1146.

Grunwald, A., Kopfmüller, J., 2012. Nachhaltigkeit, Frankfurt am Main, second ed.

Grunwald, A., Rösch, C., 2011. Sustainability assessment of energy technologies: towards an integrative framework. Energy Sustainability Soc. 1 (3) http://dx.doi.org/10.1186/2192-0567-1-3. open access.

Harremoes, P., Gee, D., MacGarvin, M., Stirling, A., Keys, J., Wynne, B., Guedes Vaz, S. (Eds.), 2002. The Precautionary Principle in the 20th Century. Late Lessons from Early Warnings. London.

Hocke, P., 2015. Nuclear waste repositories and ethical challenges. In: Wyss, M., Peppoloni, S. (Eds.), Geoethics: Ethical Challenges and Case Studies in Earth Science. Elsevier, Waltham, Massachusetts, pp. 359–367.

ITDB/WRL, 2005. Integrated Tsunami Database for the World Ocean, TSLI SD. Russian Academy of Sciences, Novosibirsk.

Jonas, H., 1984. The Imperative of Responsibility. Chicago [Original edition: Das Prinzip Verantwortung, Frankfurt am Main, 1979].

Kopfmüller, J., Brandl, V., Jörissen, J., Paetau, M., Banse, G., Coenen, R., Grunwald, A., 2001. Nachhaltige Entwicklung Integrativ Betrachtet. Konstitutive Elemente, Regeln, Indikatoren, Berlin.

Potthast, T., 2013. Geo- und Hydrotechnik inkl. Bergbau. In: Grunwald, A. (Ed.), Handbuch Technikethik. Stuttgart/Weimar.

Schepelmann, P., Ritthoff, M., Jeswani, H., Azapagic, A., Suomalainen, K., 2009. Options for deepening and broadening LCA. CALCAS - Co-ordination Action for innovation in Life-Cycle Analysis for Sustainability.

WCED – World Commission on Environment and Development, 1987. Our Common Future. Oxford.

Chapter 4

Ethics of Disaster Research

Ilan Kelman[1,2]

[1]Institute for Risk & Disaster Reduction and Institute for Global Health, University College London, London, United Kingdom; [2]Norwegian Institute of International Affairs (NUPI), Oslo, Norway

Chapter Outline

Abstract

Disaster research explores how to reduce vulnerability and disaster impacts (predisaster activities) alongside response to and recovery from disasters (postdisaster activities). Because vulnerability and disasters directly affect people and communities, disaster research is fraught with difficulties and frequently makes the difference between life and death. This chapter explores the ethics of disaster research by examining how that research might interfere with disaster-related activities and how to deal with the outcomes from disaster research. No clear-cut solutions are available. Instead, it is important to continue disaster research while being aware of the implications so that detrimental effects could be avoided while augmenting the positive consequences.

Keywords: Disasters; Humanitarian relief; Risk; Vulnerability.

OUTLINE OF THE PROBLEM

Science is often seen as being objective and neutral. Martin (1979), using pollution and resource studies, demonstrates how observed quantitative data are still subject to bias in analysis and interpretation. With qualitative data, bias is inherent and needs to be accepted as part of the scientific process. In much research, which directly affects people, the research subjects themselves can become involved in the scientific process; for instance, through "people's science" (Wisner et al., 1977) or "usable science" (Glantz et al., 1990). These methods demonstrate that science is not an objective, linear process in which

Geoethics. http://dx.doi.org/10.1016/B978-0-12-799935-7.00004-6

untainted knowledge is produced to be presented unambiguously to policy- and decision-makers. Instead, each step—including exchanges with users about problem definitions and communication techniques—is imbued with human values and judgments, leading to ethics being an integral part of the entire, interconnected science and application processes.

Some scientists resolve such ethical challenges by believing that science and scientists are and should remain above policy and political implications. Others see that viewpoint as naïve, since any communication and any interaction among people—by definition—involves politics. Still other scientists see themselves as having a duty to become involved in the societal implications of their science, going well beyond publishing in peer-reviewed, academic venues.

These ethical discussions and different perspectives appear to be particularly acute in research that has a direct and tangible impact on society. One example is disaster research, where the research results, the involvement of users, and how the results are communicated can have life and death implications—immediately and long into the future. Disaster research covers dealing with disasters, before and after they manifest. Beforehand, disaster-related activities (of which research is one) include reducing vulnerability to make people and communities safer, understanding hazards to be able to predict and analyze their behavior, and combining hazard and vulnerability to indicate disaster risk and how it might be reduced and managed. These processes are often labeled mitigation, planning, preparedness, and risk reduction. During and afterward, disaster-related activities (in which research is still important) involve emergency management, response, recovery, and reconstruction.

This chapter explores the ethics of disaster research, with disaster researchers including social scientists, physical scientists, humanists, artists, and professionals such as engineers, lawyers, social workers, doctors, and nurses. Disaster research can be disciplinary (e.g., geology and sociology), can join disciplines (e.g., geology, medicine, and engineering combining to examine how to survive inside a building during a volcanic eruption), and might be focused on solving a problem without adopting any specific discipline (e.g., how can a community improve its own disaster-related activities rather than relying on external support?). Much disaster research requires input from those who might not have extensive scientific training—although many nonetheless do—such as masons, firefighters, teachers, fishers, politicians, law enforcement officers, farmers, and plumbers.

The ethical challenges and opportunities emerging from disaster research are not resolved in this chapter. Instead, this contribution explores key questions through case-based vignettes in order to formulate some ethical questions that readers ought to be aware of in their own work. As with most ethical discussions, clear-cut or single answers are not always available. Instead, asking the right questions in the right way is as best as could be hoped for in order to acknowledge and tackle the ethical questions which arise.

For disaster research ethics, four principal questions emerging from previous studies (e.g., Kelman, 2005; Kilpatrick, 2004) are summarized here as:

1. Could carrying out disaster research interfere with disaster-related activities?
2. Could publishing disaster research interfere with disaster-related activities?
3. Should researchers take responsibility for outcomes from their disaster research?
4. Should researchers aim to create and pursue outcomes from their disaster research?

The questions merge into two clusters. First, interference with the disaster-related activities is addressed in the Section on Interference with Disaster-Related Activities. Second, disaster research outcomes are discussed in the Section on Disaster Research Outcomes. Both sections use brief case histories. Then, the Sections on Discussion and Conclusion are provided in order to propose ways forward for the practical challenge of continuing disaster research despite—or perhaps because of—the ethical issues raised.

INTERFERENCE WITH DISASTER-RELATED ACTIVITIES

Could carrying out disaster research interfere with disaster-related activities? The nature of research is to observe and to record those observations systematically for verifiable analysis and interpretation. In the midst of a disaster situation, or recovery from disaster, it is often appropriate for all available personnel to assist with solving immediate needs. Providing first aid such as staunching bleeding or using a defibrillator can save lives, as can removing rubble and setting up emergency shelters in safe locations. Anyone standing nearby in order to watch and record is one less person able to assist—and could be in the way of those trying to help.

That does not preclude researching these topics (case history 1). It is possible to be operational when needed and then, later, to use the experience, after-the-fact interviews, post hoc analysis, and field reports for scientific analysis and publication. That, in turn, has the potential for interfering with

---

Case History 1: Research on Postdisaster Shelter and Settlement

Postdisaster siting, construction, and maintenance of shelters and settlements have long been a mainstay of humanitarian relief operations. Many research questions have been investigated, from settlement layout in different contexts to postdisaster shelters for cold weather environments. For the latter, for people to survive in tents in winter conditions, the interior needs to be kept warm through sealing and heating while not suffocating the people due to too much sealing. To measure variables such as humidity, temperature, and oxygen content over time, data loggers can be set up in the tents—preferably in occupied tents to account for people's body heat and breathing.

Case History 1: Research on Postdisaster Shelter and Settlement–cont'd

The challenge is clear: Could this research interfere with the occupants' survival? When they are having trouble finding enough food each day and when they are living in a cramped, dilapidated space, will the occupants consent to researchers setting up data loggers and then coming back to collect the loggers—without providing the occupants with anything extra, such as money, which might influence their behavior and hence the data? If different types of tents are being compared, who chooses which family gets which tent? These studies help to save lives and to make the misery of postdisaster tent accommodation more survivable (Manfield et al., 2004), but the studies can interfere with helping people.

disaster-related activities. The operational teams, including decision-makers, should be informed in advance—and give their consent—that the procedures and discussions will be clinically dissected afterward. That could make them feel under pressure, thereby influencing their decisions. On the other hand, that might be positive since these decision-makers often end up in court or splashed across the media with forensic analysis of the reasons for their decisions. Knowing that scrutiny will happen might improve disaster-related decision-making while permitting more accountability (see also the Section on Discussion).

For instance, deliberating over hazard or vulnerability maps has political implications, such as for property values or allocating resources for further development and societal services. Zone delineation might be made to be as politically expedient as possible rather than relying on scientific accuracy. That could mean erring on the side of caution, to keep people out of danger zones. Alternatively, that might mean reducing the size of danger zones or shifting boundaries in order to keep as many people as happy as possible.

The second question in this category is: "Could publishing disaster research interfere with disaster-related activities?" Disasters are a newsworthy topic, frequently attracting significant attention, which can put researchers and the research process in the public's eye and the political spotlight. That attention can interfere with disaster-related activities and put others in danger (case history 2).

Tsunami and volcano evacuation plans alongside building stability in earthquake zones represent disaster research, which frequently gets attention. Publishing that work could result in political problems, if inadequacies are deemed to be found without plans to resolve them. Decision-makers might dispute the scientific findings, leading to conflict. Yet not publishing—or not publicizing—investigations that claim to identify unsafe situations could be unethical.

Case History 2: Drawing Attention to Natural Hazards

As part of "dark tourism" (tourism to sites involving death and dying), disaster tourism has been increasing in popularity (Kelman and Dodds, 2009). In addition to disaster memorials and sites of previous disasters, disaster tourism can involve visiting locations of ongoing danger such as war zones, rumbling volcanoes, and communities cleaning up after a flood. These locations are also suitable for disaster research. Scientists conducting disaster research might themselves attract tourists, putting people in danger.

In 1993, a volcanology conference was held in Pasto, Colombia, at the foot of the active volcano Galeras. A group of scientists went on a field trip, climbing up the volcano's slopes and entering the crater, despite some eruption warning signs and some scientists feeling that the trip had little scientific value. The volcano erupted while the scientists were inside the crater, killing six of them. Journalists and tourists had joined the day trip, with three tourists adding to the fatalities (see Bruce (2001)). The research on the volcano put others in danger, who had perhaps been lulled into a false sense of security by the presence of the authoritative scientists. Similarly, in 1991, during the eruptions of Mount Unzen in Japan, three volcanologists were killed along with 40 others, including journalists who had joined the volcanologists to watch the eruptions.

On September 4, 2010, an earthquake rattled Christchurch, New Zealand, followed by two aftershocks, one on December 26, 2010 and the other on February 22, 2011. The latter killed 185 people, 115 of them in the collapse of the Canterbury Television building. The subsequent inquiry found flaws in the design and approval procedure for the building's construction (Hyland and Smith, 2012). The report also described damage to the building after the September 2010 shaking, but suggested that the building had not been significantly weakened prior to February 22, 2011. Should tenants of the building have been forced to read about the damage and engineering judgments of it, so that they could make their own decision about working there? What if some employees decide that the building is safe and some decide otherwise? Publishing and publicizing any disaster research regarding the Canterbury Television building after September 4, 2010 could lead to conflict, especially since no concerns were verified by Hyland and Smith (2012). Yet not publishing such disaster research removes the choice from the people directly affected.

DISASTER RESEARCH OUTCOMES

Disaster research can interfere with disaster-related activities, as shown in the Section on Interference with Disaster-Related Activities, but it does not necessarily need to have a big influence. Instead, the research and the research publications might have limited impact on operational disaster-related activities,

Case History 3: Natural Hazards as Weapons of War

Numerous studies on weather modification (e.g., Taubenfeld, 1967) and inducing earthquakes (e.g., Ellsworth, 2013) exist. During World War II, efforts were made to use floods and tsunamis as weapons of war. In the Middle Ages, besieging armies catapulted bodies of plague victims into the surrounded communities. The success and controllability of any of the efforts is highly debatable, but publishing the analyses of the case studies could include descriptions of limitations and errors that others might learn from and improve. Is it ethical to provide solid scientific material in the public domain, which might be used to improve one's warfare capability using natural hazards?

but nevertheless affect other policies and actions, with specific outcomes linked to the research. Should researchers take responsibility for outcomes from their disaster research?

Much research exists on the human ability and choice to create, trigger, and modify natural hazards (case history 3). Ideas are easy to think of and publish, often being used in popular entertainment, irrespective of scientific accuracy or feasibility. For the research discussed in case history 3, if terrorists, governments, or the private sector pursue and succeed at creating hazards, do the scientific authors bear any responsibility for the outcomes of those hazards? After World War II, the U.S. Army and General Electric Corporation worked together on hurricane seeding, aiming to reduce storm intensity. In 1947, a seeded hurricane switched direction, crossed the shoreline into South Carolina and Georgia, and caused damage (Kwa, 2001). Blame was placed on the seeding, although it would be impossible to prove or disprove that any specific action led to the hurricane's change in direction. Do scientists who theorized hurricane modification and cloud seeding bear any responsibility? Does the responsibility rest with only those who tried to apply the knowledge?

Many scientists actively seek outcomes from their work and are supported for that by research grants. Meteorologists can have operational forecasting jobs. Much (although not all) of the volcanological research around Mount Pinatubo in the Philippines and Soufrière Hills in Montserrat was conducted after the volcanoes started rumbling, so the scientists were sent to those locations to advise on hazards, vulnerability, risk, disaster risk reduction, and disaster response (Druitt and Kokelaar, 2002; Newhall and Punongbayan, 1996).

Case History 4

Disaster diplomacy research (http://www.disasterdiplomacy.org) investigates how and why disaster-related activities do and do not influence conflict and cooperation. All case studies examined so far show that disaster-related activities have the potential for catalyzing cooperation in the short term, if a previous basis exists for

Case History 4–cont'd

that cooperation, but over the long term, other factors in war and peace dominate disaster-related factors. After an earthquake and tsunami in 2004 devastated Aceh, Indonesia, in the midst of a long-running and brutal conflict, a peace deal was reached, but the negotiations had started 2 days before the tsunami. The disaster catalyzed the peace process and supported its success, but did not create the peace.

In Sri Lanka, a conflict situation similar to Aceh persisted before the tsunami, but power brokers on all sides of the conflict had reasons for perpetuating it after the tsunami. Meanwhile, limited prior conditions inhibited the tsunami disaster ending the conflict disaster. In the end, humanitarian aid became a political flash point. After several more years of violence, the Sri Lankan Government won a military victory over its adversaries.

A disaster researcher who is asked for postcatastrophe policy advice on disaster diplomacy ends up in an ethical dilemma. Being honest would entail describing how and why disaster diplomacy fails overtly in creating peace, but might have more subtle successes in catalyzing peace, as occurred in Aceh. If decision-makers do not want peace, as in Sri Lanka, then that advice would reveal to the leader how to avoid cooperation resulting from humanitarianism. If the disaster diplomacy researcher desires both long-term peace and adequate disaster response, then being honest about scientific findings might not be feasible.

Should researchers aim to create and pursue operational outcomes from their disaster research? While it might seem shocking to even ask this question, many scientists involved in researching the hazard driver of climate change exacerbated by human activities wish to focus on only their science, without considering how their science might be used. Other climate change scientists feel the opposite. Jim Hansen, Roger Pielke, Jr., and the late Stephen Schneider have made awareness about and action on climate change as part of their personal mission, leading to bitter debates and personal acrimony when opinions differ. The role of scientists in pursuing political and operational outcomes from their original science is discussed extensively in the scientific literature (e.g., Oppenheimer, 2011; see also Martin, 1979), at times questioning how to draw the line between—and if a line should be drawn between—generating new knowledge and acting on that knowledge.

Disaster researchers are frequently involved in developing emergency management plans, recommending measures for reducing vulnerability, and modifying curricula for disaster-related education. As with climate change, the policies, actions, and values to promote on the basis of that science might not be straightforward (case history 4).

DISCUSSION

The case histories promote the questions on the ethics of disaster research, but solutions are not immediately forthcoming, partly because ways forward are contextual. Overall principles nonetheless appear from disaster ethics work, good governance in development (as well as critiques such as Sundaram et al. (2012)), and the case histories: accountability, transparency, and participation.

Accountability refers to an obligation to justify one's actions, to report or answer to others, and to be evaluated on one's choices. For example, for disaster research ethics, if a scientist or engineer gives advice on which earthquake resistance retrofitting techniques to implement, they could later potentially be held accountable if the techniques were implemented properly but the building collapsed in an earthquake.

Accountability exists for disaster research to a large degree. For instance, human subjects research has stringent rules in many countries. The Norwegian Social Science Data Services is responsible for vetting proposed procedures for projects involving human interviews or other personal data. The United States has Institutional Review Boards, which evaluate proposed protocols for human subjects research. Although these monitoring and accountability mechanisms exist with the best of intentions, they do not always fully achieve their goal. Bradley (2007) points out that insisting on anonymity might not permit a group to represent themselves or their community in the manner of their own choosing, even if dangers result from interviewees identifying themselves. Who decides how to "protect" people from themselves?

Furthermore, positive or adverse societal impacts from disaster research can occur from political processes, land use designation, policy implementation, and praxis. Should all disaster research, not just those involving interviews or other direct involvement with human subjects, be fully peer-reviewed before it is enacted? Kelman (2003, p. 121) asks "Should 'political testing'—real-time research on tinkering with human societies—undergo as stringent requirements as animal testing...or experimentation on human subjects?"

After all, accountability processes could themselves cause the harm that they seek to avoid. Citizens might demand that scientists become more involved or less involved in practical applications of research, depending on the citizens and their interests. The line between research and outcomes becomes blurred when the research process itself has clear societal outcomes. Who has the right and who has the authority to decide who is accountable to whom, how, and when?

The second overall principle is transparency, defined by the nonpartisan NGO Transparency International as "shedding light on rules, plans, processes and actions. It is knowing why, how, what, and how much" (http://www.transparency.org/whoweare/organisation/faqs_on_corruption/2/#transparency). That is, transparency aims to ensure that processes, decisions, and data are open and available to those seeking more information. For disaster research ethics, that might be putting all raw data, processing algorithms, calculations, raw results, and polished material online for climate change models. Being transparent is one way of being held accountable.

As with accountability, transparency could cause harm, as discussed above with respect to making public any flaws in evacuation plans or buildings' structures. It is also questionable whether nonscientists have the skills needed to analyze, interpret, and act on raw data and processing algorithms—although ignorance of the public might not be a suitable excuse for denying transparency to them. Yet sometimes people want to be in ignorance, as Myers (1995) differentiates between ignorance (not knowing) and ignore-ance (not wanting to know).

That raises another ethical concern: Is it enough to make disaster-related information available? If scientists analyze an emergency response plan, find flaws, and publish those findings in a scientific journal—perhaps even a popular science blog—have they done everything that they can and should? Do they have a duty to ensure that the information is acted on? Do they have a duty to avoid dramatic gestures that might lead to knee-jerk reactions? Do scientists have the skills and means to fulfill either of these duties, if they are accepted as duties?

Can transparency be implemented through codes of ethics, minimum standards, guidelines, and behavioral charters? Existing examples are "Guidelines for Professional Interaction During Volcanic Crises" from the International Association of Volcanology and Chemistry of the Earth's Interior; http://iavcei. org and the Committee on Publication Ethics (http://publicationethics.org). Meanwhile, identifying corruption in disaster-related activities has increased in importance (e.g., Lewis, 2008) recognizing, for instance, that the skills to engineer a building to withstand an earthquake exist, but those skills might not be implemented due to incentives for providing less-than-adequate work (e.g. Bilham, 2015).

The last of the three overall principles is participation, referring to full and fair involvement of research subjects in the entire research process. This principle was foreshadowed in the Section on Outline of the Problem, recognizing the importance of accepting science's subjectivity and the need to involve affected people in the disaster research process from the beginning (e.g. Bilham, 2015). This approach is often termed "participatory action research," entailing full collaboration with the people and communities being studied, so that they become research participants as well as research subjects, inspiring themselves to deal with disaster-related topics (e.g., Chambers, 1994; Cornwall and Jewkes, 1995).

The phrase's words describe how (1) those affected by the disaster research are fully **participatory** in the research, (2) **action** is needed, which can emerge from the research, to solve the disaster-related problems, and (3) the work is still **research**, so original scientific work is pursued, generating new knowledge and analysis for humanity. As with accountability and transparency, participatory action research can itself cause the problems which it seeks to avoid, by forcing participation on people (e.g., "The New Tyranny" from

Cooke and Kothari (2001)) or by perpetuating power relations and exclusion when only certain sectors in the community are permitted to participate fully.

The key is not to avoid accountability, transparency, and participation simply because they might cause difficulties. Instead, any mechanism or process requires monitoring and enforcement. It is not immediately clear who should judge or control—nor what checks and balances could exist with respect to the judge or controller. Researchers should be involved, but they cannot be the only ones with power. Nonresearchers, including community members, should be involved, but they cannot be the only ones with power. It is a balance with power distributed across all those involved.

At minimum, these questions should be asked and debated, even if the ultimate decision is that the principles cannot be made fully operational. In fact, the three principles tend to have more clarity in their definitions than in their on-the-ground usefulness for disaster research. Researchers should understand what is and is not acceptable by thinking ahead to set appropriate limits in their work—from beginning to end. During a disaster and during research, it is too late. Before starting work, researchers should be aware of their potential influence and potential responsibilities by recognizing that for disaster research, the observer is so rarely entirely divorced from the observed—as has long been known by, for example, anthropologists encompassing those who study disasters and disaster risk reduction. In most circumstances, a disaster researcher is almost inevitably a participant, even if an inadvertent one, such as for the volcano Galeras.

Because science is about thinking and most research requires planning, disaster researchers are in a strong position to recognize, anticipate, and avoid mistakes related to the ethics of their work. That suggests a responsibility to do so.

CONCLUSIONS

Disaster research is fraught with ethical difficulties. These difficulties often emerge because people's lives—including the researchers' lives—can be threatened or saved by the research's methods, results, and communication. With such difficulties, come benefits. Since the ethics of disaster research appear so starkly, with researchers being held accountable at times, the outcomes from exploring disaster research ethics can easily become engrained in the consciousness of researchers and their collaborators. If others pay attention to that, and learn from the mistakes and good practice, then spillover is feasible into other areas of geoethics.

As a consequence of the ethical discussions, disaster research should be neither stopped nor frightening. Instead, all those involved should remain aware of the challenges and opportunities in order to overcome the challenges and to make full use of the opportunities. Disaster research has and will continue to save lives and to contribute to the society. That should be permitted and encouraged while recognizing and solving the ethical challenges which surface.

REFERENCES

Bilham, R., 2015. Mmax: ethics of the maximum credible earthquake. In: Wyss, M., Peppoloni, S. (Eds.), Geoethics: Ethical Challenges and Case Studies in Earth Science. Elsevier, Waltham, Massachusetts, pp. 119–140.

Bradley, M., 2007. Silenced for their own protection: how the IRB marginalizes those it feigns to protect. ACME 6 (3), 339–349.

Bruce, V., 2001. No Apparent Danger: The True Story of Volcanic Disaster at Galeras and Nevado Del Ruiz. HarperCollins, New York.

Chambers, R., 1994. The origins and practice of participatory rural appraisal. World Dev. 22 (7), 953–969.

Cooke, B., Kothari, U. (Eds.), 2001. Participation: The New Tyranny? Zed Books, London.

Cornwall, A., Jewkes, R., 1995. What is participatory research? Soc. Sci. Med. 41 (12), 1667–1676.

Druitt, T.H., Kokelaar, B.P. (Eds.), 2002. The Eruption of Soufriere Hills Volcano, Montserrat, from 1995 to 1999. Geological Society, London.

Ellsworth, W.L., 2013. Injection-induced earthquakes. Science 341, 6142.

Glantz, M.H., Price, M.F., Krenz, M.E., 1990. Report of the Workshop "On Assessing Winners and Losers in the Context of Global Warming, St. Julians, Malta, 18–21 June 1990". Environmental and Social Impacts Group, National Center for Atmospheric Research, Boulder, CO.

Hyland, C., Smith, A., 2012. CTV Building Collapse Investigation. Hyland Fatigue + Earthquake Engineering, Auckland. Manukau and Structure Smith.

Kelman, I., 2003. Beyond disaster, beyond diplomacy. In: Pelling, M. (Ed.), Natural Disasters and Development in a Globalizing World. Routledge, London, pp. 110–123.

Kelman, I., 2005. Operational ethics for disaster research. Int. J. Mass Emerg. Disasters 23 (3), 141–158.

Kelman, I., Dodds, R., 2009. Developing a code of ethics for disaster tourism. Int. J. Mass Emerg. Disasters 27 (3), 272–296.

Kilpatrick, D.G., 2004. The ethics of disaster research: a special section. J. Trauma. Stress 17 (5), 361–362.

Kwa, C., 2001. The rise and fall of weather modification: changes in American attitudes toward technology, nature, and society. In: Miller, C.A., Edwards, P.N. (Eds.), Changing the Atmosphere: Expert Knowledge and Environmental Governance. Massachusetts Institute of Technology, Boston, pp. 135–165.

Lewis, J., 2008. The worm in the bud: corruption, construction and catastrophe. In: Bosher, L. (Ed.), Hazards and the Built Environment: Attaining Built-in Resilience. Routledge, London, pp. 238–263.

Manfield, P., Ashmore, J., Corsellis, T., 2004. Design of humanitarian tents for use in cold climates. Build. Res. Inf. 32 (5), 368–378.

Martin, B., 1979. The Bias of Science. Society for Social Responsibility in Science, Canberra.

Myers, N., 1995. Environmental unknowns. Science 269, 358–360.

Newhall, C., Punongbayan, R. (Eds.), 1996. Fire and Mud: Eruptions and Lahars of Mount Pinatubo, Philippines. PHIVOLCS and USGS, Seattle and Manila.

Oppenheimer, M., 2011. What roles can scientists play in public discourse? Eos, Trans. Am. Geophys. Union 92 (16), 133–134.

Sundaram, J.K., Chowdhury, A., Chowdhury, A. (Eds.), 2012. Is Good Governance Good for Development?. United Nations, Department of Economic and Social Affairs, New York.

Taubenfeld, H.J., 1967. Weather modification and control: some international legal implications. Calif. Law Rev. 55 (2), 493–506.

Wisner, B., O'Keefe, P., Westgate, K., 1977. Global systems and local disasters: the untapped power of people's science. Disasters 1 (1), 47–57.

Chapter 5

Toward an Inclusive Geoethics— Commonalities of Ethics in Technology, Science, Business, and Environment

Thomas Potthast

International Centre for Ethics in the Sciences and Humanities, University of Tübingen, Tübingen, Germany

Chapter Outline

Abstract

"Geoethics" is to become an integrative interdisciplinary endeavor as an upcoming field with strong links to existing approaches of application-oriented ethics, like ethics of technology, ethics of science, business ethics, environmental ethics, and political ethics. The approach of an "ethics *in* the sciences" is suggested to grapple with the challenge of interdisciplinary "mixed judgments" or "epistemic-moral hybrids" linking the necessary empirical as well as normative aspects. Three domains are the major building blocks for geoethics: (1) professional ethics of scientific and engineering experts; (2) ethical evaluation procedures of specific projects, mainly with regard to balancing the means and ends concerning risk-benefit considerations; and (3) general ethical considerations on the integrity of the earth and its geomorphological components. A matrix of relations and points to consider emerges, which may result in a method for practical deliberations

Geoethics. http://dx.doi.org/10.1016/B978-0-12-799935-7.00005-X
49

and judgments in geoethics. Ultimately, this shall provide the link to sustainable development, contributing to environmental justice considering the geosciences and their objects.

Keywords: Application-oriented ethics; Environmental justice; Epistemic-moral hybrids; Ethics in the sciences; Geoethics; Professional ethics; Sustainable development.

DOMAIN AND CHALLENGES OF GEOETHICS

As a central constituent of their cultural history, humans have systematically transformed the surface of the earth to safeguard their livelihood. By this token, the geomorphological modes of the human condition itself have been transformed. What began as one of the most basic forms of human action quickly developed into sophisticated technological systems for the secure foundation of buildings of every kind, for irrigation, drainage, and mining. The considerable consequences of these activities, which are visible in the life-world, are almost symbolic of the power of enlightened science and the related technological actions in general (Horkheimer and Adorno, 2002). Various forms of open pit mining have long been a focus of debate, as well as questions of the responsibility of civil engineers for the safety of constructions and of the reliability of disaster warnings, as in the case of the devastating earthquake of L'Aquila, Italy, in 2009. Recently, new technologies such as fracking and climate engineering widened the scope and the potential harms of geotechnical engineering and thus its ethical relevance. In that sense, geoethics is strongly driven by actors and processes of science and technology. It thus becomes a field of application-oriented ethics with a strong focus on technoscience, which is to be critically reflected and evaluated. However, thinking about holy places and the taboos of setting foot on or even altering, e.g., Ayers Rock in Australia or Kailash Mountain in Tibet, there are issues of geoethics that are much older than modern science and these aspects also have to be taken into account.

INTERDISCIPLINARY APPLICATION-ORIENTED ETHICS AND THE TÜBINGEN APPROACH TO ETHICS IN THE SCIENCES

Ethics always is practical, since it deals with human actions, and in that sense, moral practice always is an "applied" field. In that vein, the common notion of "applied ethics" might be misleading. Of course, there are fields of meta-ethics investigating the semantic structure of moral language that are quite far from life-world questions. But if it comes to normative rules or individual moral judgments, there is no algorithm of a theory to be applied, like one applies mathematics in setting up statics in construction. These considerations lead to the notion of "application-oriented ethics". It is not so much the ethics that is being applied, but rather, there are fields of applications of science and technology that lead to new ethical considerations. This view has both epistemological and ethical consequences: Science and Technology can no longer be seen as value-neutral providers of knowledge and certain means, leaving only the user and society to think about the moral dimension of their "application". Complex scientific and technological endeavors cannot be conceived of as kitchen knives, with the help of which you can cut onions as well as stab your friend.

The Tübingen approach of ethics *in* the sciences and humanities addresses the question of responsibility already in the sciences themselves, instead of considering it as a separate field for ethicists only. This program of ethics in the sciences entails an active dialogue between the sciences and humanities and the public with regard to ethical issues, e.g., concerning particular cases of geotechnologies. This approach is the most suitable to grapple with the challenge of interdisciplinary "mixed judgments" or "epistemic-moral hybrids" linking the necessary empirical as well as normative aspects. Instead of viewing the world of facts as completely separate from that of values, their intricate connections are important (Potthast, 2010).

CASES AND FIELDS OF GEOETHICS

Professional Ethics of Scientific and Engineering Geosciences

Professional ethical guidelines like Codes of Conduct from the field of geotechnical (and hydrotechnical, as it were) engineering emphasize general as well as specific aspects: generally, the responsibility for society's well-being, health, and safety is explicitly placed above the responsibility toward partial and private interests. This priority also limits the obligations of loyalty toward the employers. Protecting the integrity and the good reputation of the profession is emphasized. Furthermore, there is an obligation to inform stakeholders about possible technological, social, environmental, and other consequences of the engineering actions. This last aspect is specified for geotechnical engineers in the demand to consider the overall context of their projects and to supervise them from beginning to end, including the in situ inspection of the implementation of the technological design. Price reductions at the expense of safety are explicitly forbidden; the cooperation with other academic fields within the scope of projects is demanded. Another part of the professional ethics for engineers is the avoidance of unnecessarily definitive statements about geotechnical, geological, and environmental aspects in case of uncertainty (Brandl, 2004; ISSMGE, 2004).

Such guidelines can easily be related to current problems when buildings collapse due to the instability of the building ground. However, the question is often whether safety concerns are generally insufficiently considered (or ignored), with an eye to the consequences for the finances and reputation of the employer, or whether the implementation was flawed, deviated from the original plans, or was not controlled closely enough during construction. The latter was the case in the collapse of Cologne's City Archive on 3 March, 2009, which took place due to measures of subway construction (cf. Schmalenberg and Damm, 2010). At the same time, such a case shows the difficulty of the moral and not least the legal attribution of responsibility (cf. Hubig and Reidel, 2003 on engineers' professional ethics). More generally, issues of engineering ethics, economic ethics, as well as political ethics are all connected to the dimension of professional ethics of geoscientists and reach far beyond individual responsibility, which nevertheless is not to be neglected.

Ethical Evaluation Procedures for Specific Projects

Geoethics in a project-specific dimension primarily deals with the possible permissiveness or necessity of certain geo-/hydrotechnical measures for the—in a broad sense—improvement of the living conditions of humans. This usually takes place within the scope of an assessment of goal, means, and consequences: Does the goal of a project—in this case buildings or measures of geotechnical transformation—justify the applied technological and other means as well as the foreseeable consequences and side effects? The fundamental emphasis on the safety aspect is crucial, and it is closely linked to legal conditions, with e.g., German industrial standards (DIN) and European norms being developed and updated (cf. Ziegler, 2008). It is stressed that absolute safety is not possible due to the complexity of the substrates. However, at the same time, the desired degree of safety also depends on assessments of the necessity of the project, the question of financial and practical feasibility, and thus becomes part of a classical risk assessment. This once again brings together various domains of application-oriented ethics.

A hitherto unique case of the legal attribution of responsibility of geological expertise took place after the earthquake on 6 April, 2009, in L'Aquila in Central Italy. Six scientists and a representative of the public authorities were sentenced as members of a political-administrative commission of experts because they had allegedly ignored or downplayed the risk of an imminent heavy earthquake after the first light tremors (for details of the court case cf. Mucciarelli, 2015)). After the charge and especially after the verdict at first instance in October 2012, various national and international academics criticized the sentence because on that basis and under conditions of fundamental uncertainty, no expertise would be possible if absolute certainty of the prognoses were politically and legally demanded (The L'Aquila Trial, 2014).

Ultimately, the question of risk management of earthquakes points to a geotechnical–political problem: on which legal basis, with which acceptable financial effort in which region can and should buildings be constructed as earthquake-proof up to which seismic magnitude? "How safe is safe enough?" is not a question that engineers can answer on their own (cf. Brandl, 2004).

Mainly large-scale construction projects are, appropriately, not primarily perceived as conflicts of technology ethics because they raise a number of political and ethical questions concerning the shaping of the future by infrastructure in general. Nonetheless, the engineering dimension of the possibilities and limitations is decisive, as well. A current paradigm case for that is the construction project "Stuttgart 21", the relocation of Stuttgart's central train station below ground and the construction of a high-speed train line across the Swabian Alp in southwest Germany. One question concerns the appropriate cost-benefit balance with regard to the safety interpretation in the strength of the tunnel walls in the ground. Furthermore, possible damages to the mineral springs due to the measures of groundwater management are discussed. Uncertainties concerning the safety and cost dimensions also exist with regard to the rock of the Swabian Alp, which is particularly rich in cavities. Geoscientific expertise is one decisive factor for assessing the risks, the costs of the feasible, as well as the complexity of technological systems for underground buildings. Accordingly, the role of expertise in the political conflict is contested because ultimately, it influences the cost-benefit-risk analysis and the decision on infrastructures to a considerable extent (cf. Schlichtung S21). Ethically relevant are the question of neutrality of expertise in the political sphere, the implicit determination of assessments depending on the employer, the question of dealing with possible false-positive or false-negative prognoses and their consequences, as well as the—very controversial—question whether and how experts should explicitly comment on the political context of their expertise.

An internationally prominent example for the conflicts surrounding large-scale regulation of water bodies is the Three Gorges Dam in China. The technical question of the safety and the potential catastrophic consequences of a breach in the dam is associated with questions concerning possible negative environmental consequences (shortage of water, loss of natural high and low water regimes, and accumulation of mud and toxic materials in the reservoir; it is generally referred to experiences with the construction of the Aswan Dam on the Nile) as well as questions of social justice, which resulted mainly from forced resettlements (Parodi, 2008).

Climate engineering is used as a term for the intentional technological modification of the earth's major geochemical cycles with the main aim of counteracting the effects (but not the ultimate causes) of climate change. Climate engineering measures include various options for carbon dioxide removal (cf. Rackley, 2010) or solar radiation management (SRM). Each of these technologies has its own characteristic risks and uncertainties. To name but one example, the deployment of sulfate aerosols in the stratosphere in order to reflect solar radiation back into space entails a number of global risks such as changed precipitation patterns or potential negative effects on the ozone layer. In addition to that, the implementation of certain measures of climate engineering (especially SRM options) might divert funds and efforts from fighting the causes of climate change (The Royal Society, 2009).

Fracking has various environmental and health risks, such as pollution of groundwater and dispersal of chemicals to other layers of soil (Fossati, 2012). Discourses about fracking take place against the background of energy and planning politics and are thus connected to questions of business ethics and political ethics.

Respect for (Religions of) Nature and Dimensions of Land Ethics

Apart from technoscientific projects, the moral permissibility of certain—massive—transformations of the earth and its structures can be discussed in a more general way: This aspect is often addressed in the context of religious questions (bans on interventions in sacred mountains or sacred groves) or secular landscape and nature conservation (disfigurement through surface mining, dams, etc.). Beyond the questions of respect for people's faith in secular states, there are much broader issues of environmental justice at stake (Martinez-Alier, 2002). More often than not, indigenous people and/or socially and economically suppressed groups and classes have to carry the burden of large-scale projects like open pit mining (e.g., oil sands, Nikiforuk, 2010), whereas the gains are with private companies and industrially developed states (Ali, 2009).

In that context, it would have to be elaborated for geoethics what the dictum of Aldo Leopold's (1949, p. 217) Land Ethic can mean in concrete terms: "A thing is right when it tends to preserve the integrity, stability, and beauty of the biotic community. It is wrong when it tends otherwise." Humans with their economy and technology and the *land* as a whole should be part of this *biotic community* sensu Leopold.

OUTLOOK: GEOETHICS AND SUSTAINABLE DEVELOPMENT

The three elements (1) professional ethics of scientific and engineering experts (impartiality, incorruptibility, risk awareness, and communication), (2) risk management in a broad sense concerning specific projects, and (3) general considerations relating to environmental justice and the integrity of the earth are the crucial ethical aspects of geoethics, as well as the basis of other ethical assessments of technology (cf. Skorupinski and Ott, 2000). The connection with more comprehensive ethical conceptions of risk ethics, social ethics, environmental ethics, and political ethics is evident. With reference to geotechnical engineering, there are hardly any systematic studies and approaches concerning these aspects. A comprehensive question would be: How can the geosciences and their ensuing technologies reach an integrity of infrastructures between safety obligation, risk aversion, cost-benefit questions, and urban/landscape aesthetics?

A water ethics (cf. Llamas et al., 2009) is elaborated somewhat further. Questions of technology ethics have a firm place here, but they are often little reflected on in detail. For example, it would have to be investigated more closely in what way the juxtaposition of near-natural versus nonnatural engineering approaches can be differentiated and substantiated as a fundamental topos of ethics of nature *and* philosophy of technology. The technological systems of manipulating the earth under and above ground, including questions of safety and health, are of special ethical importance in the political–economic discussion. Should the land and the earth as such be a public property in the sense of the commons, a matter of national or supranational public services, or should they be left to the free market as a privatized good (Ostrom, 2008)? Insofar, the technological design (local, regional, and global systems) is also part of political questions of justice and human rights. This link to sustainable development as a concept of general environmental justice should be emphasized and extended to all forms of geo- (and hydro)technical engineering. It is the aim of geoethics to strengthen these connections in an interdisciplinary discourse among all academic fields and at the same time in a transdisciplinary deliberation with all members of society.

To sum up, the three domains mentioned above constitute the matrix dimensions of ethical analysis in geoethics. An integrated deliberation spelling out commonalities and specificities of geoethics in this ethical context will contribute to a realization of sustainable development, broadly understood as environmental justice for current and future generations.

REFERENCES

Ali, S.A., 2009. Mining, the Environment, and Indigenous Development Conflicts. Tucson.

Brandl, H., 2004. The Civil and Geotechnical Engineer in Society – Ethical and Philosophical Thoughts; Challenges and Recommendations. Vancouver. Online: http://www.issmge.org/images/Attachments/JML98%2804%29_HeinzBrandl.pdf (accessed 31.03.14.).

Fossati, J., 2012. The Hidden Dangers of Marcellus Shale Fracking. University of Pittsburgh Swanson School of Engineering. Online: http://www.pitt.edu/~jmf142/a3.pdf (accessed 31.03.14.).

Horkheimer, M., Adorno, T.W., 2002. Dialectic of Enlightenment: Philosophical Fragments. Stanford University Press, Stanford.

Hubig, C., Reidel, J. (Eds.), 2003. Ethische Ingenieursverantwortung: Handlungsspielräume und Perspektiven der Kodifizierung. Sigma, Berlin.

ISSMGE – International Society for Soil Mechanics and Geotechnical Engineering, 2004. Policy Document No. 1: Guidelines for Professional Practice. London. Online: http://www.issmge.org/images/Attachments/Policy%20Doc%201.pdf (accessed 31.03.14.).

The L'Aquila Trial, 2014. Documents, Articles and Comments from the Scientific Community about the Trial to the Participants to the National Commission for Forecasting and Predicting Great Risks Meeting. In: http://processoaquila.wordpress.com/ (accessed 31.03.14.).

Leopold, A., 1949. A Sand County Almanac. Oxford University Press, New York.

Llamas, M.R., Martínez-Cortina, L., Mukherji, A. (Eds.), 2009. Water Ethics – Marcelino Botín Water Forum Santander 2007. Taylor & Francis, Leiden.

Martinez-Alier, J., 2002. The Environmentalism of the Poor. A Study of Ecological Conflicts & Valuation. Edward Elgar Publishing, Cheltenham & Northampton.

Mucciarelli, M., 2015. Some comments on the first degree sentence of the "L'Aquila trial". In: Wyss, M., Peppoloni, S. (Eds.), Geoethics: Ethical Challenges and Case Studies in Earth Science. Elsevier, Waltham, Massachusetts, pp. 205–210.

Nikiforuk, A., 2010. Tar Sands – Dirty Oil and the Future of a Continent. Revised edition. Greystone, Vancouver.

Ostrom, E., 2008. The Challenge of Common-Pool Resources. Online: http://www.environment-magazine.org/Archives/Back%20Issues/July-August%202008/ostrom-full.html (accessed 31.03.14.).

Parodi, O., 2008. Technik am Fluss – Philosophische und kulturwissenschaftliche Betrachtungen zum Wasserbau als kulturelle Unternehmung. Oekom Verlag, München.

Potthast, T., 2010. Epistemisch-moralische Hybride und das Problem interdisziplinärer Urteilsbildung. In: Jungert, M., Sukopp, T., Romfeld, E., Voigt, U. (Eds.), Interdisziplinarität. Theorie, Praxis, Probleme. Wiss. Buchgesellschaft, Darmstadt, pp. 173–191.

Rackley, S.A., 2010. Carbon Capture and Storage. Burlington & Oxford.

The Royal Society, 2009. Geoengineering the Climate – Science, Governance and Uncertainty. RS Policy document 10/09, issued September 2009. Online: https://royalsociety.org/~/media/Royal_Society_Content/policy/publications/2009/8693.pdf (accessed 31.03.14.).

Schlichtung S 21. Informationsseiten der Schlichtung Stuttgart 21. Online: http://www.schlichtung-s21.de (accessed 31.03.14.).

Schmalenberg, D., Damm, A., 2010. Geschlampt, gefälscht – Ursachenforschung. Schaden von mehr als 700 Millionen Euro. Special issue of the Kölner Stadtanzeiger on 3 March 2010.

Skorupinski, B., Ott, K., 2000. Technikfolgenabschätzung und Ethik. vdf Hochschulverlag, Zürich.

Ziegler, M., 2008. Sicherheitsnachweise im Erd- und Grundbau. In: Witt, K.J. (Ed.), Grundbau-Taschenbuch, Teil 1: Geotechnische Grundlagen. Ernst & Sohn, Berlin, pp. 1–42.

Chapter 6

Humans' Place in Geophysics: Understanding the Vertigo of Deep Time

Telmo Pievani

Department of Biology, University of Padua, Padua, Italy

Chapter Outline

Abstract

We define the phenomenon of "denialism in science" (denial of scientific evidence without any support), its multiple facets, and its influential cognitive roots. Biological evolution and geophysical evidence are preferred topics for deniers in scientific issues. The case history of peculiar denialism in Canadian conservative government is outlined and interpreted as an example of insidious manipulative strategy against the acceptance and diffusion of scientific evidences. Earth sciences and evolutionary biology share a set of counterintuitive features, which can explain the psychological success of denialism in public debates. The roots of these counterintuitive features are briefly presented. The popular appeal of pseudoscience and denialism makes it clear that receiving solid teaching of the value of scientific thought and the scientific method early in life is of fundamental importance.

Keywords: Creationism; Denialism in science; Philosophy of earth sciences; Teleology.

DENIALISM IN SCIENCE

In this chapter, we argue that evidence and explanatory models in earth sciences, like in other life sciences, contain counterintuitive features. A counterintuitive fact, concept, or reasoning, in philosophy of science, is something contrary to

Geoethics. http://dx.doi.org/10.1016/B978-0-12-799935-7.00006-X

intuition or everyday life expectations. It could be, for instance, a scientific theory that contradicts naïve observations, gut feelings, and our spatiotemporally limited intuitions. Counterintuitive ideas suggest that the relationship between science and common sense is a nonlinear one: common sense nurtures science, but science is able to reach conclusions that obvert common sense.

The counterintuitive features of life and earth sciences, running counter common sense, could explain, according to recent evidence in cognitive sciences (Girotto et al., 2014), the long-term persistence in the population of the denial of scientific evidence. There exists strong hostility toward supported research programs and established and refined models (such as anthropic role in climate warming and neo-Darwinian theory of evolution). Also, religious motivation, economic interests, and psychological constraints (among others, our teleological stance and ethics of strict proximity) could explain (but not justify) the popular appeal of such antiscientific movements.

We define here "denialism in science" as the public's refusal to accept empirically corroborated scientific evidence, like biological evolution by natural selection or the geological history of planet Earth, and scientific consensus, like climate change as a global process due to human activities. Any form of denialism (i.e., historical) is clearly related to social and political issues, like religious fundamentalism spreading in Western countries and conservative movements, but here we concentrate mainly on its philosophical and cognitive features.

It could be argued that denialism is an expression of trivial irrationalism, opposed to the use of plausibility and rationality. For a philosopher of science, however, the allegedly simple contradiction between the self-evident rationality of science and the obscurity of lazy superstition is not the whole story. We propose here to consider denialism in science as based on much more influential cognitive roots, on the ambiguities of the demarcation of sciences, and on the counterintuitive results of some scientific research (Pievani, 2012a). Adding the ingredients of cunning politics and economic interests, the recipe is insidious.

Case History

As a case study, we briefly outline the recent conservative campaign in Canada that involves both the bases of geophysics and the idea of the relentless evolution of our planet. According to recent polls (2008),[1] Canada has an average good level of acceptance of evolution (classical "creationism" is followed by only 22% of Canadians). It implies that the antiscientific campaigns are better addressed if not openly declared in public debates, in order to avoid an immediate rejection.

The Canadian conservative government seems strongly influenced by what some call "evangelical religious skepticism," a new mixture of religious fundamentalism and pragmatic denialism of anthropic climate change and its effects, as suggested by the following facts. Canada has pulled out of the Kyoto agreement, without any alternative national plan to battle carbon pollution. The historical

Case History—cont'd

Canadian Environmental Assessment Act and the Fisheries Act are questioned now by the government. The expansion of excavations for oil sands in Alberta is causing major concerns among environmentalist international associations for their ecological impact. Furthermore, in 2013, there has been a public debate about the fact that the National Research Council should focus more on practical, directly commercial science and less on basic science that may not have business applications.

According to the opinion reported by Canadian Professor of Science Broadcast Journalism at Carleton University and President of the Canadian Science Writers' Association, Kathryn O'Hara, in *Nature*, September 2010, "The signs were there in spring last year, when press reports revealed that climate scientists in the government department Environment Canada were being stymied by Harper's compulsive message control. Our researchers were prevented from sharing their work at conferences, giving interviews to journalists, and even talking about research" (O'Hara, 2010). Even the discovery of a colossal flood that occurred in northern Canada at the end of the last ice age has been censored: a strange reticence with regard to the geosciences. Federal scientists and government workers are warned not to talk with the media about these issues without permission of the media-relations offices. The obstructionism refers mainly to evolution, climate, and health studies. *Nature* in March 2012 officially requested that it was "time for the Canadian government to set its scientists free" and respect the duties of accountability, integrity, and openness with the press ("Frozen out," *Nature*, 483:6).

Prime Minister (since 2006) Stephen Harper does not speak about his religious beliefs and refuses to answer media questions about it. Press agencies have speculated about whether he belongs to the Christian and Missionary Alliance, an evangelical Protestant church with two million members. If he did, he might endorse the "evangelical climate skepticism," which doubts climate change and champions burning fossil fuels. According to this view, environmental policies would be a threat against economic freedom and individual liberties.

Another clue for this general lack of transparency may be the apparent reluctance to accept the fact of evolution as demonstrated by the following incident. In March 2009, the Federal Science Minister, a Christian and former chiropractor, Gary Goodyear, promoter of federal funding cuts to researchers, in an interview with "The Globe and Mail" refused to say whether he believes in evolution, because it is a religious matter. Faced with protests from many Canadian scientists, he subsequently stated that he does accept evolution, but offered a very confusing and twisted representation of it. ("We are evolving, every year, every decade. That's a fact. Whether it's to the intensity of the sun, whether it's to, as a chiropractor, walking on cement versus anything else, whether it's running shoes or high heels, of course, we are evolving to our environment."[2]) This will have done little to dispel any doubts about his beliefs.

We can see here the pragmatic flexibility of denialism in science. It can be used to placate political supporters, while not openly opposing the majority's view. A denial can be hidden in operational decisions, justified by economic interests, and exercised by controlling press releases. By this technique, a negative reaction by the electoral consensus that a public and aggressive fundamentalism would entail

Continued

is avoided or minimized. Evolution and geophysics are always in the eye of the storm. The Canadian conservative party's manipulative strategy could be a Trojan horse for a successive popular campaign, and should not be underestimated.

[1] The complete poll by Angus Reid Strategies is available here: http://www.angusreidglobal.com/polls/33181/canadians_choose_evolution_over_creationism. For a comment by the National Center for Science Education: http://ncse.com/news/2008/08/polling-creationism-canada-001375.
[2] "Power Play: Gary Goodyear responds to criticisms." CTV News. March 17, 2009.

DISCUSSION: COUNTERINTUITIVE SCIENCE AND OUR TELEOLOGICAL MINDS

Politics has its well-known rules, as well as the imperatives of economic agendas and interests. Much less well known are the rules of the persuasiveness of some ideas with respect to others. Human minds are equipped with propensities and preferences due to our evolutionary history. Like biological evolution and common descent, earth sciences have counterintuitive features for persons non literate in science. We put them in three nested levels of perception and then we try to briefly explain why we consider them counterintuitive:

1. The *general perception of deep time*, with its disconcerting consequence on some people's idea of the (no longer central) humans' place in nature.
2. The *indirect perception of historical contingency*, due to the crucial role of geophysical factors in biological evolution, including human evolution (Pievani, 2012b).
3. The *direct perception of unpredictability* (earthquakes, volcanic eruptions, tsunamis, hurricanes, floods, and other natural hazards) and of being in a system out of our complete control, due to ecological and climate instability, and violent expressions of geophysical forces. It is a desire of humans to be in control and be able to predict the future, which is to a large measure impossible. That means, a wish is denied us, but it is not a desire like anyone else: it is a wish concerning the direction of the history of a system, that is, an intuition about the degree of predictability of history.

According to a philosophical analysis, these are three wounds to those who hold a terrestrial narcissism. A prevalent attitude among people is that they would like to be, but are not, masters in their house. The clash between human time and geological deep time changes irreversibly the perception of our place in history. The scientific explanation of the patterns of deep time requires a more costly cognitive investment with respect to intuitive feelings. A teleological Intelligent Design (ID) (both referring to general biological evolution and humans' place in nature) is much more psychologically comforting and politically profitable, due to our natural attitudes discovered in a growing amount of experiments in cognitive and developmental psychology. We argue that these results prove the counterintuitive quality of previous statements 1–3.

Convergent data, coming from developmental psychology, evolutionary psychology, anthropology, and neuroscience, suggest a biological predisposition of our minds, even in the earliest phases of development, to distinguish inert entities (like physical objects) and entities with psychological features (like living agents) instinctively (Bloom, 2004), and to attribute or, incidentally, hyperattribute intentions and purposes to animate and inanimate objects, producing "teleofunctional explanations" of the natural world (Keleman, 2003). The discovery of the cognitive bases of this promiscuous teleology—so similar to the folk teleology of Voltaire's Dr Pangloss in the novel "Candid" debunking teleology with respect to natural disasters—could explain the natural propensity to find the animistic justifications of natural events, or those based on invisible intelligent designers, psychologically and emotionally satisfying. Thus, a natural disaster is intuitively believed, mostly by people deprived of scientific education, to be a message from some entity, possibly a punishment for past sins. A personalized "Nature" becomes a loving mother or a cruel stepmother depending on the circumstances. And rarely do we consider that a "natural" disaster has possibly a human component in terms of responsibility and lack of foresight.

The widespread diffusion of cognitive detectors of causality and causal agents also in other animals (Vallortigara, 2008) and the presence of inferences about hidden causes in early childhood (Saxe et al., 2005) suggest that it could be a mental habit rooted in adaptive specializations. To suppose rapidly and economically that a purposeful agent with projects and aims is hidden behind the complexity of natural phenomena, rather than blind mechanic processes indifferent to human fate, has been an effective evolutionary strategy for species under predation (Guthrie, 1993). As a result, even today, humans seem affected by an hypertrophy of the system dealing with animated objects (Boyer, 2001). We tend to attribute desires, intentions, and projects to natural objects, whereas they do not exist. The mental world of William Paley's, in which a Creator of the Universe is derived seemingly by logic, the world of a natural theologian, is still today a place where many find a comforting refuge. Paradoxically, humans are inclined to think this way by adaptive natural evolution.

Of course, this does not imply that all humans are teleological fundamentalists, or teleological naïve thinkers, because contemporary cognitive psychology and evolutionary psychology teach us that our mind is equipped by natural attitudes intended as natural precursors and not as deterministic innate instincts. A natural precursor is able to make some beliefs easier than others ("easier" meaning more intuitive, the first choice among others in a given situation), due to ancient evolutionary reasons. Nevertheless, early education, nurture, and cultural and social influences are sufficiently powerful for readdressing or suppressing such phylogenetic biases (Girotto et al., 2014). On the other hand, the fact that someone, scientifically trained in a secular world, has no teleological biases (or she/he thinks so) does not demonstrate that the natural precursors discovered and repeatedly cross-checked in experimental, developmental, and cognitive psychology do not exist.

The evidence that the teleological propensity and the hyperattribution of mental states are not stupid or childish human attitudes, but a mental activity crucial for the functioning of our minds, does not belittle the fact that some people can misuse these attitudes badly on many occasions, like when they deny the validity of a corroborated scientific program following instead fallacious but intuitively amusing arguments. To be conscious of the evolution of our adaptive or exaptive cognitive constraints is a tool for dealing with them in a more careful and rational way. The evolutionary explanation of their emergence is not a justification by nature. The understanding of a behavior as product of the biological evolution of our species does not imply that the behavior itself is forever written on the stone. Nevertheless, we should consider the intentional misuse of these mental habits carefully, with respect to the counterintuitive nature of many scientific explanations.

Our intuitive ideas about the "nature of history," as Stephen J. Gould pointed out, are under discussion here (Gould, 1989). The point is that the evolutionary and geophysical explanations—in their mix of functional, structural, and historical inferences, with at least three robust inflows of chance (random mutations, random genetic drifts, and contingent ecological macroevents)—are deeply ateleological. They show that history has no preestablished directions and designed purposes. The dependence of our evolution as African hominins on external (and frequently accidental) circumstances and processes, like the Great Rift Valley formation and the Pleistocene climate oscillations, does not mean, on the other side, that human evolution occurred exclusively "by mere chance." Our natural history is rather an entanglement of functional factors (produced by selective pressures), structural constraints, and historical contingent events. This complex interplay between random events and lawlike regularities (Eldredge, 1999; Gould, 2002) respects all the parameters of a counterintuitive scientific result, with respect to our intuitive teleological attitudes. *Homo sapiens* is an improbable and tiny branch at the end of a luxuriant tree of species (Tattersall, 2012; Cavalli Sforza and Pievani, 2012). The massive contingency of human evolution means that particular events, or apparently meaningless details, were able to shape irreversibly the course of natural history. Other worlds were possible; other forms of the present were achievable.

A study of denialistic arguments shows that they are carefully suited for those minds that continue to be attracted to teleological beliefs, and for the intuitive refusal of accepting chance and contingency in causal explanations by parts of the population. Furthermore, all such unfair psychological advantages are powered by the privileged position of any unorthodox minority in a public debate. For these reasons, without any intention of justification, we should admit (as in *Nature*, April 2005, about the diffusion of ID and neo-Creationism in North American University campuses, Brumfiel (2005)) that neo-Creationism is not only the manifestation of blind religious fundamentalism but also an ideological campaign able to fit the intuitions of a large number of people, and perfect to be manipulated as a political instrument of consensus. Therefore, such pseudosciences are a powerful cocktail in our societies.

What can we do? We have to rely on human plasticity and on the effects of early learning. Several studies in psychology of thought show that education could make the difference in the development of the human mind and adult propensities toward science. Early scientific education in geoethics could make a real difference, promoting for instance an "ethics of foresight," the awareness of the interdependence of global factors in our evolution, and the philosophical and psychological acceptance of our fragility in the history of Earth.

HOW CAN ONE ANSWER DENIALISM IN SCIENCE?

Pseudoscience is different from plain nonscience: it is a camouflaging of science and an abuse of its methods (Kitcher, 1982), pushing cognitively and psychologically persuasive and deceptive inferences. In such a situation—a powerful mix of psychological appeal, cognitive constraints, demagogic communication, and political support—is it still useful to oppose arguments (like in Dawkins (2006), or in Coyne (2009)) based only on simple evidence of "truth" and on a shared naïve definition of what is science and what is not?

In 2007, the rapid spread of the creationist interreligious movements over Europe produced a formal critical response by the Council of Europe ("The Dangers of Creationism in Education"). In its document, the European Committee on Culture, Science and Education stated that the teaching of creationism in schools cannot be considered as an expression of freedom of thought, but only as a form of religious fundamentalism, which threatens both the freedom of science and the freedom of teaching: "There is a real risk of serious confusion being introduced into our children's minds between what has to do with convictions, beliefs, ideals of all sorts and what has to do with science. An 'all things are equal' attitude may seem appealing and tolerant, but is in fact dangerous" (statement n. 7; resolution 1580/2007, Doc. 11375).

Therefore, how can we answer the creationist arguments in this new political and cultural context? A reactive, defensive, and spot-by-spot policy of rebutting seems not enough (Pievani, 2008). Science is an open way of thinking, with common rules; it has a public role in our societies; and, as a matter of principle, any dissent is potentially useful. If ID were a good alternative "school of thought" about the "explanation of life"—as former US President George W. Bush said in August 2005—it should explain all the empirical bases of current neo-Darwinian research program,[3] it should explain something more, and it should do all that using laws and factors different from neo-Darwinian ones.

---

3. We mean by neo-Darwinian "research program" (using the terminology proposed by epistemologist Imre Lakatos) the core of any current evolutionary explanation, based on variation and selective processes, plus all the advancements, revisions, and extensions of this core, from paleontology to molecular biology (Pievani, 2012c).

This is not the case. As many scientists and philosophers analytically pointed out, giving ID the benefit of doubt, we can easily check that there is no empirical evidence at all, just negative results (about the origins of life or the Cambrian explosion) confusing an impossibility *de facto* with an impossibility *de iure*, and ignoring scientific explanations already available. We can also check that in ID works there are no good inferences at all, but many logical fallacies concerning the relationships between the concepts of chance, probability, complexity, and information (Pievani, 2013b). Hence, we do not find any useful controversy, because ID does not help to outline any potential weakness of the neo-Darwinian frame and its objections cannot be falsified.

So far, we have seen the negative side of the debate. But, in a positive way, useful for the outreach of evolutionary research: What makes science different from pseudoscience? Science is discovery, not only theories and frames of concepts. A process of human creativity, a mix of method and imagination, and a marvelous adventure of the mind, such as the one of the young geologist Charles Darwin described day by day in his early Transmutation Notebooks (1836–1844). Furthermore, scientific theories and ideas evolve, whereas ID still consists of the same arguments William Paley used in 1802. The theory of evolution, like any evolving theory, is flexible. There is no "Darwinian orthodoxy" and today's theory is quite different from the formulation of 1859 and from that in the XX century, although still consistent with the basic neo-Darwinian processes. Some evolutionists are beginning to talk about a possible new "extended evolutionary synthesis," where a neo-Darwinian extended core of the program (variation, selection, genetic drift, and macroevolution) is surrounded by new hypotheses concerning the plurality of rates of change, the plurality of levels and units of change, and the plurality of patterns of change deriving from the different trade-offs between biological structures and functions, constraints and plasticity (Pievani, 2012c). While we spend time in debates with creationists, a lot of interesting discoveries are accumulating in evolutionary science, exciting to popularize.

Is this strategy enough for public awareness? Probably not, because denialism in science is a political and cultural campaign, manipulating and transcending any contents, and neo-Creationism is psychologically attractive, whereas science is frequently counterintuitive. Nevertheless, what we need is to clarify that evolution is not the door to atheism. Scientific (methodological) naturalism is different from metaphysical (ontological) naturalism. So, we have to do much more than "answer." We need a positive message about the beauty of freedom of inquiry, a message able to create consensus among professionals concerned with science education.

As a methodological example (Pievani and Serrelli, 2008), we could set friendly and interactive educational situations (hands-on and minds-on) in which we could propose positive, and not discouraging, messages coming from the awareness of our historical contingency. We have had an extraordinary opportunity to be here, within deep time. We would not be here if the Earth was not unstable, always continuing its history. Biological and geophysical evolution teaches us that we are not special, thus we are not alone in this adventure. The construction of our niche depends on us, and the process (being accurate or inaccurate) will have good or bad countereffects on our well-being.

Mostly, we have to communicate science as a rational process of discovery and intersubjective control, and not simply as a congeries of hypotheses and results. Values are deeply entrenched and inherent in science, but values must be rationally argued in the continuously ongoing conversations between scientists, and between science and society. Rational arguments need to be supported by shared data; otherwise, it will never be possible to find a common ground of discussion. Therefore, geoethics is not possible (just as bioethics is not possible) without a methodological agreement upon the indispensability of rational argumentation and the sharing of public, corroborated, and reliable scientific data. Different biases and worldviews are not sufficient to transform science into a relativistic enterprise, where all points of view are equivalent per se. Science is an irreversible process of criticism and growth of knowledge, fostered by the plurality of points of view (which is different from their relativity) sharing a common set of rules.

CONCLUSIONS

Denialism in science is a philosophical, cognitive, and political challenge. We should take advantage of it for learning how to better communicate what science is and how results are derived methodologically. The role of scientific literacy, at popular and diffused levels, is crucial, as well as the role of communication of science intended in a participative and not in a paternalistic and unidirectional way. When we talk about the survival of human populations and settlements, as is the case in geoethics, the maximum of transparency and comprehensibility is a moral imperative, starting with the basic distinction between the prediction of specific events and the prediction of risks. However, it is not enough. Science has also to be communicated as a shareable, hands-on and minds-on, experience, avoiding the idea of a linear transfer from who knows to who does not.

For these reasons, a philosophy of earth sciences is welcome. For instance, this analysis could also show a philosophical difference between geoethics and ecologist movements. In environmental ethics, we aim at protecting and preserving natural equilibrium and biodiversity, personally considering nature as an evolved process or as the creation. In these praiseworthy efforts, we undertake the psychologically impossible task to pass from an anthropocentric approach to an ecocentric approach. The task is impossible because in any environmental ethics, *H. sapiens* decides to self-assume (for duty, for passion, for guilt, and for indirect advantage) the role of the rescuer of the planet.

There is an altogether reassuring psychological component in this mission. On the other hand, in geoethics, we are in front of the disconcerting scientific fact that the Earth's surface—instable, unpredictable, and living—was not made for us. Therefore, it is not our property, but a set of common received goods enabling our survival accompanied by a set of common received risks enabling our extinction (i.e., volcanic supereruptions and climate changes) (Pievani, 2013a).

Considering the human losses, the economic costs, and the collective social impacts of natural events that occur in situations in which *Homo* so-called *sapiens* placed itself in conditions of high risk, the duty to survive maintaining as best as we can our ecological niche and territory for the next generations is both a scientific and a humanistic achievement. We are not the glorious rescuers of the planet; we are part of it, and a nonnecessary part. It is interesting that "adaptation" (along with mitigation) is becoming the keyword in climate change discussions: Again and again, we have to adapt to ever changing ecological niches the alteration of which we contributed. But in evolution success is not guaranteed. This risky contingency, and the evolutionary humility that follows, suggests that we have no right to stop the human experiment in progress on the third planet of the Solar System.

The counterintuitive contents of scientific evidence show that science, like democracy, is not an irreversible advancement of humanity. Science and democracy need continuous care, supervision, and alertness. As the Council of Europe stated in its resolution against creationism in 2007 (point n. 12): "Our modern world is based on a long history, of which the development of science and technology forms an important part. However, the scientific approach is still not well understood and this is liable to encourage the development of all manner of fundamentalism and extremism. The total rejection of science is definitely one of the most serious threats to human and civil rights."

REFERENCES

Bloom, P., 2004. Descartes' Baby: How the Science of Child Development Explains What Makes Us Human. Basic Books, New York.

Boyer, P., 2001. Religion Explained. Basic Books, New York.

Brumfiel, G., 2005. Intelligent design: who has designs on your student's minds? Nature 434, 1062–1065.

Cavalli Sforza, L.L., Pievani, T., 2012. *Homo sapiens*. The Great History of Human Diversity. Codice Edizioni, Turin.

Coyne, J.A., 2009. Why Evolution Is True. Viking Press, New York.

Dawkins, R., 2006. The God Delusion. Mariner Books, London.

Eldredge, N., 1999. The Pattern of Evolution. W.H. Freeman and Co., New York.

Girotto, G., Pievani, T., Vallortigara, G., 2014. Supernatural beliefs: adaptations for social life or by-products of cognitive adaptations? Behaviour 151, 385–402. special issue "The Evolution of Morality" (Ed. by Churchland, P., De Waal, F., Parmigiani, S., Pievani, T.).

Gould, S.J., 1989. Wonderful Life. The Burgess Shale and the Nature of History. Norton, New York.

Gould, S.J., 2002. The Structure of Evolutionary Theory. Belknap-Harvard University Press, Cambridge, MA.

Guthrie, S., 1993. Faces in the Clouds: A New Theory of Religion. Oxford University Press, Oxford.

Keleman, D., 2003. British and American children's preferences for teleo-functional explanations of the natural world. Cognition 88, 201–221.

Kitcher, P., 1982. Abusing Science. The Case against Creationism. The MIT Press, Cambridge, MA.

O'Hara, K., 2010. Canada must free scientists to talk to journalists. Nature 467, 501.

Pievani, T., 2008. What makes science different: how to answer neo-creationistic "arguments". Ecsite Newsl. autumn (76), 2–3.

Pievani, T., November 2012a. Denialism. What is "real" in public debates today? The case of evolution. Phainomena XXI 82/83 ("Selected Essays in Contemporary Italian Philosophy", Ed. by Verč, J.), Ljubljana, pp. 145–156.

Pievani, T., 2012b. Geoethics and philosophy of earth sciences: the role of geophysical factors in human evolution. Ann. Geophys. 55 (3), 349–353. http://dx.doi.org/10.4401/ag-5579.

Pievani, T., 2012c. An evolving research programme: the structure of evolutionary theory from a Lakatosian perspective. In: Fasolo, A. (Ed.), The Theory of Evolution and Its Impact. Springer-Verlag, New York, pp. 211–228.

Pievani, T., November 2013a. The sixth mass extinction. Anthropocene and the human impact on evolution. Rend. Lincei Sci. Fis. Nat.http://dx.doi.org/10.1007/s12210-013-0258-9. special issue "Anthropocene".

Pievani, T., 2013b. Intelligent design and the appeal of teleology: structure and diagnosis of a pseudoscientific doctrine. Paradigmi 1-2013, 151–164.

Pievani, T., Serrelli, E., October 2008. Education in evolution and science through laboratory activities. Evol. Educ. Outreach 1 (4), 541–547. http://dx.doi.org/10.1007/s12052-008-0072-5.

Saxe, R., Tennenbaum, J.B., Carey, S., 2005. Secret Agents: Inferences about hidden causes by 10- and 12-month-old infants. Psychol. Sci. 16 (12), 995–1001.

Tattersall, I., 2012. Masters of the Planet. Palgrave Macmillan, New York.

Vallortigara, G., 2008. Animals as natural geometers. In: Tommasi, L., Peterson, M.A., Nadel, L. (Eds.), Cognitive Biology: Evolutionary and Developmental Perspectives on Mind, Brain and Behavior. The MIT Press, Cambridge, MA.

Section II

GEOSCIENCE COMMUNITY

Chapter 7

Research Integrity: The Bedrock of the Geosciences

Tony Mayer
President's Office, Nanyang Technological University, Singapore

Chapter Outline

Abstract

Research integrity (RI) is fundamental in all areas of scientific endeavor. As scientists, we rely and build on the scientific record and if this is being deliberately distorted, then it not only results in the waste of resources, especially time, and may divert research into dead ends but it also erodes public confidence in science. Policies are now in place both to address misconduct and, through educational programs, to promote the responsible conduct of research. The frequency of misconduct is discussed and four famous cases in geosciences spanning the twentieth century are described. Recent developments in the formulation of RI policy are outlined culminating in the Singapore Statement, one of the key foundational documents in the evolution of RI policy. The four principles and 14 responsibilities of researchers and their institutions are given. While there is a need for education and training and for institutions to create the best conditions to promote good research, ultimately, responsibility for the good conduct must lie with the individual. There can be no other way.

Keywords: Research integrity (RI); Responsible conduct of research (RCR); Singapore Statement; World conferences on RI.

INTRODUCTION

What is research integrity (RI) and why should it matter in the geosciences or in any other scientific endeavor?

Human knowledge, especially science, advances incrementally. As Newton said "we stand on the shoulders of giants" (Turnbull, 1959: Correspondence of

Geoethics. http://dx.doi.org/10.1016/B978-0-12-799935-7.00007-1

Isaac Newton). So any blemish in the scientific record may set future researchers off in the wrong direction resulting in a waste of precious resources of time,
effort, and finances not to mention the likely demoralization of researchers, dis-
illusion in the public perception of research, and the resulting loss of reputation.
So RI and the responsible conduct of research (RCR), or good research practice
(GRP) as it is sometimes known, really does matter and is not an "add-on" or
bureaucratic invention.

Why do people, who have been trained to be dedicated scientists with a
commitment to search for the truth, commit research misconduct? Is the motive
to try to gain professional acknowledgment or to cover up inadequacies? One
may speculate to no avail as each case has its own peculiarities. Fortunately, the
frequency of serious research misconduct, usually defined as fabrication and
falsification of data and plagiarism (FFP), is relatively rare while behaviour
below that expected in RCR is far more widespread than most people in the
research community have always assumed was the case and while asserting that
there is not a problem of misconduct in research. In fact, of the cases of serious misconduct—FFP—the first two are relatively rare while plagiarism is the
most common of the serious breaches of integrity. Fabrication and falsification
are the two most damaging to the scientific record while plagiarism is a form
of "reputational theft" and creates an unnecessary multiplication of papers and
cluttering of bodies of knowledge.

Thus, the best prevention is to ensure that all researchers understand the
principles of RCR/GRP and receive training and mentoring in the early stages
of their careers. Research institutions have the responsibility to provide such
training and to create an environment where misconduct is unlikely to happen
and to have rigorous procedures for dealing with misconduct when it occurs.
This policy has the threefold aim of

> Providing for the development of a **greater professionalism** in the conduct
> of research,
> **Reducing risk** by providing an environment in which breaches of integrity
> should be rare, and
> Cultivating **RCR** in all its aspects.

In her complementary chapter, "Formulating the American Geophysical
Union's Scientific Integrity and Professional Ethics Policy: Challenges and
Lessons learned," Linda Gundersen shows the steps taken by one of the World's
leading Earth sciences learned societies to develop and create a policy that can
govern behavior into the future. Before proceeding further, it is necessary to distinguish between ethics and integrity. According to the Oxford English Dictionary, integrity is the "quality of having strong moral principles, the state of being
whole" while ethics is the moral principles governing or influencing conduct.
In most definitions, integrity includes ethics, and breaches of ethical norms are
considered as research misconduct.

So, how frequent are the cases of research misconduct and less than acceptable RCR behaviour?

In a study by of American researchers Titus, Wells, and Rhoades, published in *Nature* in 2008 (Titus et al., 2008), of the 2212 researchers surveyed, 201 instances of likely misconduct were observed over a 3-year period, which translates as three incidents per 100 researchers per year occurring in all disciplines and research endeavors. Examples quoted by Titus, Wells, and Rhodes et al. include

- "A post doc changed the numbers in assays in order to 'improve' the data."
- "A colleague duplicated results between three different papers but differently labeled data in each paper."
- "A coinvestigator on a large, interdisciplinary grant application reported that a postdoctoral fellow in his laboratory, falsified the data submitted as preliminary data in the grant. As principal investigator of the grant, I submitted supplementary data to correct the application."
- "A colleague used Photoshop to eliminate background bands on a Western blot to make the data look more specific than they were."

Note: The Western blot is a means of identifying antibody proteins by gel electrophoresis and is normally displayed through imagery.

Other surveys show that poor RCR may occur in around 30% of all researchers, including taking shortcuts in their experimental procedures and "massaging" data (being highly selective in the data used) and not reporting "outliers" or using Photoshop tools to change imagery.

Over the past decade, there has been a large increase in the numbers of researchers in the world and, especially, in the number of papers produced, a substantial number of which are multiauthor works with authors drawn from several institutions and working in international partnerships. In its report "Knowledge, Networks and Nations" (2011) (RSL Policy 03, 2011), the Royal Society of London reported that in the period between 2002 and 2007, the global spend on research increased by nearly 50% to $US 1145.7 billion/year, the number of researchers went up by 25% to 7.1 million/year, and the number of publications increased by nearly one-third to 1.58 million/year. Despite the economic crisis of 2008, there has been a continuing rise in research investment, in the total number of researchers in the world and their research output as measured by publications.

Taking the figures for misconduct and applying them to the new volume of research activity shows that instances of research misconduct are probably more common than we think and may be a serious impediment to research across the globe.

Furthermore, as researchers become more involved in public debates on issues of concern such as "fracking" or the use of genetically modified organisms, the public has to be able to trust the scientific record on which public policy may be based.

SOME FAMOUS GEOSCIENCE CASES

While the geosciences probably experience the same levels of misconduct and poor practice as other endeavors, there have been some *causes célèbres* over the past century.

The first was the case of Piltdown man, which took over half a century to be shown as a fake. Of course, today, with DNA analysis at one's disposal, this would have been shown up quite quickly, but, at the time, there was no way of determining that this "discovery" of a new hominid was false.

A skull, purportedly showing an early hominid, was reported by an amateur paleontologist, Charles Dawson, from a site in Sussex, UK.

Parts of the skull were deliberately stained to make them look old and other fossil animal remains were also planted at the site to provide an "authentic" environment to the find. The "finds" were verified by Arthur Smith Woodward, Keeper of the Geological Department of the British Museum and Teilhard de Chardin also reported the "find" of a hominid tooth at the site. One can only guess at the motivation for this fraud, which was not exposed until 1953, more than 40 years after the first reports were made. It certainly led the study of paleo-anthropology along a false trail for many years. It seems that the one motive may have been the search for "professional acknowledgment and respectabil-ity" by the amateur, Dawson, but what was the motivation by an acknowledged expert such as Woodward (Weiner, 1955)?

Another famous case was that of Jacques Deprat, a paleontologist in the early twentieth century, who trained at the Sorbonne before working on the paleontology of French Indo-China (now Cambodia, Laos, and Vietnam) He had been acknowledged as a brilliant young geologist, both in France and on the world stage, before setting out to work in the Service géologique de l'Indochine. Roger Osborne (Osborne, 1999), in his book about the case, shows that the accusation of fabrication—in this case, the "planting" of trilobites from Bohe-mia—was complicated by personal and professional rivalries coupled with the class-ridden attitudes of the time. The final evidence of fraud emerged when Deprat went back to his field area, under supervision, and failed to find the trilobites. Following his disgrace as a geologist, Deprat ended up as a writer (*Les Chiens aboient*, published under the pseudonym of Herbert Wild was an attempt at self-justification) and mountaineer. Richard Fortey, in his review of the Osborne book (in the London Review) (Fortey, 1999), comments that in searching for a reason for the fraud, one can only surmise Deprat seemed to have "felt the need to gild his achievements with a false cosmetic."

Similarly, Vishwa Jit Gupta, a faculty member of Panjab University, India, was also accused of "planting" fossils—this time in the Himalayas. Once again we are dealing with a paleontologist of genuine academic repute. This time, the "offending" fossils were conodont fauna from North America, which "found" their way to northern India. Despite the findings of an investigation by a former Chief Justice of the Sikkim High Court which concluded that Gupta was guilty and following an investigation by the Geological Survey of India, the University retained Gupta on its staff although not allowing him to teach paleontology to students. This particular case, which concerned many papers in the scientific record, is considered to be possibly one of the most extensive instances of malpractice in the whole scientific record, not just in the geosciences. Clearly, the motivation was to further enhance a professional reputation. This was revealed in an article in *Nature* published in 1989 and subsequent correspondence (Talent, 1989, 1990; Gupta, 1989; Ahluwalia, 1989; Bhatia, 1989; Srikantia, 1996).

Finally, in these *causes célèbres*, there is the case of John Heslop Harrison, a renowned professor of botany from Durham University, UK, who had devoted himself to the study of the island of Rhum in the Hebrides in Scotland. He was undoubtedly an excellent field biologist who identified Arctic species in rare localities on the island. Yet he also fraudulently claimed the identification of non-Arctic species in support of a theory that the Western Isles of Scotland had not been subject to glaciation. Unusually, this is a case of someone intent on proving his theory rather than for self-aggrandizement, even though the theory was highly speculative and contrary to accepted wisdom (Sabbagh, 2001).

While these are all historic cases, the evidence from surveys such as those by Titus, Wells, and Rhoades (see above) and reports that are to be found in science journals and on Web sites such as *Retraction Watch* show that the problem of research misconduct continues within the scientific community.

THE WORLD CONFERENCES ON RI AND THEIR OUTCOMES

Following some well-documented cases of scientific malpractice at the end of the twentieth century, including that of Jan Hendrik Schoen in the Bell Laboratories, a brilliant physicist who fabricated data (Report of Bell Lab Inquiry) RI moved to the center stage. Following these scandals, the scientific "establishments" in both Germany and the United States were shocked and, consequently took remedial action that resulted in the development of their policies and procedures to counter the breaches of RI and setting benchmarks in standards. In the United States, there had been Congressional action as early as 1985 following some well-documented cases in that country, largely in the biomedical field. This eventually resulted in the consolidation of activities into the Office of Research Integrity (ORI) in the Department of Health and Human Services in 1992 and covering not only the National Institutes of Health but also universities in receipt of grant funding. The National Science Foundation also established a unit dealing with integrity within the Office of the Inspector General of that organization. Also in 1992, the U.S. Academies of Science published a landmark document "Responsible science: ensuring the integrity of the research process" (Responsible Science, 1992).

In Europe, Germany had led the way in establishing policies for its grant agency (Deutsche Forschungsgemeinschaft). In 2000, in response to a request from its Member Organisations, the European Science Foundation (ESF) published a study reviewing what had been done to date across the world to combat research fraud and recommended actions, focusing on funding agencies and academies of sciences (ESF Policy No 10, 2000).

In 2005, Professor Nick Steneck, University of Michigan, then at ORI, concerned at the lack of broad international agreement on definitions and actions regarding research misconduct, initiated discussions into the creation of a forum for discussion between the USA and Europe. In partnership with the ESF, this evolved into a proposal for a world conference. At the same time, responding to the problems of RI within its own research system, Japan (with Canada) proposed that the Organization for Economic Cooperation and Development Global Science Forum should investigate the issue of RI in the international context. Thus, the First World Conference on RI brought both these activities together in a meeting in Lisbon in September 2007 (ESF Policy No 30, 2008). This was the first occasion at which there was a truly global sharing of experiences and problems and recognition of the importance of RI at a time when there was a step change in the quantum of research being performed across the world. One firm conclusion was that there should be a continuation of such global exchanges.

The result was a Second World Conference on RI, which was held in Singapore in 2010 (Mayer and Steneck, 2011). Concerned that such meetings should be more than just a sharing of experiences and an exchange of views, the organizers proposed that the Conference should create a lasting mark on global research. The result was the Singapore Statement on Research Integrity—a document agreed by consensus of the participants that sets out four principles and 14 responsibilities, which should underpin all research. The idea was to create a succinct statement on a single page of paper that could be attached to a wall in every research laboratory worldwide in many translations. This has provided a "foundational document" for what followed. The Singapore Statement is discussed in more detail below.

A Third Conference followed in 2013, this time in Montréal, at which a further statement has emerged, this time addressing all the issues and problems that may arise in cross-boundary research collaboration and which recognized the increasing trend of collaborative and interdisciplinary research. Again, the Montréal Statement has emerged as a consensus document from the participants.

Further world conferences are planned and the they have established themselves as a forum for debate on RI, each time with an evolving focus from definitions and examples to the creation of statements of principles and on toward education, mentoring, and training (Montréal Statement).

THE SINGAPORE STATEMENT

Formulated by a small drafting group consisting of M. Anderson, N. Steneck, and T. Mayer, the Statement was adopted as a consensus document of the 350 participants from 55 countries who attended the meeting in 2010.

The purpose of the Statement is "...intended to challenge governments, organizations and researchers to develop more comprehensive standards, codes and policies to promote research integrity both locally and on a global basis."

The principles and responsibilities summarized in the Statement provide a foundation for more expansive and specific guidance worldwide. Its publication and dissemination are intended to make it easier for others to provide the leadership needed to promote integrity in research on a global basis, with a common approach to the fundamental elements of responsible research practice.

The Statement is applicable to anyone who does research, any organization that sponsors research, and any country that uses research results in decision-making. Good research practices are expected of all researchers: government, corporate, and academic.

In its Preamble, the Statement recognizes the universality of the principles of integrity by stating, "The value and benefits of research are vitally dependent on the integrity of research. While there can be and are national and disciplinary differences in the way research is organised and conducted, there are also principles and professional responsibilities that are fundamental to the integrity of research wherever it is undertaken." The four principles on which the Statement is based are "**Honesty** in all aspects of research; **Accountability** in the conduct of research; **Professional courtesy and fairness** in working with others; and **Good stewardship** of research on behalf of others."

Of course, the ultimate responsibility to behave with integrity has to rest with the individual and thus it is no surprise that the first responsibility is addressed to the individual researcher. This is fundamental to all policies that deal with RI. It has to be one's own moral values that govern behavior and we know that even the best researchers, including Nobel Laureates, sometimes may be tempted to take shortcuts or ignore inconvenient results.

The 14 responsibilities set out in the Singapore Statement are as follows:

1. **Integrity**: Researchers should take responsibility for the trustworthiness of their research.
2. **Adherence to regulations**: Researchers should be aware of and adhere to regulations and policies related to research.
3. **Research methods**: Researchers should employ appropriate research methods, base conclusions on critical analysis of the evidence, and report findings and interpretations fully and objectively.
4. **Research records**: Researchers should keep clear, accurate records of all research in ways that will allow verification and replication of their work by others.
5. **Research findings**: Researchers should share data and findings openly and promptly, as soon as they have had an opportunity to establish priority and ownership claims.
6. **Authorship**: Researchers should take responsibility for their contributions to all publications, funding applications, reports, and other representations

of their research. Lists of authors should include all those and only those who meet applicable authorship criteria.

7. **Publication acknowledgment**: Researchers should acknowledge in publications the names and roles of those who made significant contributions to the research, including writers, funders, sponsors, and others, but do not meet authorship criteria.

8. **Peer review**: Researchers should provide fair, prompt, and rigorous evaluations and respect confidentiality when reviewing others' work.

9. **Conflict of interest**: Researchers should disclose financial and other conflicts of interest that could compromise the trustworthiness of their work in research proposals, publications, and public communications as well as in all review activities.

10. **Public communication**: Researchers should limit professional comments to their recognized expertise when engaged in public discussions about the application and importance of research findings and clearly distinguish professional comments from opinions based on personal views.

11. **Reporting irresponsible research practices**: Researchers should report to the appropriate authorities any suspected research misconduct, including FFP and other irresponsible research practices that undermine the trustworthiness of research, such as carelessness, improperly listing authors, failing to report conflicting data, or the use of misleading analytical methods.

12. **Responding to irresponsible research practices**: Research institutions as well as journals, professional organizations, and agencies that have commitments to research should have procedures for responding to allegations of misconduct and other irresponsible research practices and for protecting those who report such behavior in good faith. When misconduct or other irresponsible research practice is confirmed, appropriate actions should be taken promptly, including correcting the research record.

13. **Research environments**: Research institutions should create and sustain environments that encourage integrity through education, clear policies, and reasonable standards for advancement, while fostering work environments that support RI.

14. **Societal considerations**: Researchers and research institutions should recognize that they have an ethical obligation to weigh societal benefits against risks inherent in their work.

Many of the responsibilities address professional behavior of researchers such as those dealings with authorship, acknowledgments, and peer review. Then there is the responsibility to inform and explain to the public while avoiding slipping into an advocacy role, sometimes in areas outside one's own expertise. One important responsibility is that addressed to institutions (Responsibility 13). Everyone is aware of the increasing pressures on researchers to produce papers and patents quickly. There are also strict time limits set for projects, including postgraduate and postdoctoral studies and so the pressure to take shortcuts and possibly breach integrity are there. Therefore, it is important that research institutions themselves take responsibility for RI.

In her complementary chapter on the development of the American Geophysical Union's RI policy, Linda Gundersen shows how it has used the Singapore Statement and other foundational documents to create its policy. It is no surprise that the elements of this policy are similar to the principles and responsibilities contained in the Singapore Statement (AGU, 2013).

Finally, and possibly, the most controversial of the responsibilities is that referring to "societal considerations." There was substantial debate at the Second World Conference as to whether such issues should be included in a statement of RI. In the end, a majority agreed to its inclusion. This is an important responsibility in the geosciences, especially when environmental matters are at stake. Many activities in the geosciences will have societal implications including such matters as exploitation of mineral resources (including hydrocarbons), the use of new and possibly controversial techniques such as hydraulic fracturing ("fracking"), waste disposal of all sorts, including nuclear waste and contaminated land, engineering geology for major developments, especially new industrial sites, and coastal defenses. Therefore, it is important that geoscientists pay serious attention to the societal implications of their research and ensure that there exists a public awareness and understanding of the issues.

The Singapore Statement is now a key "foundational" document and was one of those used in the formulation of the Statement of Principles by the Global Research Council in May 2013 (Statement of Principles, 2013). For the first time, the research funding agencies worldwide accepted that they have a responsibility alongside the research institutions themselves. This includes ensuring that all those in receipt of research funding follow good research practice and that the funding agencies and the institutions should also promote education and training in this area. Such agencies have only recently started to adopt policies that require their institutional "clients" to have in place rigorous RI policies and procedures.

The Montréal Statement, which takes the Singapore Statement as its starting point, builds on it and develops responsibilities in collaborative research, whether this be across disciplines, between institutions, in partnership projects between the public and private sectors, or internationally. It reflects the new trend that has been seen in an increasing level of collaborative research.

One common theme in these statements and in policies generally is the acknowledgment that researchers must be educated and trained in good research practice. This is not a static topic, especially with the evolution of new technologies. It is now such a simple thing to "copy and paste" using a computer that there is a danger of both deliberate and inadvertent plagiarism. This necessitates training programs so that researchers are fully aware of all the pitfalls in terms of quoting from other's work, with correct citing, and referencing. There are now packages available to check for plagiarism such as *Turnitin* and *iThenticate*, but they do have constraints as they report textual matching that could produce a high matching figure, especially in review papers even where all the citations have been done correctly. Therefore, it is insufficient just to rely on machine-checking. Photoshop and similar packages allow both for the legitimate enhancement of imagery to illustrate a phenomenon or to aid interpretation, but also provide a mechanism for distortions and falsifications. Doubtless, there will be further technological developments, which will need to be taken into account. The message here has to be transparency. Whatever process in data manipulation has been performed requires documenting and reporting so that others can understand and, if necessary, repeat the manipulations.

Correctly, much emphasis is now placed on the collection and archiving of data and its accessibility for others. Laboratory data should not be a problem if good practice is followed with the maintenance of properly kept laboratory notebooks. In the past, these have been kept in hard copy; but, with an increasingly common requirement to make data available for lengthy periods (many journals insist that research records are retained for 10 years post the publication date), the use of electronic notebooks is rapidly becoming the norm.

For the Earth scientist, while the field notebook still remains essential, new technologies mean that all records and observations, imagery, and measurements can be recorded electronically and easily archived.

CONCLUSIONS

It cannot be stressed enough that the key to maintain RI has to be through education and training at an early stage in a researcher's career plus learning by good example and mentoring by one's seniors plus constant vigilance on infringements. This final point can create difficult choices when one may be torn between friendship and working relationships and the need to report infringements and bad practice in which the need to protect the scientific record is paramount.

In the end, though, this has to be an individual responsibility and we all owe it both to ourselves and the community at large to conduct our research in the best possible manner following accepted good research practice.

REFERENCES

Ahluwalia, A.D., September 07, 1989. The peripatetic fossils: part 3. Nature 341, 13–15.

American Geophysical Union, 2013. AGU Scientific Integrity and Professional Ethics.

Bhatia, S.B., September 07, 1989. The peripatetic fossils: part 4. Nature 341, 15. http://www.cehd.umn.edu/olpd/MontrealStatement.pdf.

ESF Policy Briefing No 10. Good Scientific Practice in Research and Scholarship 2000. European Science Foundation.

ESF Policy Briefing No 30 (with ORI). Research Integrity: Global Responsibility to Foster Common Standards 2008. European Science Foundation.

Fortey, R., November 25, 1999. Did the self-made man fake it with Bohemian fossils? London Rev. Books 21 (23), 31.

Gupta, V.J., September 07, 1989. The peripatetic fossils: part 2. Nature 341, 11–12. http://www.ithenticate.com/and http://turnitin.com/.

Mayer, T., Steneck, N., 2011. Promoting Research Integrity in a Global Environment. World Scientific. ISBN:978-981-4340-97-7.

Turnbull, H.W., 1959. In: The Correspondence of Isaac Newton, vol. 1. p. 416.

Osborne, R., 1999. The Deprat Affair: Ambition, Revenge and Deceit in French Indochina. 240 pp. ISBN:0 224 05295 0.

Report of Bell Lab Inquiry: Report of the Investigation Committee on the Possibility of Scientific Misconduct in the Work of Hendrik Schön and Coauthors http://www.alcatellucent.com/wps/DocumentStreamerServlet?LMSG_CABINET=Docs_and_Resource_Ctr&LMSG_CON-TENT_FILE=Corp_Governance_Docs/researchreview.pdf. http://www.researchintegrity.org.

Responsible Science: Ensuring the Integrity of the Research Process, vol. 1, 1992. NAS Press.

Policy Document 03/11. Knowledge, Networks and Nations 2011. Royal Society of London. ISBN:978-0-85403-890-9.

Sabbagh, K., 2001. A Rum Affair: The True Story of Botanical Fraud. Da Capo Press. ISBN:10: 0306810603.

Srikantia, S.V., February 10, 1996. The Himalayan fossil fraud and its aftermath. Curr. Sci. 70 (3), 198.

Statement of Principles for Research Integrity, May 2013. Global Research Council, Berlin.

Talent, J.A., April 20, 1989. The case of the peripatetic fossils. Nature 338, 613–615.

Talent, J.A., February 01, 1990. The peripatetic fossils: part 5. Nature 343, 405–406.

Titus, S.L., Wells, J.A., Rhoades, L.J., June 19, 2008. Repairing research integrity. Nature 453, 980–982.

Weiner, J.S., 1955. The Piltdown Forgery. Oxford University Press.

Chapter 8

Formulating the American Geophysical Union's Scientific Integrity and Professional Ethics Policy: Challenges and Lessons Learned

Linda C. Gundersen[1] and Randy Townsend[2]

[1]*U.S. Geological Survey, Reston Virginia;* [2]*American Geophysical Union, Washington, DC, USA*

Chapter Outline

Abstract

Creating an ethics policy for a large, diverse geosciences organization is a challenge, especially in the midst of the current contentious dialogue in the media related to such issues as climate change, sustaining natural resources, and responding to natural hazards. In 2011, the American Geophysical Union (AGU) took on this challenge, creating an Ethics Task Force to update their ethics policies to better support their new Strategic Plan and respond to the changing scientific research environment. Dialogue with AGU members and others during the course of creating the new policy unveiled some of the following issues to be addressed. Scientific results and individual scientists are coming under intense political and public scrutiny, with the efficacy of the science being questioned. In some cases, scientists are asked to take sides and/or provide opinions on issues beyond their research, impacting their objectivity. Pressure related to competition for funding and the need to publish high quality and quantities of papers has led to recent high-profile plagiarism, data fabrication, and conflict of interest cases. The complexities of a continuously advancing digital environment for conducting, reviewing, and publishing science

Geoethics. http://dx.doi.org/10.1016/B978-0-12-799935-7.00008-3

has raised concerns over the ease of plagiarism, fabrication, falsification, inappropriate peer review, and the need for better accessibility of data and methods. Finally, students and scientists need consistent education and encouragement on the importance of ethics and integrity in scientific research. The new AGU Scientific Integrity and Ethics Policy tries to address these issues and provides an inspirational code of conduct to encourage a responsible, positive, open, honest scientific research environment.

Keywords: American Geophysical Union; Ethics; Ethics policy; Responsible science; Scientific integrity.

INTRODUCTION

The American Geophysical Union (AGU) is a nonprofit, professional scientific organization with over 62,000 members worldwide. It is dedicated to advancing the Earth and space sciences for the benefit of humanity through its scholarly publications, conferences, and outreach programs. Its membership created a new Strategic Plan, released in 2010, that contains a number of core values including "Excellence and integrity in everything we do" and "Benefit of science for a sustainable future." To ensure its ethics policy supports the aspirations of the new Strategic Plan and to provide critical tools to respond to the increasingly complex issues of ethics and integrity that scientists face, the AGU convened an Ethics Task Force in the fall of 2011. The Ethics Task Force was asked to review the current state of AGU's scientific ethical standards and update AGU's protocols and procedures for addressing ethical violations. The new AGU Scientific Integrity and Professional Ethics Policy (AGU, 2013) was approved by the AGU Board and Council in 2012 and released in 2013.

In 1992, the National Academy of Sciences (NAS) published its landmark report "Responsible Science Volume 1: Ensuring the Integrity of the Research Process," (National Research Council, 1992) which included the recommendation that the federal government and research institutions "should adopt a single consistent definition of misconduct in science that is based on fabrication, falsification, and plagiarism." This definition forms the core of the current Federal Research Misconduct Policy and many of the scientific integrity codes found in universities and scientific institutions in the United States, including the previous AGU Misconduct in Science policy. In the 20 years since that report, much has changed in the research environment. Research funding is more competitive and is often outcome-driven, focused on solving societal problems. Instantaneous news, coupled with the politicization and sensationalism of dialogue on controversial societal issues that require scientific solutions such as climate change, future energy strategies, and environmental health have created a complex environment for publishing and communicating science. Scientists may find themselves the target of intense public scrutiny, both positive and negative, where it is difficult to maintain objectivity and be free of conflict of interest (Steneck, 2013). Researchers may need to defend all aspects of their research and methods, and the integrity of science may be threatened through unwarranted attacks and criticism (American Association for the Advancement of Science, 2011).

New complex digital tools, the large magnitude of data and analyses generated by advanced instruments and computers, and the ability to instantaneously access and publish science have revolutionized research, but at the same time created opportunities for the integrity of science to be weakened. These digital tools also make it easier to plagiarize, fabricate, or falsify data, and without a comprehensive digital infrastructure for the sciences, it is more difficult to reproduce and confirm results (National Research Council, 2009a). The desire for instant access to data and results also poses a concern to the integrity of the peer review process. In the last decade, numerous scientific organizations and the federal government have called for the reexamination of scientific and ethical policies, open access of data and publications, and a digital infrastructure to share data and models (Obama, 2009; Union of Concerned Scientists, 2009; World Conference on Research Integrity, 2010, 2013; American Association for the Advancement of Science, 2013; Holdren, 2013). The NAS recently convened a new Committee on Responsible Science with expected publication of a report in 2014.

BUILDING THE AGU SCIENTIFIC INTEGRITY AND PROFESSIONAL ETHICS POLICY

Charge to the Task Force and Foundational Documents

In November 2011, the AGU Ethics Task Force convened their first meeting with the following charge:

- Review the current state of AGU's scientific ethical standards in the geophysical sciences and those of other related professional/scholarly societies.
- Based on this knowledge, update AGU's protocols and procedures for addressing violations of its ethical principles.
- As appropriate, revise and augment AGU's current ethical principles and code of conduct for AGU meetings, publications, and interactions between scientists with their professional colleagues and the public.
- Propose sanctions for those who violate AGU's ethical principles.
- Consider whether AGU should adopt a statement of ethical principles as a condition of membership or for participation in certain activities of the Union. If so, develop a recommendation on how the principles would be applied to AGU members and or participants in AGU activities.

The 13-member AGU Ethics Task Force included representatives from AGU committees and an AGU Past President, university and federal research scientists with experience in developing ethics and science integrity policies, and one representative from industry and the Center for Ethics at Illinois, each. The Task

Force examined AGU's then-current ethical standards and science misconduct policy and found abundant relevant guidance in various places throughout the AGU Web site with which to begin building a new comprehensive policy.

After considerable discussion with members, committee representatives, editors, and officers of the AGU, it became evident that the new policy should be comprehensive, encompassing both professional ethics and scientific integrity and directly address scientists, editors, authors, reviewers, and AGU volunteer leadership. Members of AGU and the Task Force felt strongly that the new policy should be positive, encouraging, and aspirational, but also contain a detailed process for handling allegations of scientific misconduct, with clear lines of authority at the highest levels. The Task Force examined the numerous new federal science integrity policies, including those from the Department of Interior (DOI, 2011) and the National Oceanic and Atmospheric Administration (NOAA, 2011). The Task Force used publication codes of conduct from the Committee on Publication Ethics (2011a,b), the Council of Scientific Editors (2012), Taylor and Francis UK (2006), and the World Association of Medical Editors (2009). Ethical codes for research involving human subjects laid out in the Nuremburg Principles (International Committee of the Red Cross 1950) and Belmont Report (1979) were also foundational. The most influential guidance came from the Second World Conference on Research Integrity (2010), which produced the "Singapore Statement" (e.g., Mayer, 2015), a comprehensive code of scientific ethics created with input from scientists and ethics specialists worldwide. After considerable discussion, the Task Force decided to start with the structure and general text of the Singapore Statement and adapt it to create a general code of conduct for all AGU members.

Structure of the Policy

The AGU Scientific Integrity and Professional Ethics Policy (AGU, 2013) has six distinct parts: an explanatory Prelude, a Code of Conduct and Professional Ethics for all members, an expanded Code of Conduct for Volunteer Leaders, a discussion and definition of Scientific Misconduct, Ethical Guidelines for Publication of Scientific Research, and explicit Processes for Allegations of Scientific Misconduct.

The Code of Conduct and Professional Ethics for all members is comprised of a Preamble, seven Principles, and 13 Responsibilities. The Preamble states:

> *The value and benefits of research are dependent on the integrity of the research and researcher. Members will place quality and objectivity of scientific and scholarly activities and reporting results ahead of personal gain or allegiance to individuals or organizations. Members as scientists and AGU members will act with the highest level of professional ethics and scientific integrity.*

The Task Force wanted to emphasize the individual responsibility of each researcher in contributing to science. Since science is built upon a system of trust and upon previous research, it is essential that each scientist strives for excellence, integrity, and honesty. The general standards of professional ethics for members are provided in the seven Principles:

Excellence, integrity, and honesty *in all aspects of research*

Personal axzccountability *in the conduct of research and the dissemination of the results*

Professional courtesy and fairness *in working with others*

Unselfish cooperation *in research*

Good stewardship *of research on behalf of others*

Legal compliance *in all aspects of research, including intellectual property*

Humane approach *in evaluating the implications of research on humans and animals*

The 13 Responsibilities outline the requirements of each individual for maintaining integrity within the research environment, including responsibilities regarding the scientific process, research records, publishing, peer review, conflict of interest, public communications, working with others, creating a positive research environment, reporting irresponsible research practices, and ethical responsibilities that scientists have to society, humans, animals, and the environment.

The expanded Code of Conduct for AGU Volunteer Leaders addresses the unique responsibilities that AGU volunteers have as leaders to uphold ethical and professional standards of conduct when participating in AGU affairs and representing AGU in an official capacity. Because they are the public face and ambassadors for AGU and the geosciences, Volunteer Leaders have a responsibility to hold themselves to the highest standards of professional behavior and avoid any behavior that could damage the reputation and credibility of AGU and the geosciences.

The Ethical Guidelines for Publication and Scientific Research are one of the most critical parts of the policy because they cover the final step of the scientific process that ensures the integrity of science. In the Overview for this section, the AGU Scientific Integrity and Professional Ethics Policy states:

AGU aspires to select and publish, through peer review, the highest quality Earth and space science research. To achieve this, the peer review process must be objective, fair, and thorough. The ethical basis for this aspiration is absolute trust and honesty among editors, authors, researchers, reviewers, and funding agencies. Decisions about a manuscript should be based only on its importance, originality, clarity, and relevance to the journal's scope and content.

This statement is supported in the policy by extensive responsibilities during the review and publication process for editors, reviewers, authors, and the AGU as an organization. These responsibilities are designed to enhance and protect the editorial, peer review, and publishing process from commercial, personal, political, or other nonscientific influences.

The AGU Scientific Integrity and Professional Ethics Policy contains a broad definition of scientific misconduct that includes any violation of the codes of conduct and professional ethics as provided in the AGU policy as well as the formal definition of research misconduct from the Federal Policy on Research Misconduct. The Task Force had extensive discussions on the advantages and disadvantages of broadening the definition of scientific misconduct beyond the federal policy definition of plagiarism, fabrication, and falsification to include questionable research practices and ethics. Ultimately, it was agreed that the codes of conduct in the policy represent behaviors critical to ensuring scientific integrity and responsible science.

The final section of the policy, Processes for Allegations of Scientific Misconduct, contains explicit instructions for filing, receiving, and responding to allegations, procedures for conducting investigations, recommending sanctions, and other actions that can be taken in response to an outcome, and the conditions of appeal and dismissal. Allegations of scientific misconduct may be submitted to AGU by anyone when the alleged action is directly connected to a program operated under the direction of the AGU, including its publications, presentations, and meetings. A person filing an allegation is referred to as the complainant. When an allegation received by AGU also involves federally funded research and meets the federal definition of research misconduct, AGU will follow the reporting requirements of the Policy on Federal Research Misconduct. The allegation process also contains numerous points of review as well as opportunities for anyone who has an allegation filed against them (known as a respondent) to respond and appeal during the process. Strict confidentiality is maintained throughout the process.

This section of the policy also establishes an AGU Ethics Committee that reports to the AGU Board of Directors and is chaired by the AGU Past President. The Ethics Committee is responsible for investigating allegations that are deemed by the Chair of the Ethics Committee to be substantial and in need of further inquiry. The Ethics Committee membership consists of representatives from AGU committees and one editor. Members may be added to a specific investigation to provide subject matter expertise pursuant to the allegation at the discretion of the Chair.

The Board of Directors has the final authority to determine what actions will be taken, if an allegation of scientific misconduct is found to be substantiated and appeals of decisions go directly to the Board. Sanctions include a broad range of actions that vary in severity from written reprimands to suspension of membership, expulsion from AGU, rejection of a manuscript, notification to the home institution, and most severely, a public notice regarding the misconduct.

CHALLENGES AND LESSONS LEARNED IN EVALUATING AND RESOLVING SCIENTIFIC MISCONDUCT ALLEGATIONS

Although the AGU Scientific Integrity and Professional Ethics Policy contains detailed directions for evaluating and resolving misconduct cases, it is clear that every case is unique and requires careful review to determine if an investigation is warranted. Any allegation filed under this policy requires the name of the complainant as well as a rigorous explanation of any supporting evidence of why the allegation constitutes misconduct under the policy. This requirement provides a filter for unwarranted allegations. Further, initial review by the Chair and the requirement of strict confidentiality allows for dismissal of an allegation before notifying or damaging the reputation of an innocent member. Another consideration is that scientists do make honest mistakes, which need to be corrected and prevented, but are not misconduct. The greatest difficulty in dealing with an allegation arises when all parties are uncooperative or have no confidence in the process. It is critical that correspondence be professional and for due process and confidentiality to be maintained.

In resolving allegations and determining sanctions, it is important to consider the impact of the misconduct and determine if a scientist who has committed misconduct can recover, and in the future, contribute important science. One of the biggest challenges is encountered when an allegation affects a group, laboratory, or whole institution. This is when organizations need to step in to help make sure that the misconduct of one scientist does not affect the careers of many scientists.

Since implementing the new policy, the AGU has received 18 formal allegations of misconduct, ranging from plagiarism and authorship disputes to scientific misrepresentation. All but two of the allegations have been resolved before having to utilize the Ethics Committee. Because the policy requires a detailed explanation of the allegation, AGU was able to readily understand the nature of the misconduct and work with both the complainants and respondents to resolve the allegations. Some of the allegations submitted did not follow instructions and needed extensive follow-up to be resolved.

Whenever possible, it is preferable for an ethics or integrity issue to be resolved among the complainants and respondents first, before going to a formal allegation, or before going to the Ethics Committee. For example, the AGU received a formal allegation of misconduct involving a meeting abstract submission. The allegation, received a week before the meeting, claimed that names of some coauthors were omitted from the list of authors. As the AGU began to gather supporting and refuting information, it was revealed that the involved parties had agreed to meet and discuss author attribution but scheduling conflicts prevented this meeting from taking place before the abstract was submitted to the AGU. The AGU was able to work with the complainants and respondents to resolve the issue and concluded that the author omission was the result of an unintentional mistake and the complainant withdrew the allegation.

Because the policy and AGU ethics Web site is explicit in its processes and what to expect, AGU was able to quickly refer complainants and respondents to these resources to answer questions. AGU was able to adhere to the policy for most cases; however, they found that the time frames for some parts of the process needed to be adjusted to accommodate the complexity of the issue. Some cases were particularly difficult, for instance, the AGU collaborated with the home institution to investigate the context and merit of a formal allegation claiming authorship on a paper published in an AGU journal. This unique allegation required a thorough examination of research participation and international intellectual property issues that dated back 10 years.

IMPORTANCE OF EDUCATION AND A POSITIVE RESEARCH ENVIRONMENT

The AGU Ethics Task Force made a number of recommendations for implementing the new policy. Among them were

1. Provide opportunities for members to discuss this new policy, hold symposia on scientific integrity and professional ethics, and provide workshops on integrity and ethics.
2. Provide special sessions, workshops, and training aimed specifically at students and early career scientists.
3. Create a system for collecting, tracking, recording, and archiving relevant information from the allegation and publications processes related to scientific misconduct.
4. Publish annually for the members a summary of statistics regarding the information collected above to demonstrate AGU use of the policy and help the science community better understand and be aware of scientific misconduct in the community.
5. Work with sister societies to reinforce these policies across the international scientific enterprise.
6. Consider requiring the reading and agreeing to this policy as a condition of membership.

The purpose of these recommendations are to begin building a positive, aware environment of the importance of professional ethics and scientific integrity, provide education, invite dialogue, and, through awareness, prevent or reduce future misconduct. The AGU has implemented most of the recommendations, including providing a comprehensive ethics Web site, dedicated ethics e-mail, and internal system of archiving and tracking allegations and requiring acknowledgment of the ethics and scientific integrity policy upon joining and renewing membership. The AGU has held several ethics events at their annual Fall Meeting including an Ethics Town Hall in 2012, an Ethics Workshop in 2013, and an ethics dialogue with students in 2013. The AGU is also considering how it might make public the statistics and general information on the misconduct cases it has processed. Recently, the AGU participated in the American Geosciences Institute Leadership Forum on Ethics and was one of 20 societies who agreed on a consensus statement asserting the importance of ethics in the geosciences and proposing that the geoscience societies develop a common Code of Ethics (http://www.agiweb.org/gap/events/LF13/index.html).

Understanding the effectiveness of education and sanctions on reducing scientific misconduct is challenging and complex (Weed, 1998; NRC, 2002). There are some studies available on how much misconduct exists, but it is difficult to extrapolate the results since the majority of these studies focus on the biological/medical community (Titus et al., 2008; Fanelli 2009; Fang et al., 2012; Mayer, 2015). However, as Wolpe points out in his recent article (Wolpe, 2013),

What sociological research has shown us for almost a century is that well-meaning people, when placed in an environment that puts pressure on them to behave in certain ways, will tend to do so. If the environment encourages integrity, transparency, and fidelity to the work, people will tend to behave that way. If the environment encourages results at any cost, a me-first attitude, or cutthroat competition, people will generally conform their behavior to that mode of thinking as well.

Research organizations, academic institutions, and professional societies should proactively encourage an environment of professional ethics and scientific integrity. Students and scientists should receive consistent education and training on the importance of ethics and integrity in scientific research. There are substantial education resources available. One source of excellent education materials and guidance on ethics and scientific integrity is the Office of Research Integrity of the National Institutes of Health (NIH) (http://ori.hhs.gov/). Another source is the National Research Council's handbook "On Being a Scientist," which provides case studies and activities for constructive dialogue on ethics and integrity issues. Section 7009 of the America Creating Opportunities to Meaningfully Promote Excellence in Technology, Education and Science Act (42 U.S.C. § 1862o-1) requires each institution applying for financial assistance from the National Science Foundation (NSF) for science and engineering research or education to describe in its grant proposal a plan to provide appropriate training and oversight in the responsible and ethical conduct of research to undergraduate students, graduate students, and postdoctoral researchers participating in the proposed research project. Education resources and guidance for this requirement for grant proposals at the NSF are available from http://www.nsf.gov/bfa/dias/policy/rcr.jsp. Both the NIH and NSF are currently funding research into how effective such education is in mitigating scientific misconduct.

CONCLUSIONS

The science environment has changed dramatically in the last 20 years and continues to change. Accordingly, professional ethics and scientific integrity policies need to be responsive to this changing environment, not only responding to the traditional scientific misconduct defined by plagiarism, falsification, and fabrication, but also providing codes of conduct that inspire scientists to act ethically and with integrity within the fast paced, competitive, contentious, digital world in which we live. Policies should be created by listening to the users of the policy and accounting for all that may be affected. Comprehensive processes and dedicated resources are needed to successfully implement an ethics policy. Capturing professional ethical behaviors such as honesty, striving for excellence, supporting a healthy sustainable earth, valuing the benefit to society of our work, respecting all living things, unselfish cooperation, mentoring others, sharing data, and avoiding conflict of interest situations are critical in creating a healthy proactive science environment. Teaching and valuing responsible scientific practice, objectivity, sound record keeping, accessibility of data, and promoting honest and fair peer review are among the important features needed for scientific integrity to flourish. Science organizations need to take responsibility for promoting and providing positive science environments where ethics and integrity are valued above competition and reward. Education in ethics and sound scientific practice is critical for students and young scientists.

REFERENCES

American Association for the Advancement of Science, 2011. Statement of the Board of Directors Regarding Personal Attacks on Climate Scientists. Available from: http://www.aaas.org/sites/default/files/migrate/uploads/0629board_statement.pdf.

American Association for the Advancement of Science, 2013. News: Scientific Integrity and Transparency. Available from: http://www.aaas.org/news/aaas-participates-capitol-hill-discussion-scientific-integrity-and-transparency.

American Geophysical Union, 2013. AGU Scientific Integrity and Professional Ethics Policy. Available from: http://ethics.agu.org/policy/.

Committee on Publication Ethics, 2011a. Code of Conduct for Journal Editors. Available from: http://www.publicationethics.org/files/Code_of_conduct_for_journal_editors_Mar11.pdf.

Committee on Publication Ethics, 2011b. Flowcharts for Cases of Suspected Misconduct. Available from: http://www.publicationethics.org/resources/flowcharts.

Council of Scientific Editors, 2012. White Paper on Promoting Integrity in Scientific Journal Publications. Available from: http://www.councilscienceeditors.org/wp-content/uploads/entire_whitepaper.pdf.

Department of Health Education and Welfare, 1979. The Belmont Report on Biomedical Research. Available from: http://www.hhs.gov/ohrp/humansubjects/guidance/belmont.html.

Department of Interior, 2011. Scientific and Scholarly Integrity Policy. Available from: http://www.doi.gov/scientificintegrity/index.cfm.

Fanelli, D., 2009. How many scientists fabricate and falsify research? A systematic review and meta-analysis of survey data. PLoS One, http://dx.doi.org/10.1371/journal.pone.0005738. Available from: http://www.plosone.org/article/info%3Adoi%2F10.1371%2Fjournal.pone.0005738#pone-0005738-g006.

Fang, F.C., Steen, R.G., Casadevall, A., 2012. Misconduct Accounts for the majority of Retracted scientific publications. Proc. Natl. Acad. Sci. U.S.A. 109 (42) http://dx.doi.org/10.1073/pnas.1212247109. Available from: http://www.pnas.org/content/109/42/17028.

Federal Policy on Research Misconduct. Available from: http://www.aps.org/policy/statements/federalpolicy.cfm.

Holdren, J.P., 2013. Memorandum for the Heads of Executive Departments and Agencies, Subject: Increasing Access to the Results of Federally Funded Scientific Research. Available from: http://www.whitehouse.gov/sites/default/files/microsites/ostp/ostp_public_access_memo_2013.pdf.

International Committee of the Red Cross, 1950. Principals of International Law Recognized in the Charter of the Nuremburg Tribunal and the Judgment of the Tribunal. Available from: http://www.icrc.org/ihl.nsf/INTRO/390?OpenDocument The Nuremburg Principles.

Mayer, T., 2015. Research integrity: the bedrock of geosciences. In: Wyss, M., Peppoloni, S. (Eds.), Geoethics: Ethical Challenges and Case Studies in Earth Science. Elsevier, Waltham, Massachusetts, pp. 71–81.

National Oceanic and Atmospheric Administration, 2011. NOAA Administrative Order 202-735D on Scientific Integrity. Available from: http://ethics.iit.edu/ecodes/node/5088.

National Research Council, 1992. Responsible Science. Ensuring the Integrity of the Research Process, vol. I, The National Academies Press, Washington, DC.

National Research Council, 2002. Integrity in Scientific Research: Creating an Environment that Promotes Responsible Conduct. The National Academies Press, Washington, DC.

National Research Council, 2009a. Ensuring the Integrity, Accessibility, and Stewardship of Research Data in the Digital Age. The National Academies Press, Washington, DC.

National Research Council, 2009b. On Being a Scientist: A Guide to Responsible Conduct in Research, third ed. The National Academies Press, Washington, DC.

Obama, B., 2009. Memorandum for the Heads of Executive Departments and Agencies, Subject: Scientific Integrity. Available from: http://www.whitehouse.gov/the-press-office/memorandum-heads-executive-departments-and-agencies-3-9-09.

Steneck, N.H., 2013. Responsible Advocacy in Science: Standards, Benefits, and Risks. written for the Workshop on Advocacy in Science American Association for the Advancement of Science. Available from: http://www.aaas.org/report/report-responsible-advocacy-science-standards-benefits-and-risks.

Taylor and Francis UK, 2006. Ethical Guidelines to Publication of Scientific Research. Available from: http://www.tandf.co.uk/journals/pdf/announcements/tmph_guidelines06.pdf.

Titus, S.L., Wells, J.A., Rhoades, L.J., 2008. Repairing research integrity. Nature 453. Available from: http://www.nature.com/nature/journal/v453/n7198/pdf/453980a.pdf.

Union of Concerned Scientists, 2009. Scientific Integrity and the Presidential Transition. Available from: http://www.ucsusa.org/scientific_integrity/solutions/big_picture_solutions/transition-main-page.html.

Weed, D.L., 1998. Preventing scientific misconduct. Am. J. Public Health 88 (1).

Wolpe, P.R., 2013. Scientific work in a changing environment. Off. Res. Integr. Newsl. 22 (1). Available from: http://ori.hhs.gov/images/ddblock/dec_vol22_no1.pdf.

World Association of Medical Editors, 2009. Publication Ethics for Medical Journals. Available from:http://www.wame.org/resources/ethics-resources/publication-ethics-policies-for-medical-journals.

World Conference on Research Integrity, 2010. Singapore Statement on Research Integrity. Available from: http://www.singaporestatement.org/statement.html.

World Conference on Research Integrity, 2013. Montreal Statement on Research Integrity in Cross-Boundary Research Collaborations. Available from: http://www.wcri2013.org/doc-pdf/MontrealStatement.pdf.

Chapter 9

Ethical Behavior in Relation to the Scholarly Community: A Discussion on Plagiarism

Stefano Solarino
Istituto Nazionale di Geofisica e Vulcanologia, Genova, Italy

Chapter Outline

Abstract

The action of using someone else's production, ideas, or research without acknowledging the source and then claiming credit for them is known as plagiarism. Plagiarism is in principle a moral offense; it is not always an illegal action but certainly is an ethical complex case. Copying without permission or stealing someone else's work violates the standard codes of scholarly conduct and ethical behavior in professional scientific research. Science may be somewhat more prone to plagiarism in that scholars exchange ideas and proposals more frequently than artists, for example. Workshops, scientific meetings, and round tables are places where primitive and preliminary researches are presented and discussed before their official presentation. This chapter discusses plagiarism and self plagiarism and tries to point out the role of the scientific community toward this kind of scientific misconduct.

Keywords: Copyright infringement; Ethical conduct; Plagiarism.

INTRODUCTION

Plagiarism is committed every time that literary, scientific, and artistic works are made public, on own accord or not, without permission or acknowledgments of the authors, or better of the rights' owner, by printing, publishing, playing them, or adopting more subtle forms.

Geoethics. http://dx.doi.org/10.1016/B978-0-12-799935-7.00009-5

Although plagiarism is in principle an offense, it is not always an illegal action. Copying ideas or proposals that are not registered and protected under a copyright is often not considered a crime (theft or stealing) in civil or penal codes worldwide. In principle, a copyright is established anytime an original work is fixed in a tangible medium of expression, for instance, as a word processing document, a Web page, a digital photograph, or a recorded podcast, but only an official registration of the work ensures the chance of a public claim in case of infringement. Sometimes, there is a fine line between common and personal, improved knowledge and it is not easy to prove plagiarism, even in cases of research or productions protected by copyright.

However, plagiarism is always an unethical action, whether or not a copyright is established or registered, and in science, plagiarism is listed as one of the main scientific misconducts. The problem of plagiarism is therefore an ethical complex case, where recognizing the will to copy or steal may make the difference, although the action itself is always a violation of the standard codes of professional scientific research.

In this chapter, I discuss the aspects of plagiarism in science, although most of the observations described here can be applied to other nonscientific fields. However, natural or physical science is somewhat more prone to plagiarism in that scholars exchange ideas and proposals more frequently than artists, for example. In fact, differently to what happens in other disciplines, preliminary scientific ideas or results are often discussed among scientists prior to publication. Sometimes, new ideas are disseminated in the form of preliminary reports. Workshops, scientific meetings, and round tables are places where primitive and preliminary researches are presented and discussed before their official presentation. In addition, the peer-review system makes original work available to other scholars, who may develop similar studies taking inspiration from the manuscripts under review or simply copy parts of them. Finally, scientific research is often the result of studies conducted by a group of scholars and academics collaborating in search of a specific target or result. In these cases, it is often difficult to recognize the contribution of each scientist and it is even more difficult to assess which intuition or idea, and then which author, made the final result successful. The presentation of formerly obtained results by a single scientist or together with a different research group is thus always up for a debate about plagiarism. If no clear agreements about the rights have been a priori established among the scientists developing the original study, the latter remains a property of the whole group and should not be used or published by the single researcher. In most of these cases, only ethics can discriminate between correct and inappropriate behavior.

This chapter tries to discuss plagiarism; it is not meant to be a complete treatise on the topic because it has many different societal and legal aspects that cannot be taken into account in a short paper and moreover I do not have enough experience to deal with them all. The phenomenon is then merely analyzed under the point of view of ethics. In particular, the topics treated here can be summarized as follows:

- Definition of plagiarism in science, with particular attention to geosciences
- The legal aspects of plagiarism (for copyrighted artwork and research, in case the work is paid for)
- The various levels of plagiarism
- The complex problem of assessing plagiarism in science. Some specific issues will be discussed—self-plagiarism
- The role of ethics and the scientific community

DEFINITION OF PLAGIARISM

Plagiarism is a scientific misconduct: in the academic world there is a consensus on this statement. The action of copying without permission or stealing someone else's work violates the standard codes of scholarly conduct and ethical behavior in professional scientific research. Moreover, the words "cheating" or "stealing" already imply a negative and immoral action.

Although the concept of plagiarism is intuitive and the meaning behind the word is innate, a unique sentence to define it is not easy to achieve. There are a number of possible definitions, each one reflecting a different aspect of the same concept.

In a simple way typical of dictionaries, one definition of plagiarism arbitrarily chosen is "to steal and pass off (the ideas or words of another) as one's own: use (another's production) without crediting the source" (Merriam-Webster, an Encyclopaedia Britannica company, http://www.merriam-webster.com/dictionary/plagiarize). This definition only describes the action, it does not include the special kinds of plagiarisms that will be discussed later in this chapter, namely, self-plagiarism and text duplication (the definition reads only about another's production, ideas, or words), and does not consider the ethical aspects.

Another interesting definition, which may be found on the Web page http://wpacouncil.org/node/9, reads "In an instructional setting, plagiarism occurs when a writer deliberately uses someone else's language, ideas, or other original (not common-knowledge) material without acknowledging its source." This definition has a more ethical issue (the word deliberately), but apparently it only refers to writers and to "the instructional setting." This reference is noteworthy, because I will later describe how plagiarism is common among students.

A more complete definition can be found on Wikipedia (http://en.wikipedia.org/wiki/Scientific_misconduct): "(plagiarism) is the appropriation of another person's ideas, processes, results, or words without giving appropriate credit. One form is the appropriation of the ideas and results of others, and publishing as to make it appear the author had performed all the work under which the data was obtained. A subset is *citation plagiarism*—willful or negligent failure to appropriately credit other or prior discoverers, so as to give an improper impression of priority. Sometimes it is difficult to guess whether authors intentionally ignored a highly relevant cite or lacked knowledge of the prior work." This definition is quite complete, although it lacks the reference to self-plagiarism.

Overall, we could define plagiarism as the immoral, unethical, and, under certain circumstances, illegal actions of using, intentionally or not, own or someone else's production, ideas or research without crediting the source, with the aim to appear as the author or owner of the production.

THE LEGAL ASPECTS OF PLAGIARISM

An author transfers the rights of his/her work to a third party through a contract that also defines, if applicable, the financial matters. In most cases, a payment is due to the author for the transfer of copyright and sometimes also royalties are established and paid proportionally to the "success" in terms of sales of the work. This is common in the arts, with slight differences between the various expressions. In music, for example, the artist by transferring the copyright loses any control on the final use of the artwork and is not allowed to make copies of it, but he/she gets some remuneration for every media sold (in general a fraction of the amount cashed by the "new" owner of the rights) and for every public performance (concert, radio or TV broadcasts, and soundtrack) through specific agencies that collect a fee from the spectators or the theaters.

Often in science, the publication of a scientific work, and then its transfer of copyright, is on a "free from payment" basis, except in special cases. The author does not get any reimbursement if the copyright passes to the publishing company. The author loses the property of figures and tables included in the manuscript and is not allowed to make copies, while the publishing company may sell the work (for example, including the article in a book or a journal) to cover the publishing expenses and to get profit. Any request of copying or printing copyrighted material from third parties must be addressed to the publishing company as the final owner of the rights. The company may then decide to allow the material to be used for free or upon reimbursement.

However, authors are granted "scholarly rights," which allow them to use their articles for noncommercial purposes. They are then allowed to use their own publications at a conference, meeting, or for teaching purposes, to use in a subsequent compilation of the author's works or include in a thesis or dissertation, or to reuse portions or extracts from the article in other works.

Within the umbrella of these "scholarly rights," other colleagues may use short sentences from the original work and report them in their articles provided that the original authors are mentioned together with the complete references and the citation does not provide economic benefits to the user. If exactly identical to the original, the copied lines must be included in quotation marks to highlight that the source is published somewhere else.

From the ethics point of view, there are no big differences between the cases of free or paid transfer of copyright: violation of the property is always a fault no matter what kind of contract is involved. However, it becomes a crime, with legal consequences, when economic aspects are involved. While it is in general tolerated if it carries no economic loss, the use of copyrighted material is worldwide considered illegal and prosecuted when the responsible gains from the action. In the academic environment, there are no financial issues, as already stated, and the plagiarism becomes a mere ethical issue.

THE DIFFERENT FORMS OF PLAGIARISM

Plagiarism may be carried out at various levels. In principle, anytime a scholar claims credit for someone else's work or ideas or uses/copies parts of someone else's work without acknowledging the source, a plagiarism is carried out. The severity of the action, both from a legal and an ethical point of view, depends on many factors: the "amount" of copied material, the "importance" of the material in a context, and the form with which the plagiarism is carried out.

In science, plagiarism of an entire research project is quite uncommon. On the one hand, the scientific system requires the results to be justified by original data and to be obtained in consecutive steps. It is then difficult to prove to have achieved similar (possibly copied) results with different data sets or skipping intermediate steps. On the other hand, the publishing policy requires scientific results to be peer reviewed by experts who are in principle capable to recognize how much of the results are founded on the method and especially on the data discussed in a scientific article.

The arduousness to copy entire research results not surprisingly makes recognizing plagiarism more subtle. In most cases, only parts of a research project are copied, often changing the original sentences to mask their source. In these cases, it is much more difficult to assess the commitment of plagiarism. In a random level of severity, plagiarism may take the form of (http://www.ethicsed. org/programs/integrity-works/pdf/Plagiarism.pdf)

- failing to put a quotation in quotation marks
- giving incorrect or no information about the source of a quotation
- changing words but copying the sentence structure without giving appropriate credit
- copying words or ideas from someone else without giving credit
- turning in someone else's work as one's own

Some scientists wrongly believe that rephrasing a sentence taken from someone else's work is allowed and frees the user from quoting the original source. But changing the words of a text is not sufficient to avoid plagiarism. The fault in plagiarism is to retain the original idea or result, no matter what kind of alteration is made in the sentences describing it. Most of the time then, searching for plagiarism becomes a complicated job to find both copied and masked original content.

In a peer-reviewed journal, the reviewers should be experienced enough to correctly recognize the original author of the studies they are about to judge, provided that these reviewers are adequately chosen and familiar with the topics. However, there are exceptions to this point. Sometimes reviewers have a broader experience and, when asked to review specific or innovative studies, may not be able to assign the origin of a research result to the right author.

There exists software (e.g., Duplichecker, the plagiarism checker) designed to search for copied material, which can help in finding frauds. The software is designed to cross-check a manuscript over a large database of digital publications and to detect pairs, estimating a percentage of similarity between the two texts in what we can call the quantitative part of the detecting process. The result of this search for duplicates must then be checked with the reference section of the investigated article to identify which entry is not listed and then plagiarized. The whole operation may take some time if the threshold to detect copies is set to low percentage. A thorough description of the detecting tools is out of the scope of this chapter and will not be given here. The reader may find information by searching the Web. Most of these programs in their free versions have a limited functionality and require a registration or the payment of a fee for full and unlimited use. They may fail in finding duplications if texts have been rearranged and they may not work properly with languages other than English or when cross-checking between different languages, especially when different characters are used, like in Russian or Chinese. Finally, they can be run on digital sources only: thousands of printed books and theses have never been scanned and may be used for cheating. These detecting tools are not often provided to the editors and the reviewers, and thus the adoption of any checking tool is a free choice of the person responsible for the peer-reviewing process.

If the results are presented by a scholar as his own in an environment different than science (TV or radio broadcasts, newspapers, magazines) often no proof, scientific explanation, or hypothesis is required to ascertain the ownership, making plagiarism easier. Finally, some journals are not based on the peer-review system and no assessment of intellectual property right (human or virtual) is performed.

DETECTING PLAGIARISM

As shown in the previous sections of this chapter, it is not easy to detect plagiarism in science. Every day hundreds of scientific articles are published, in printed or electronic form. Even assuming that plagiarism may be performed on a whole text (for example, complete copy of a manuscript with slight to significant text rearrangements), for a human it would be impossible or at least hard to discover it. A reviewer may be suspicious when reading a sentence about a finding or result not adequately supported by the data, but it is impossible, or at least difficult, to get back to the original source. It may be possible to detect self-plagiarism, described in the next chapter, if the reviewer gets suspicious about titles of similar articles by the same author, but in case of self-plagiarism there is a further difficulty in assessing whether a work is a clone because it has to be estimated if and how much the latest work is "improved." This aspect will be discussed later.

On the other hand, the automated tools for searching for duplicate texts may fail if the manuscript is translated in/from a different language, if the original source is not in electronic format or simply not available on the Web (some articles can only be accessed via a subscription and cannot be browsed automatically), if only very short sentences of the original work are copied. Sometimes even a "smart" rephrasing of the original text may be not detected as a copy by the detection software.

Many articles have been published about plagiarism and its ethical aspects (Gowrinath, 2012; Gelman and Basbøll, 2013 to quote two recent ones). Most of these articles deal with single case studies or with the particular situation of the single country. Conversely, only few studies on the statistics of occurrence of the phenomenon of plagiarism in science have been conducted and published.

In these studies, the number of likely-to-be-recognized "pairs" has been estimated using standard software for duplication search or anonymous interviews in particular research fields. According to these estimates, the duplication of manuscripts or parts of them can be quantified in a fair fraction of the published work. For the results of a study on Medline, the database of biomedical publications, see Long et al. (2009) or Errami and Garner (2008). Here, I just underline that according to these studies, the number of suspect duplicates ranges from 2 to 9 per 1000 publications (about 421 potential copies on 62,000 analyzed publications), with a steady increase from 1975 to 2005, proportional to the increasing number of published papers.

In a previous work (Sorokina et al., 2006) conducted on arXiv, an open-access archive of mathematics, physics, computer science, biology, and statistics papers, the authors detected some 500 cases of likely plagiarism and over 1000 cases of likely mild plagiarism in a set of data consisting of about 287,000 publications for the period 1991 to 2005. In all the above described cases, some duplicates are from the same author publishing twice his/her own work alone or in cooperation. I therefore spend a few words about self-plagiarism.

SELF-PLAGIARISM

Self-plagiarism is defined as a scholar reusing parts of his or her own articles and publishing them in another journal without citing the original study. Another common form of fraud is when a scholar duplicates texts and content from earlier papers and quotes the original. The two forms of cheating are not substantially different and the terms self-plagiarism and text duplication are sometimes considered synonyms. An interesting treatise on self-plagiarism and on the guidelines to recognize it is given in Bretag and Mahmud (2009).

If it is not common, as stated above, to copy entire manuscripts published by other scientists, it is conversely fairly common to do it with one's own work. It is frequent to read similar articles from the same authors in different journals. In most cases, the two articles are submitted to two different journals at the same time. In the first step of the submission process, authors are required to confirm to both editors that the manuscript is unpublished and has not been submitted to another journal. According to some points of view, both manuscripts are original in that they are not accepted for peer review yet and this supposedly circumvents the rule and the authors believe they have not made a false declaration and cannot be accused of plagiarism.

This kind of action is unfair because it wastes editor's time in case both journals would accept the contribution. The authors should decide in which journal to publish. But it is of course even less ethical if the authors decide to publish both articles, which is actually technically possible because no publishing company or software can detect clones on their way to publication.

The double submission and acceptance clearly shows a will of the authors to commit fraud attempting to increase their reputation. The visibility and reputation of individual scholars are linked to the number of publications through indexes. The citation metrics just counts the number of published papers and their citations, no matter what the content or the exclusive/innovative character of the papers is. Two similar or clone articles will count as two distinct entries in the computation of metrics indexes.

Since the citation metrics is sometimes used to hire or fund scientists, as discussed in Solarino (2012), the rating scheme based on the number of published articles may augment the number of cases of plagiarism.

There exists no comprehensive statistics on self-plagiarism, but the little information available shows that the phenomenon is not negligible, as the significant number of retracted articles and the increased attention of the publishing companies toward the problem are confirming (Reich, 2010). The number of cloned articles dropped after the adoption of a software to detect fraudulent manuscripts.

However, on the threshold to consider an article as self-plagiarized there is no consensus. The scientific community agrees that each scientific paper should be an original contribution. In principle, it is accepted that two articles may contain similar parts (e.g., the description of the method), but the other sections (results, discussion, and abstract) should present new results. It is not straightforward to estimate, if the slight differences that sometimes characterize two papers are relevant enough to consider them as separate and distinct manuscripts.

The scientific method envisages the chance to develop research in several steps. Each of these improvements may be slight. It is well accepted that a research team presents in several, consecutive meetings improvements of results on the same topic, often using the same data set or an update of them. But it cannot be accepted with publications: only significant evolution of the original work should be published, and a subsequent article should be kept on hold until new material is available. It is not just an ethical matter; since the authors passed the copyright of the previous manuscript to the publishing company, they are not cheating themselves, but a third party, that could possibly start legal procedures against them.

THE ROLES OF ETHICS AND THE SCIENTIFIC COMMUNITY

Plagiarism is a global and frequent phenomenon, probably larger than documents and statistics estimate. The recent widespread dissemination of electronic documents via the Internet made plagiarism easier. Its occurrence will probably rise in the future, especially if no detecting policy and means will be adopted. The Internet has several advantages compared to traditional bibliographic sources; it is fast, easy to browse, and always available. The Web is, among other things, a huge library. It is then becoming more common for scientists to obtain information and data from Web sources instead of from printed books, with the additional convenience to get what they need already in an electronic format.

Having the Internet available as a source of bibliographic texts has introduced unprecedented problems in the fundaments of ethics. They basically are "technical" questions, but they directly reflect the moral aspects of science. For example, one of the most common sources for information on the Web is Wikipedia. When searching the Web, this site is often at the top in the list of results. The information contained within a Wikipedia page is not always fully reliable; however, this topic is out of the scope of this chapter and is discussed in Solarino (2014). Many scientists and nonscientists get hints and data from Wikipedia, whose pages are compiled by several authors in a long process of "quality, reliability, and completeness" improvement. To avoid plagiarism and to deal with the codes of professional scientific research, one should then cite the right authors of the copied sentences, but these are not easy to identify (their names are published on the "modify source" page, sometimes under nicknames, and in some cases the single sentence is the result of the joint contribution of more volunteers).

In this case, quoting the Web page appears to be the appropriate solution, but it adds an additional difficulty because a Web address cannot be considered a "standard" bibliographic reference (in some journals Web pages do not have a proper rule to be quoted or are not allowed yet) and it is also a form of "authorized" plagiarism because it is similar to quote, in a nonvirtual environment, the publishing company rather than the author. The same problem applies to blogs, articles, and other nonreferenced sources on the Web. These technical problems may induce the authors to omit the reference, and to commit plagiarism.

Despite the recognition of the problem of nonreferenced use of other authors' material, scientists of many countries are fighting against plagiarism without any intervention from their governments. There are exceptions. In the United States, for example, a statutory body against scientific misconduct is established. The Office of Research Integrity takes care of the offenses to research integrity and implements activities and programs to teach the responsible conduct of research, promote research integrity, prevent research misconduct, and improve the handling of allegations of research misconduct; it develops policies, procedures, and regulations related to the detection, investigation, and prevention of research misconduct and the responsible conduct of research (http://en.wikipedia.org/wiki/Office_of_Research_Integrity).

In all other countries, only the principle of ethics and the standard codes of professional scientific research can be claimed to defeat plagiarism. There is a lively debate going on about all types of scientific misconduct in the research environment and the society, in devoted meetings (e.g., The World Conference on Research Integrity) or in the literature (Kalleberg, 2007, 2010).

These precepts should equally apply to any level of a scholar's career, but are often neglected by graduate or PhD students compiling their dissertations. With some notable exceptions in instructional structures where strict principles to detect, document, and report plagiarism are established together with the associated disciplinary procedures, the plagiarism carried out by students is not always detected and sometimes even tolerated by universities, colleges, and advisors. The reason for students to copy from an original work is intuitive. They do not have enough experience for writing a scientific work on their own, do not have enough skill for doing that in a sound manner, and especially do not have much time to devote to their "first" (and sometimes only, few of them will continue a career in the academic environment) article, being often in the rush to respect the deadlines. On the other hand, even the point of view of advisors is easy to understand: by allowing students to copy from an existing work, their duty is reduced to check the homogeneity of the various sections of the thesis rather than assisting the student. This way they can save time. Moreover, although the thesis is an official document, it will be hardly checked for duplicates (if no policy to avoid that is in place) and in most cases will be disseminated to a limited number of readers (just the members of the supervisory committee). However, these are not by any means convincing motivations for tolerating plagiarism: students of today will become scholars of tomorrow. If they are not requested to adhere to the ethical codes in the beginning of their career, they will likely be tempted to behave similarly in the rest of their scientific activity.

Since 2009, an Internet-based plagiarism-prevention service named Turnitin specifically designed for instructional institutions is available. A description of the software is out of the scope of this chapter, and will not be treated here. The service is similar to other duplicate checkers; universities and high schools buy licenses to submit essays to the Turnitin Web site, which checks the documents for copied content. The results can be used to identify similarities to existing sources, to warn students on how to avoid plagiarism and improve their writing (http://en.wikipedia.org/wiki/TurnItIn). The service is adopted by some 10,000 educational institutions, located in 126 countries. In 2012, a multilingual translation technology has been implemented in order to let Turnitin to identify content potentially plagiarized from publications in other languages. The main goal of the service is to detect duplications, but it also has the aim to help students understand how plagiarism may take form.

Among students plagiarism turns out to be common, and according to the statistics provided by Turnitin regards about 14–20% of the analyzed worldwide student population. The scientific community must cooperate, in any country, to educate students to avoid plagiarism and to convince them that copied material equals poor scholarship because it simply presents someone else's work instead of their own.

Educating students is the lower level action to prevent plagiarism, but there are other activities to be undertaken by the publishing companies and the scientific community. Publishing houses could provide reviewers with ad hoc facilities to detect duplicates. This solution has probably been already adopted in some fields of science or medicine, but it is not common. I have mentioned in the previous sections that in some cases, the software for duplication detection works quite well. Any of the already existing services, accessible via a password from the Web site of the publishing company, would be probably enough to detect big portions of nonoriginal material. To avoid an increase in expenses, publishing houses could instead design their own software and could limit the search to the field of study of the article to be checked. This search for duplicates would be necessary only in the first stage in order to warn authors that plagiarism can be detected and to prevent scholars from copying. However, in this case, a careful choice of the threshold to discriminate between plagiarism and fair, correct use of sentences taken from other sources must be established. Moreover, detecting copied material must be followed by careful comparison of the proposed reference section in the article to estimate the amount of plagiarized material, which will increase the time for the completion of the reviewing process. An automated "alert" system could be implemented in the manuscript submission process in order to report to the reviewers how much of the manuscript they are going to review is already published. If this amount exceeds a fixed value, the reviewers could then cross-check the reference section. Even in case of nonplagiarism, this would help avoiding publishing a review paper, intended as an article where the fraction of novel data and results is low, if not declared as such.

The automated system could complement the quality check of figures, already implemented in many automatic manuscript submission platforms, as an additional control to warn the authors about weird amount of copied material and allowing them a chance to improve the manuscript before the submission by deleting nonreferenced material or adding the proper bibliography.

Publishing houses have of course also the chance to retract a manuscript, should the plagiarism be proven beyond a doubt. This is a strong action that must be carefully considered. The consequences of the action may vary: it does not occur frequently and there is no consensus and not much information about what happens to the retracted paper. For example, there may be problems in deleting the entry from the citation score of the scholar or, due to its electronic publication, the paper may remain available to the scientific community.

The described actions are technical solutions to an issue that has to do with moral behavior, so the strongest defense can be provided by ethics. The scientific community must cooperate in improving the moral and ethical inclination of scientists by making it clear that plagiarism is a dual dishonest action. It provides scholarly poor articles to the plagiarist and hurts authors who are not adequately referenced.

Moral issues in science are often considered a second priority problem or even a taboo and are rarely discussed in meetings and workshops, while it would be probably worth to hold devoted round tables, blogs, and discussions on the topic. Scientists, with few exceptions, are not required to sign or adhere to a code of conduct like it happens in private or public associations, which probably makes us scientists believe that moral and ethical problems are second order. It is my personal forecast that ethics in science will have a hard time in the next future, but it is probably not too late yet to intervene.

REFERENCES

Bretag, T., Mahmud, S., 2009. Self-plagiarism or appropriate textual re-use? J. Acad. Ethics 7, 193–205.

Errami, M., Garner, H., 2008. A tale of two citations. Nature 451, 397–399.

Gelman, A., Basbøll, T., 2013. To throw away data: plagiarism as a statistical crime. Am. Sci. 101 (3), 168.

Gowrinath, K., 2012. Plagiarism in scientific research. Narayana Med. J. 1 (2), 49–51.

Kalleberg, R., 2007. A reconstruction of the ethos of science. J. Classical Sociol. 7 (2), 137–160. ISSN:1468-795X.

Kalleberg, R., 2010. The ethos of science and the ethos of democracy. In: Calhoun, C. (Ed.), Robert K. Merton: Sociology of Science and Sociology as Science. Columbia University Press, pp. 182–213. ISBN:9780231151122.

Long, T.C., Errami, M., George, A.C., Sun, Z., Garner, H.R., 2009. Scientific integrity. Science 323, 1293–1294.

Reich, E.S., 2010. Self-plagiarism case prompts calls for agencies to tighten rules. Nature 468, 745.

Solarino, S., 2012. Impact Factor, Citation Index, H-Index: are researchers still free to choose where and how to publish their results? Ann. Geophys. 55 (3), 473–477.

Solarino, S., 2014. Geoethics and communication 2: Ethics, the notably absent from the Internet. In: Lollino, G., Arattano, M., Giardino, M., Oliveira, R., Peppoloni, S. (Eds.), Engineering Geology for Society and Territory - Volume 7. Education, Professional Ethics and Public Recognition of Engineering Geology, pp. 39–45. doi: 10.1007/978-3-319-09303-1_7.

Sorokina, D., Gehrke, J., Warner, S., Ginsparg, P., 2006. Plagiarism detection in arXiv. In: Clifton, C.W., Zhong, N., Liu, J., Wah, B.W., Wu, X. (Eds.), Proceedings of the Sixth International Conference on Data Mining ICDM, pp. 1070–1076.

Section III

THE ETHICS OF PRACTICE

Chapter 10

When Scientific Evidence is not Welcome...

Paul G. Richards

Lamont-Doherty Earth Observatory of Columbia University[1], Palisades, NY, USA

Abstract

This chapter concerns interactions in the 1980s between the technical community and the Reagan administration. Reports to the US Congress from the executive branch of government had stated that the Soviet Union was "in likely violation" of the bilateral Threshold Test Ban Treaty (TTBT), which had been negotiated between the USA and the USSR in 1974 and which imposed a limit of 150 kilotons for the explosive yield of any underground nuclear weapons test conducted by these two countries after March 1976.

The TTBT had led to the need for making estimates of explosive yield, and several methods came into use of which the most prominent was based on seismology. In the specialized work of estimating explosive yield from analysis of the strength of seismic signals, the expert community became convinced that it was inappropriate to claim that the Soviets were cheating on this arms control treaty.

In a revealing TV interview, Assistant Secretary of Defense Richard Perle stated with reference to the opinion of seismologists on the size of the largest Soviet tests, "It's not a question of scientific evidence. It's a question of scientists playing politics."

This chapter discusses Mr Perle's claim here, from a personal perspective: why I became involved; briefly, what was the scientific evidence; and, most importantly, to what degree was it true that scientists were playing politics? Mr. Perle stated, "I didn't much care what their answer was. It doesn't have any profound bearing on our policy."

To maintain a good professional reputation in the face of allegations of being biased because of a policy issue, an expert must be diligent to establish key facts and to defend them as bulwarks that may influence policy, with the potential to discredit policy makers who mischaracterize them. Seismological methods for measuring the size of the largest Soviet tests were endorsed by the results of two special nuclear explosions conducted in 1988 for which intrusive methods of monitoring were allowed.

The TTBT was eventually ratified in 1990 (by President G.H.W. Bush). Seismology continues to play a role in policy debates with reference to the much more important Comprehensive Test Ban Treaty of 1996.

1. This is Lamont-Doherty Earth Observatory Contribution number 7823.

Geoethics. http://dx.doi.org/10.1016/B978-0-12-799935-7.00010-1

Keywords: Underground nuclear explosions; Yield estimation; Nuclear arms control; Seismic monitoring.

Policy issues that have a strong technical component, rooted in the geosciences, include responses to the prospect of climate change, national security decisions on dual use technologies,[2] and options for nuclear arms control where there is an underlying need for consensus on technical aspects of monitoring capability. This chapter describes the treatment of a specific arms control policy issue that arose in the Reagan Administration (1981–1989). The technical component was very simple. It boiled down to an understanding of the relationship between the energy released by an underground nuclear explosion (UNE) and the P-wave magnitude of the seismic signals generated by the explosion. And because of the forcefully expressed opinions of a person speaking for the Administration, we are able to see in this case, very clearly, an unwelcome reaction to a conclusion that was strongly held by the technical community, and also some of the consequences of how this particular technical issue played out in the policy arena.

In the 1980s, the US Congress required annual reports from the executive branch of government on the Union of Soviet Socialist Republics' compliance with arms control agreements. Some of these reports stated that the Soviet Union was "in likely violation" of the bilateral Threshold Test Ban Treaty (TTBT), which had been negotiated between the United States and the Union of Soviet Socialist Republics in 1974 and which imposed a limit of 150 kilotons (kt)[3] on the explosive yield of any underground[4] nuclear test explosion conducted by these two countries after March 1976.

Many methods of estimating nuclear explosive yield have been developed using remote observations—going back to the very first nuclear test explosion, of July 1945 (Trinity), in the atmosphere. Nuclear testing moved underground in later years and estimates of explosive yield began to be made using seismological methods. And then with the TTBT it became necessary to interpret yield estimates in the political context of assessing compliance with a formal arms control treaty.

The basic seismological observations were not seriously in dispute and were as follows: The largest underground explosions conducted by the United States at the Nevada Test Site (NTS), at yields that were reported by US agencies to be somewhat less than the 150 kt threshold, had seismic magnitudes of about

---

2. "Dual use" in this context means, military and civilian, for example, with reference to satellites.
3. In this context, a kiloton is an energy unit, originally taken as the energy released by exploding 1000 tons of TNT, but today the formal definition is 1 kiloton = a trillion calories (~4.2 trillion joules). So, 1 g (of TNT equivalent) is 1000 calories.
4. In 1963, the Limited Test Ban Treaty, sometimes called the Atmospheric Test Ban Treaty, had banned nuclear testing in the oceans, in the atmosphere, and in space. It did not ban nuclear testing underground.

(a) original, STS   **(b)** original, NTS   **(c)** corrected, NTS

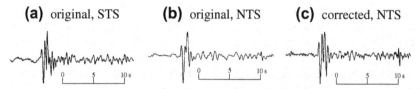

FIGURE 1   Here are three seismograms, showing about 15 s of the seismic signal at the station code named as EKA in Scotland, all with the same timescale. Seismograms are recordings of ground motion at a particular site as a function of time. In each case, the wave has traveled to EKA through the Earth's deep interior from an underground nuclear explosion several thousands of kilometers away. (a) is the signal at EKA, as originally recorded from an underground nuclear explosion at the Semipalatinsk Test Site (STS). (b) is the signal at EKA, as originally recorded from an underground nuclear explosion at the Nevada Test Site (NTS). (c) is a corrected version of (b), as described in the text. *Adapted from Douglas (1987).*[7]

5.6–5.7. The Soviet Union after March 1976 conducted its largest underground tests mostly at the Semipalatinsk Test Site in Kazakhstan, at magnitudes that steadily attained higher and higher values over a few years, and that by the early 1980s were up at around magnitude 6.1.[5] Since magnitude scales are logarithmic on a base of 10, this difference meant that the amplitude of signals recorded from the largest Soviet underground explosions were about $10^{0.4}$ or $10^{0.5}$ larger than the signals recorded from NTS. Since this factor is approximately 2.5–3, an assumption that the magnitude–yield relationship was the same for the Nevada and Semipalatinsk test sites led straightforwardly to estimates of the yield for the largest Soviet tests that were roughly three times greater than the largest tests in Nevada.[6]

But from the early 1970s to the mid-1980s, a growing body of evidence emerged from seismology that the magnitude–yield relationship was not the same for the two test sites. For example, there was the observation that for stations on shield regions, at distances of several thousands of kilometers from Nevada or Kazakhstan, the signals from Soviet explosions had significantly higher frequency content than those from US explosions. This feature is described in Figure 1, which shows three seismograms. The first signal, at a station in Eskdalemuir, Scotland, with code name EKA, is from an underground nuclear explosion on June 30, 1971, of magnitude 5.9, at the Semipalatinsk Test Site. The second signal, also recorded at EKA, is from an underground nuclear explosion in granite on June 2, 1966, of magnitude 5.6, at the Nevada Test Site.

5. Several different magnitude scales are in use by different agencies, and typically they report slightly different values for the same seismic event, whether it is an earthquake or an explosion. The different values arise because of the use of slightly different measurement techniques and use of data from different seismographic networks.
6. If it were further concluded that the yields of deployed Soviet nuclear weapons were three times larger than had been estimated, then there would be consequences for an evaluation of the effectiveness of a Soviet first strike.
7. Douglas, A., Differences in upper mantle attenuation between the Nevada and Shagan River test sites: can the effects be seen in *P*-wave seismograms? Bull. Seismol. Soc. Am. 77, 270–276, 1987.

Although the two test sites are at comparable distances from the recording station in Scotland, these seismograms are significantly different. The first, with its signal from Kazakhstan, contains higher frequencies than the second, with its signal from Nevada. From this and a broad range of other evidence, it is concluded that beneath the test site from which the higher frequencies are not observed, there must lie an attenuating region.

Furthermore, we can quantify the amount of this attenuation by modeling the way in which the original seismograms differ across the whole spectrum of frequencies. When the high-frequency components are restored to make the attenuated signal (from Nevada) match the unattenuated one, the outcome is as shown in the third seismogram of Figure 1.

The amount of the correction needed to make the seismograms (a) and (c) in Figure 1 look similar turns out to have an effect on the amplitude of the signal in the band of frequencies where the magnitude is measured. In this case, it was found that the effect of attenuation beneath the Nevada Test Site had reduced the size of the signal (b) recorded at EKA by about 0.3 magnitude units.

Many studies of this type have been done, using sources and seismographic recording stations all over the world.[8] Several other lines of argument[9] have also pointed to the conclusion that Nevada signals were being attenuated more than was the case for signals from Semipalatinsk. When this extra attenuation was quantified, the Union of Soviet Socialist Republics, like the United States, appeared to be observing a yield limit, which was about 150 kt, in its underground nuclear testing program.

The first evidence for differences in attenuation began to emerge in the early 1970s.[10] The general subject of how to estimate yield by seismological means was vigorously pursued for 15 years, culminating, as we shall see, in 1988. Within the US research community, which was well funded from the 1960s to the 1980s by Department of Defense agencies seeking to improve monitoring capability, there developed a widening acceptance of the conclusion that teleseismic signals from sources in the tectonically active western United States (including NTS) were attenuated more than was the case for sources in stable continental regions such as northern Kazakhstan (including the Semipalatinsk Test Site).

I did not participate directly in this research but was well aware of it through contacts at Caltech (my grad school in the late 1960s) where my advisor was

---

8. For a detailed study of the method summarized in Figure 1, see Der, Z., McElfresh, T., Wagner,R., Burnetti, J., 1985. Spectral Characteristics of P waves from Nuclear Explosions and Yield Estimation. Bull. Seismol. Soc. Am., 75, 379–390. Also note some important errata, given in the same year at vol. 75, 1222–1223.

9. Another piece of evidence was the seismic magnitude of the 80-kt US nuclear test LONGSHOT conducted in the Aleutians, for which the magnitude was significantly greater than would be expected on the basis of magnitude-yield relations for a nuclear test in Nevada.

10. For example, Filson, J., Frasier, C.W., 1972. Multisite estimation of explosive source parameters, J. Geophys. Res. 77, 3303–3308.

Charles Archambeau, and at the Lamont-Doherty Earth Observatory of Columbia University (which has employed me since 1971) where I have followed the work of my colleague, Lynn Sykes (more on Archambeau and Sykes, below). From 1971 to 1984, my career followed a conventional academic track that had little to do with public policy issues, but then in 1984, I took a leave of absence for a year and became a visiting scholar at the US Arms Control and Disarmament Agency, specifically in a unit that participated in writing the President's annual Report to Congress on Soviet compliance with arms control treaties. For me, this was a fascinating opportunity to see professional worlds very different from my own—academia—and to see how they dealt with highly specialized technical issues, which in one or two cases I was encouraged to understand (most importantly, yield estimation using seismological methods). For the most part, I found a willingness to listen to technical arguments and to bring relevant information to bear on the subject at hand. But sometimes, there was hostility, and in making this point we can examine the openly expressed views of a senior policy maker, the Assistant Secretary of Defense for International Security Policy, Richard Perle. His views were televised on May 9, 1986, by KRON, the National Broadcasting Company affiliate in San Francisco, as part of its evening news broadcast in a segment prepared by KRON's "Target 4" unit, when Perle was interviewed by an experienced investigative reporter, Rollin Post.

Let me give excerpts from this interview, which is available on YouTube,[11] and which is (perhaps unfortunately) framed as "Perle versus the experts" who in this case include two seismologists from academia, Charles Archambeau and Lynn Sykes, and Willard (Jim) Hannon of Lawrence Livermore National Laboratory.

Introduction: *The Administration says, that the Soviets have tested weapons more powerful than agreed to in 1974. But have they? … Our investigative unit spent the last month looking into it. They found the Administration is ignoring evidence that the Soviets never cheated….*

Rollin Post: *What you are about to hear, now, is a story of political ideology and scientific truth, and what happens when the two come into conflict. It's also a story of Soviet-American tensions and nuclear weapons. The key players in this drama are an Assistant Secretary of Defense named Richard Perle, and a group of seismologists who have spent much of their careers working for the government.*

The issue; has Richard Perle, because of his distrust of the Soviets, deliberately rejected and suppressed evidence?…

[Cut to Perle: *Baloney. It's not a question of scientific evidence, it's a question of scientists playing politics!*]

---

11. The interview is available as an 8-min clip on YouTube, via http://www.youtube.com/watch?v=MfjG1hE0Qfg (or, search YouTube on Richard Perle + experts).

Post, voice over (with images of underground nuclear tests, and of Nixon and Brezhnev signing the bilateral TTBT): *The argument concerns scientific testing of nuclear weapons. In 1974, the United States and the Soviet Union agreed to a treaty that said each side could test nuclear weapons as long as the explosion didn't exceed a hundred and fifty kilotons....*

Perle: in a clip from the MacNeil-Lehrer Report /1986 March 24/ *I've looked carefully at the evidence, and have concluded, as President Reagan did, that there is significant evidence that the Soviets have violated the hundred and fifty kiloton threshold.*

Post: *This has been the position of the Administration since 1983, a position formed and spearheaded by Assistant Secretary of Defense Richard Perle. But it has also caused a rebellion among the very scientists that the Defense Department relies on to estimate the size of the Soviet tests.*

So deep has been the controversy that last year the Defense Department secretly convened three groups of experts who monitor Soviet tests. They are seismologists, the same people who measure the size of earthquakes.

Target 4 has interviewed three of those seismologists, and they all said the Soviet Union has not violated the 1974 test ban treaty.

One of them is Dr. Charles Archambeau, who has been monitoring the Soviets since the Eisenhower Administration.

[Cut to Archambeau: *At present, there is no evidence that the Soviets have tested over 150 kilotons—none whatsoever.*]

Another is Professor Lynn Sykes, who negotiated the test ban treaty for the Nixon Administration: *The treaty itself, it states, that...neither country shall test above 150 kilotons. And I have no evidence that indicates to me that the Soviets have done that.*

Post: *Dr. Willard Hannon heads seismic monitoring for the Lawrence Livermore National Laboratory*

[Cut to Jim Hannon: *I don't believe the evidence supports a militarily significant violation.*]

The man who ultimately receives this advice is Richard Perle. His position is far different.

Perle: *The best experts available, spent years studying this, and came to the conclusion that it was likely the Soviet have violated the 150 kilotons...*

Post: *There are seismologists...who have written, for instance, Dr. Charles Archambeau—I think you know...him—wrote "at present, there is no evidence the Soviet have tested over 150 kilotons, none whatsoever."*

Perle: *Well, with all due respect...he's wrong. There's lots of evidence. He may not be persuaded on the basis of the evidence, but to say there's no evidence is flatly wrong.*

Post: *Dr. Sykes, of Columbia, says that the Soviet Union has not violated a 150 kilotons limit of the threshold treaty as alleged. He's wrong?*

Perle: *He's entitled to his opinion. He's a professor sitting up at Columbia.*

Post: *But he is an authority on the issue, isn't he?*

Perle: *Well, all the seismology enables you to make an estimate of the yield of the event... Even by that standard alone, there is evidence that suggests the Soviets have violated it, your experts, notwithstanding.*

Post: *They were your experts at one time. They worked for the Defense Department.*

Perle: *There is other evidence, as well...of a sensitive and classified nature.*

Mr. Perle would only say, that the other evidence involves satellite and electronic surveillance...but Target 4's investigation learned that Mr. Perle had already convened a panel of experts which looked at this other evidence, and rejected it.

Post: *You had a Defense Intelligence Agency study, not very long ago, for precisely that, to evaluate seismic and non-seismic ways of determining what the Soviets are doing, and they still came up with the conclusion that seismology, was the most accurate way.*

Perle: *They came to the conclusion that...of the many ways of estimating yield, seismology was the single most important—and I happen to agree with that.*

Post: *Even after acknowledging seismology as the best indicator, Mr. Perle then rejects it.*

Perle: *I didn't much care what their answer was. It doesn't have any profound bearing on our policy.*

Post: *Target 4's also uncovered evidence that Mr. Perle improperly tried to manipulate intelligence agencies in a biassed direction. Example, a Perle letter to the Air Force when its intelligence unit asked seismologists to advise on Soviet tests. According to sources who have seen the letter, it said: "The intelligence community is trying to undermine the Administration's position. My Department will control this area." I asked Perle about the letter:*

There was a letter I understand, that you wrote on April the 15th, of last year, to General Fats who is the Air Force Assistant Chief of Staff, I think...

Perle: *I don't remember the exact words of the letter, but my concern, as I've expressed throughout this interview, which is that...we have tended—I think wrongly—to exclude the non-seismic evidence that bears on the estimation of the yield of Soviet tests.*

Post: *When you cut through these disagreements you find that the real issue may be more personal and political than scientific.*

Perle: *They're all seismologists! They're a bunch of seismologists feathering their own nests. Well, look, seismologists have dominated this field since the beginning, it's how they make their living. The day that it is concluded that we can get along without attributing the importance to seismology that we do, some of these fellows are going to be looking for jobs.*

Archambeau: *The scientific opinion is close, to, I'd say to unanimous. Right now Mr. Perle finds it extremely difficult to find any scientist that will defend the DOD/Perle position. And that's because, there just aren't any, that believe it.*

Sykes: *I think one view that is often put forth…is that arms control agreements are not in the best interest of the United States. That the Soviets will cheat. And then attempt to have a self-fulfilling prophecy, by coming up with a procedure, an incorrect one, that indicates to them that the Soviets have cheated.*

I am writing about this interview almost 30 years later, and still find it fascinating for the direct expression of views. The title of this chapter is "When scientific evidence is unwelcome…" and for the present example, we can be more complete: "When scientific evidence is unwelcome, shoot the messenger!" (In this case, the seismologist.)

The scientists see the technical issue as clear cut (especially Charles Archambeau, with his statement "… there is no evidence that the Soviets have tested over 150 kilotons—none whatsoever"), whereas in Mr. Perle's view: "It's not a question of scientific evidence, it's a question of scientists playing politics!"

In practice, policy makers must act in situations where knowledge is incomplete. But this is not an excuse to make up one's own reality. Was Mr Perle right, in saying that scientists were playing politics here?[12] To the extent that a scientist personally becomes convinced that a certain technical conclusion is correct and is being misrepresented, is there not an obligation—if the issue is important—to push harder? This, to me, is the central ethical issue. One must avoid the temptation to embarrass a policy maker who may not be aware of the strength of the evidence against his or her position, but surely there is a responsibility to stand up for technical conclusions if they are being misrepresented.

Though not the most important of arms control treaties, the TTBT received extensive attention in the late 1980s in part as a bargaining chip between the Reagan Administration and the US Congress, which had several members wanting to see the TTBT ratified and then see progress on negotiating a Comprehensive Test Ban Treaty. The Reagan Administration agreed to submit the TTBT for Senate advice and consent to ratification, but then stunned senior Senators, after making that submission of the TTBT, by taking the position that even if the Senate voted its consent to ratification, the President would not ratify it, at least until its verification provisions were improved.

Most remarkably, with the appointment of Mikhail Gorbachev as General-Secretary in the Union of Soviet Socialist Republics, a breakthrough was achieved on the issue of TTBT verification, when a pair of "Joint Verification Experiments" (JVEs) were conducted in the summer of 1988 according to new protocols negotiated between the two superpowers in the preceding months. These experiments were a significant part of the opening up of the Soviet Union, and presumably would not have happened if there had not been challenges concerning TTBT compliance. One of the experiments was a special nuclear explosion carried out by the United States on the NTS, and the

---

12. In March 1987, he resigned from his position in the Department of Defense.

other was a reciprocal explosion by the Union of Soviet Socialist Republics on the Semipalatinsk Test Site. It was agreed ahead of time that these two explosions would be in the yield range between 100 and 150 kilotons and that each side could bring monitoring equipment right on to the other side's test site to monitor the host country's nuclear explosion close up. This agreement was remarkable in view of the decades of history of operations at the superpowers' nuclear test sites, some of the most highly protected and secretive installations in the world, where explosions larger than the totality of all World War II explosions were conducted, in part as an expression of Cold War competition.[13] Special equipment was installed for the two JVEs to measure the strength of each blast deep underground as a hydrodynamic shock wave,[14] using sensors in a vertical hole going down hundreds of meters, parallel to the shaft in which the nuclear device was emplaced. The monitoring system was thus deployed at distances going down to just a few tens of meters away from the exploding device.

From such near-in underground measurements, and an analysis to estimate the volumetric extent of the hydrodynamic shock wave, this nonseismic method applied to the JVE at Semipalatinsk on September 14, 1988, provided a yield estimate deemed accurate and that was indeed in the range 100–150 kt. Stations around the world provided measurements teleseismically, giving a seismic magnitude of 6.1 and thus comparable to the largest magnitudes of Semipalatinsk explosions since 1976, indicating that they too had been conducted in a way that respected the 150 kt limit of the TTBT. The reciprocal JVE had been conducted earlier at the NTS, on August 17, 1988, with a Russian team making its own close-in measurements of the shock wave from a large US underground nuclear test that was intended to be in the range 100–150 kt. (According to many news reports, the yield of this explosion somewhat exceeded 150 kt. Timerbaev[15] and news reports have given it as 180 kt.)

The TTBT eventually received the Senate's advice and consent in 1990, and it was ratified by President George H. W. Bush. The momentum/progress on that treaty may well have helped in the complicated process of reaching an agreement 6 years later on the text of the Comprehensive Nuclear Test Ban Treaty, finalized in 1996, about 40 years after such a ban first began to be discussed in the 1950s.

The JVEs of 1988 were a massive experiment[16] that enabled TTBT ratification and thus entry-into-force and that proved what seismologists had long

---

13. See, e.g., "A Review of Nuclear Testing by the Soviet Union at Novaya Zemlya, 1955–1990" by Khalturin, V.I., Rautian, T.G., Richards, P.G., Leith, W.S., 2005. Sci. Global Sec. 13, 1–42. More than 200 megatons of nuclear explosive energy was released in 130 tests at this site.

14. At the shot point of an underground nuclear explosion, rock is vaporized and a hydrodynamic shock spreads out, eventually slowing to become a linear elastic P-wave. The volume of rock within which the shock travels faster than the P-wave, is proportional to the explosion yield.

15. See http://www.pircenter.org/kosdata/page_doc/p1650_1.pdf.

16. The cost of the US share was about $28 million.

worked to demonstrate, namely, that for UNEs of around 150 kt at the Semipalatinsk Test Site the resulting seismic signals were about 0.4 magnitude larger than those from UNEs of around the same yield at the NTS.

In part, I suspect that the hostility shown by some members of the Reagan Administration to the TTBT, and, in a different way, to seismologists, was a reflection of their policy views on the much more important Comprehensive Test Ban Treaty. From this perspective, agreeing to TTBT was indeed a slippery slope that led to a CTBT.

Coming back to the issues raised by Rollin Post in his interview with Richard Perle: what can an expert do, to maintain a good professional reputation in the face of allegations of being biased because of a policy issue? Mainly, my advice is to be diligent to establish key facts, and to defend them if attacked. But do not become vulnerable by making a technical mistake, and do not become embittered over time if your views are not accepted. Just keep on pushing, if you are sure you are right.

Every person has to make his or her own choice as to the level of activism in pushing a point of view, and here I suppose that Mr Perle could have been right, in a limited sense, in saying this a question of scientists playing politics. I worked for years on technical issues surrounding the TTBT, and felt the need to get out the story that technical arguments were being treated in some quarters with contempt. So I was very glad to have had an opportunity to work with the people who prepared Rollin Post (whom I never met—nor have I met Mr Perle). Rollin Post certainly elicited some remarkable statements from Mr Perle.[17]

In my opinion, it does not work very well to claim that certain policy makers are behaving unethically. More important, is to monitor the behavior and the activities of the person for whom one most directly has responsibility, namely, oneself; and to stand up for the views of the expert community especially when it appears that those views are disparaged.

So then, was it ethical to set up Mr Perle by working with a very competent TV interviewer? I had heard secondhand of Mr Perle's views from several people, and it seemed important to bring these views out for a wider audience.

I am comfortable with having worked to place his opinions on the record.

---

17. "I didn't much care what their answer was. It doesn't have any profound bearing on our policy."..."They're all seismologists! They're a bunch of seismologists feathering their own nests. Well, look, seismologists have dominated this field since the beginning, it's how they make their living. The day that it is concluded that we can get along without attributing the importance to seismology that we do, some of these fellows are going to be looking for jobs."

Chapter 11

M_{max}: Ethics of the Maximum Credible Earthquake

Roger Bilham

CIRES and Department of Geological Sciences, University of Colorado, Boulder, CO, USA

Chapter Outline

Abstract

The prediction of earthquakes or their forecast is associated with uncertainties in timing that are generally not easy for the public to assimilate. Estimates of the maximum credible earthquake (M_{max}) in a region, which excludes considerations of time, may also result in public concern. In one view, it is indefensible to withhold findings of relevance to public safety, yet in another the release of this information may cause alarm, or set in motion a chain of circumstances that may unsettle other scientists, cause public anxiety, or financially depress sectors of the economy. Two examples are discussed in India where the revelation of international scientific discussions of M_{max} by news media both alarmed the public sector, and taxed the civic authorities tasked with responsibilities for the mitigation of earthquake risk. Since the scientist is fundamentally responsible for the disclosure of sensitive information to news media, to what extent is the scientist also morally responsible for the subsequent chain of events? In the USA the National Earthquake Prediction Evaluation Council (NEPEC) provides a possible model for the delivery of sensitive

Geoethics. http://dx.doi.org/10.1016/B978-0-12-799935-7.00011-3

information to the public avoiding the potential pitfalls of sensationalism. A recent evaluation of M_{max} in the New Madrid Seismic Zone, for example, provided a consensus view of diverse opinions already known to the public. It is not clear, however, that a NEPEC confirmed assessment of M_{max} newly derived by a team of scientists would be handled by the world's press with any greater sensitivity than in the past.

Keywords: Ethics; Jaitapur; Kashmir; M_{max}; NEPEC; New Madrid Seismic Zone; Nuclear power plant; Seismic hazard.

INTRODUCTION

How does one educate a society endowed with abundant information that often requires expert opinion if it is to be correctly evaluated? In particular, how does society identify the expert opinion that provides the most plausible interpretation of a future that will affect its members? The diversity of opinions and interpretations in news media and the Internet bring with them no absolute stamp of credibility. To the casual browser of the Internet, the crafted writings of a crackpot can carry the same weight as the considered opinion of a group of scientists. However, even the most uninformed visitor usually accesses some minimal but critical tools for sifting information. Schwarz and Morris (2011) distinguish four types of credibility invoked by those attempting to distinguish reliable from unreliable information on the Internet:

1. *Presumed credibility* based on knowledge of the source (e.g., United Nations statistics) or the supposition of its trustworthiness from domain identifiers like ".gov."
2. *Surface credibility* is derived from the balance of professional content and the absence of slang, superfluous adjectives, and obvious errors in grammar or page construction.
3. *Earned credibility* influenced by a site's historically dependable information (e.g., BBC news vs a personal blog).
4. *Reputed credibility* based on recommendations from others or public acknowledgment of their integrity in the form of international or other recognized awards.

These qualifiers apply equally to Web-based, television, or newspaper sources. In the Internet as well as in news media, *surface credibility* can often mislead, concealing hidden agendas or biased points of view. Surface gloss in such cases may subdue other estimates of credibility. It is for this reason that professional presentation of information is identified as a fundamental method for establishing what others may perceive to be credible (Fogg et al., 2003). The importance of, and varied forms for, establishing credibility in its numerous guises were recognized in World War II in the development of the discipline of deception known as disinformation (Jones, 1978).

Scientific findings may be judged by each of the four criteria for credibility listed above. Indeed, scientists themselves use identical criteria, if not to evaluate the work of their peers, to decide whether to invest time in reading an article closely for its potential contribution to scientific advance. This chapter addresses problems inherent in informing the public of scientific findings, and the ethical dilemma that can accompany the transfer or nontransfer of sensitive information.

RIGHT AND WRONG: GUT DECISIONS AND APPROACHING DANGER

At this point, it is necessary to digress from issues of credibility briefly to a consideration of issues of right and wrong. A subsequent discussion of earthquakes and ethics requires a working definition of moral behavior if we are to interpret some behaviors as appropriate and others as unacceptable.

There is some evidence to suggest that our species is predisposed at birth to distinguish certain types of behavior that form the basis of human ethics (Hauser, 2006). From experimental observations of children faced with simple moral decisions, Hauser concludes that many decisions are based on evolutionary hard-wired behavioral patterns that underlie our notions of good and bad. For example, game-playing experiments showed that 90% of all participants automatically would save five hitchhikers from a runaway tram, by deflecting a tram to a branchline where only one victim would be killed (cf. Figure 1(b)). He concludes that humans have inherited these early behaviors because ancestors who did not exhibit them were less adapted to survive.

A loose interpretation of Hauser's bimodal tests is illustrated graphically in Figure 1. An approaching out-of-control tram approaches a switch (we shall later consider the tram as metaphor for an earthquake). In the figure, a participant is asked to decide between inaction or diversion of the trajectory of the oncoming wagon. In Figure 1(a), he must decide between saving members of his species at the expense of another. In Figure 1(b), his decision is to save many people at the expense of a single person. In Figure 1(c), he is disposed to save his family and relatives (the white-flag team) at the expense of strangers

(a)   **(b)**   **(c)**

FIGURE 1   Instinctive moral reactions inherited by humans. Under gamelike conditions, a person must decide which way to deflect on approaching danger. (a) Survival of the species (the cow represents an alternative species), (b) survival of the majority (90% of Hauser's participants chose to save the group and to sacrifice the individual), and (c) survival of the kinship group (family members carry white flags).

(the black-flag team). All the decisions have outcomes that are unpleasant, but unless one is a psychopath, the decisions are obvious—to switch the dangerous wagon to a path that minimizes the perceived effects of an approaching disaster. The decisions, once made, are black-and-white choices. Not to decide is to decide. There is no half-way option. It is perhaps no accident that "yes-or-no" information can be easily assimilated and communicated to society, and uncertainty less so.

Another example of a hard-wired response inherited by humans is one ingeniously explored by Darwin (1871)—our human propensity to blush with shame. Blushing signals to a group that an individual admits personal discomfort with a recent or pending antisocial transgression. On finding that blushing was a common trait in many cultures, Darwin concluded that shame offered an evolutionary advantage for our ancestors (Boehm, 2012). Darwin also considered that a tribe whose members included many who were patriotic and courageous would be victorious over other tribes, thus ensuring the survival of certain complex behaviors that we now categorize as virtuous. We commonly encounter advanced forms of shame nowadays in the form of political cover-ups. Less common is the embarrassment of scientists who are reluctant to abandon pathological scientific findings, such as, for example, the proposition of cold fusion (Close, 1992).

This chapter is not the place to explore the thousands of books and articles that have been written on the subject of the paradox of moral behavior (cf. Gaur, 2015). It is sufficient to note that human evolution has provided us with certain basic tools that assure the survival of the species, the advantage of the tribe, and the survival of the group at the expense of the individual (Figure 1). In this chapter, I shall naïvely assume that the complex ethical behavior of individuals or groups of individuals in society are founded upon a small number of inherited survival instincts that give rise to actions we consider morally correct.

EARTHQUAKE PREDICTION AND FORECAST, AND A VULNERABLE PUBLIC

Most scientific findings published between the covers of academic journals are of interest only to other scientists. However, occasionally something surfaces that is an important result, a secure and insightful foundation on which to advance future science, or a discovery that may impact society.

The study of earthquakes provides numerous opportunities for the identification of information of importance to the public. For example, the research of historical evidence for the power and reach of former destructive earthquakes in a region has implications for the required resistance of new construction to future shaking. Less interesting to the public, but of great utility to the earthquake engineer responsible for public housing might be the identification of locations where earthquake shaking can be amplified. The discovery that buildings may have just escaped failure in a previous earthquake with short shaking duration, and that they may not be so lucky in a future one of longer duration, would also be noteworthy. But by far the most important of public earthquake issues concerns the forecast or prediction of the time, magnitude, and location of future earthquakes.

The specific time, magnitude, and location of future earthquakes cannot at present be predicted; although, there have been times when scientists have been optimistic that prediction might be possible (Hough, 2010). For some seismologists, this optimism remains (Jordan et al., 2011), but attempts to predict earthquakes from the observation of physically observable changes that have preceded some earthquakes have been shown not to be reproducible ubiquitously, and have now mostly been discarded.

Given that at present damaging earthquakes cannot be predicted, an educated public should be predisposed to discredit any announcements of the time, place, and magnitude of future earthquakes. However, the public is in general unaware of the current inability of scientists to predict an earthquake for tomorrow, next week, or next year. Moreover, even among seismologists, the announcement of a pending earthquake would not be discredited without learning first the reasoning adopted for its prognostication, because there is always the possibility that the announcement of a new prediction has been based on a potentially successful new discovery. Thus it is recognized that earthquake predictions are likely to surface from time to time, and that assessments of their credibility will require evaluation.

For these evaluations, the public seeks guidance from that small segment of the seismological community who have taken a special interest in the study of the physical, historical, and statistical processes that precede earthquakes, gleaned from previous patterns of earthquake sequences. In 1980, at a time when the sensitivity and potential pitfalls of earthquake prediction were recognized, the United States formed the National Earthquake Prediction Evaluation Council (NEPEC), a group of a dozen experts designated to advise the director of the United States Geological Survey (USGS) on timely warnings of potential future disasters. Similar bodies exist in many countries of the world. Their task in the United States is not to undertake earthquake prediction research but to evaluate the credibility of the warnings and findings of others.

Earthquake prediction is a special (and currently unsuccessful) case, of a successful discipline known as earthquake forecasting. It is quite common for seismologists to evaluate the approximate location and probable magnitude of an earthquake with a qualified statement of when it will occur. Uncertainties in the forecast are quantified through statistical probability—a numerical estimate of the chances of specified earthquake occurring within a specific time interval. As such, the forecast typically has a time aperture of many decades. In practice, such forecasts have great benefit for society because they provide engineers and city planners with sufficient time to outline and implement building codes for a region prior to the occurrence of an earthquake. A successful short-term earthquake prediction (were it possible) in a city of unreinforced buildings would not prevent the destruction of infrastructure and dwellings; although, lives could be saved through emergency mobilization of the public.

The ethics of three well-known cases of earthquake prediction are discussed by Sol and Turan (2004). The first two of these predictions were initiated by U.S. individuals and the third by the U.S. scientific community. All three predictions were/are considered failed predictions by the scientific community, and Sol and Turan (2004) argue the consequences to the general public of all three undesirable. The reasons for failure are well known (Hough, 2010): the 1981 Brady/Spence prediction for Peru, the 1990 Iben Browning prediction for the New Madrid region of the United States, and finally the 2004 Parkfield earthquake that occurred later than expected on the San Andreas Fault. The authors discuss the ethical dilemma associated with the issuance of predictions justified by the need to alert people to impending danger, and contrast them with the consequences of predictions that result in distress of local populations. They conclude that "the publication of unwarranted hypotheses that are of immediate concern cannot be justified by appealing to a principle of absolute freedom of speech," since they lead to public anxiety. They furthermore reject the notion that the resulting heightened awareness of earthquake risk attending these predictions is beneficial, since this violates an ethical principle that individuals should not be used as merely a means to an end. Figure 1(b) illustrates the instinctive opposite, the probability that in an evolutionary sense the common good may be preferable to the distress of the individual. Dolce and di Bucci (2015) examine practical solutions to this dilemma.

The complexities invoked in considering the ethics of earthquake prediction borrow from the basic moral instincts evolved in humans mentioned in the previous section. The earthquake is the tram and the scientist, or group of scientists, potentially control, if not a switch, a megaphone to alert those in the path of the tram of its approach. Let us dismiss for a moment our current inability to predict earthquakes (which amounts to not knowing the timetable of the trams in Figure 1). The scientist convinced of a prediction must decide whether to alert the people to get out of the way of the expected tram, or to remain silent. Clearly, if he is convinced that an earthquake is approaching, most people would agree he has a moral duty to alert the people involved. However, if his hypothesis is wrong, the outcome is that he may unnecessarily alarm the people in the path of the tram. Again most people would agree that he has done the right thing in announcing that in his view, the people are better off out of the way.

It is only when we introduce some real-life complexities into a well-meant prediction can we compromise the moral predicament of the predictor. Let us imagine that in the path of the tram are senior citizens with heart ailments sensitive to fright, or that the rail crosses a bridge over a raging torrent requiring those threatened by the approaching tram to leap to possible death. In this case, the outcome of a failed prediction may be worse than not making the prediction, and most people would consider it morally questionable. Moral aspects of such decisions are considered by Kelman (2015).

Earthquake predictions are rare, and currently not credible, and will not be considered further in this chapter. More important, but less perilous, are the moral consequences of forecasting and discussing scenarios for future damaging earthquakes.

A HYPOTHETICAL CASE OF NUCLEAR POWER INVOLVING ETHICAL DECISIONS

Let us imagine that a nation has decided to construct a large nuclear power plant somewhere in its country. Physicists, engineers, and economists would be involved in the selection of the type of power plant from a short shopping list of national and/or international nuclear power plant vendors. The selected design and vendor combination would be no doubt one chosen with a good safety record based on levels of credibility selected by government officials responsible for power production.

The cost of the power plant is a decisive factor, and a simple cost–benefit analysis usually governs the final decision. Thus the total construction/operating cost must be capable of delivering power at rates lower than the anticipated cost per kilowatt of electricity from other means (kWH/$), for the anticipated duration of the plant. Various hidden costs might be subdued or excluded in this decision. Should the societal consequences of nuclear power versus coal-fired plants be considered? That is, should one consider the undesirable loss of livelihood to coal miners, or should the undesirable global consequences of carbon emissions be balanced against the costs and dangers of storing nuclear waste indefinitely? Should worst-case meltdown conditions or the cleanup costs of a major accident be included in the cost–benefit analysis? Should construction costs include decommissioning costs at the end of the design life of the plant? Often such considerations are apportioned to subcommittees designated to decide and provide information to a higher-level decision-maker, thereby absorbing moral decisions at a level remote from the decision to proceed.

At some fundamental level even the kWH/$ cost of the power plant is not a fatal issue that opponents and proponents may raise in discussions. On the one hand, opponents can argue that nuclear power must be avoided at all costs since a severe accident may result in undesirable mutations to the human race and other species and on the other hand, proponents can argue with equal conviction that the replacement of coal-fired plants with nuclear power will ensure fewer climatic changes and assure the survival of species. A third path invoking renewable energy is omitted in the discussions considered here, which with certain caveats would avoid both these undesirable outcomes.

Let us return to the decisions faced by a government body attempting to solve the need for future power. Having decided on the type of reactor, the next step in the process would be to choose a location for the plant. Some economic and physical considerations are common to all types of steam-generating plants: the need to be located sufficiently close to the user community that transmission costs are reasonably low, and the need to be close to abundant water for cooling purposes essential in the conversion of heat into electrical power. Special safety considerations are naturally considered for nuclear power plants. Unlike coal, plutonium is toxic to life, and hence the most important problem to address is to minimize the accidental exposure of people to toxic and radioactive release. Favored solutions are to place the power plant far from large population centers, and downwind and downstream from any agricultural regions upon which local populations depend. The avoidance of natural disasters that might compromise plant stability is equally important: coastal tsunami, flooding, landslides, future earthquakes, or volcanic eruptions.

The need for a go/no-go decision is easy where branch decisions are black-and-white, but often this will not be the case. For example, the placement of a nuclear power plant where population density is low does not mean that populations in the selected site will be zero. The decision is analogous to the decision to divert the tram to the path with fewer casualties (Figure 1(b)). This is an effortless decision, predicated by, that in the unlikely event of a radioactive leak, placement of the plant in a remote area spares the majority at the expense of the minority who live closer. A moral decision becomes only more difficult when secondary issues of concern to the minority close to the power plant are considered. Examples might be that in the course of normal operation (the accident-free operating status envisaged by proponents), the increase in temperature of local cooling water may damage the local ecology, threaten an endangered species, or reduce fish populations. Again the disadvantages to the local population when weighed against the advantages to the majority will result in a possibly uncomfortable (but inevitable) decision in favor of the majority.

The consideration of future natural disasters, whose occurrence near a planned nuclear power plant may introduce an unforeseen accident, requires evaluation by a broad spectrum of scientific experts (meteorologists, oceanographers, hydrologists, volcanologists, and seismologists) whose written statements typically confirm self-evident, noncontentious physical conditions near a plant. The issue of future seismicity and associated coastal tsunami generation is often less transparent. Earthquakes are transient, largely invisible phenomena that occur irregularly. Not all earthquakes occur at plate boundaries and it is almost impossible to say that an earthquake has never occurred, or will never occur at a location of interest. Recourse in such circumstances must be made to the calculated probability of an earthquake based on several considerations, many of which may be uncertain.

The possible occurrence of one or more future earthquakes is not an issue that will normally prevent the construction of a nuclear power station; although, in some nations the proximity and amplitude of potential intense shaking can condemn a site as unsuitable should certain prescribed limits be approached or exceeded. No existing nuclear power plant has ever been destroyed directly by shaking near, or even above, its design shaking limits. The Fukushima reactor shut down safely which is located approximately 100 km from a magnitude 9 earthquake only to be flooded 50 min later by a tsunami. For most nuclear power plants, the possibility of future violent shaking merely requires the introduction of various additional safety elements in the construction of the plant. These elements will, of course, add to the cost of the plant and its kWH/$ economics, and hence it would be naïve to imagine that the prospect of future seismicity has little relevance to the viability of a chosen plant location. Doubling the construction cost would not necessarily double the cost of electricity, but it would most certainly make it less competitive. To render a nuclear power plant safe during shaking, an engineer requires an estimate of the probability of future shaking, and quantitative estimates of the amplitude and duration and frequency of probable extreme ground motions during the design life of the power plant. For this, he/she consults a seismologist.

Quantifying Future Earthquakes where Few Have Occurred Historically

Seismologists have established a number of methods to investigate future seismicity (see Wyss, 2015a,b). Where earthquakes are sufficiently frequent and the historical record sufficiently long, it is possible to recognize patterns indicating the probable recurrence intervals between damaging earthquakes. In some locations, however, the history of a region is short compared to the time between damaging earthquakes. For example, the historical record could be 200 years long, with intervals of a 1000 years between earthquakes. The absence of a damaging earthquake in such cases does not necessarily imply that an earthquake cannot occur. It may mean than an earthquake has been brewing for 999 years and could occur one year after plant completion.

In such circumstances, it is necessary to extend the historical record backward, from hundreds of years to hundreds of thousands of years using geological evidence for the slip on earthquake faults in the area. Through excavation geologists can quantify the amount of offset of Earth's surface along faults and determine the timing of these offsets (Meghraoui and Atakan, 2014). Various mechanical constraints apply to earthquake faults: the longer the fault—the larger the possible earthquake; the closer the fault—the higher the possible accelerations; the less eroded the fault scarp—the more recent the historical earthquake. One question that is particularly important concerns whether a fault that has not slipped in an earthquake for a very long while, is more, or less, likely to slip again. If it could slip again, it is defined as a *capable* fault. In 1996, the definition of a *capable* fault adopted in the United States for the purposes of nuclear site investigations was that the fault has to have slipped once in the past 35,000 years, or recurrently in the past half million years. In other countries, different definitions apply, and there has been a recent tendency to realize that a simple numerical definition for all earthquake regions may be inappropriate. The International Atomic Energy Agency (IAEA, 2010) has suggested that periods as long as the Pliocene (2.6–5.3 MyBP) should be considered where earthquakes are infrequent. A recent proposal in Italy is to consider a fault capable if it has slipped in the past 0.8 Myr in Italy's extensional terrain, and capable if it has slipped in the past 2.6 Myr in Italy's compressive tectonic region (Galadini et al., 2012).

Guidelines for the proximity of capable faults are also prescribed in some detail. No capable fault of length x km can be located closer than distance y km to a nuclear power plant, where x and y are varied in accordance to anticipated shaking from the largest credible earthquake that can occur on a fault with those specified dimensions. Extreme examples are x = 1 km where y = 1 km, and x = 40 km where y = 200 km. The largest credible earthquake in a region is known as M_{max} and is discussed in a following section.

Hence it is reasonably simple for a nuclear agency to request a report from a competent group of seismologists and geologists to check off all the requirements for safety required by nuclear construction engineers. Having determined the probability for slip of each mapped fault during the design life of the plant, Earth scientists are no longer involved in the planning process. Based on the reports of the probability for shaking intensity, frequency, and duration, the engineer designs the plant and provides the cost estimate to the planning team.

Societal Pressures and Moral Principles

In the above site evaluation process, it will be noted that neither the seismologist nor the geologist is confronted with moral decisions. His or hers is a strictly professional commitment to do the best possible job with the tools available. The acquisition of geological data may be subject to different interpretations, but if these are discussed transparently, it is often possible to place numerical uncertainties upon them that can provide engineers with probabilities for design purposes.

In practice, however, other pressures may be at work within the planning committee. For example, if the cost of the project will be prohibitive if it is designed for accelerations exceeding 0.5 g compared to modest accelerations of 0.2 g, evidence supporting these high accelerations will be examined with microscopic detail, in the hope of detecting any flaw in the data or reasoning that has led to this economically undesirable conclusion. From the point of view of the planning team, two interpretations may be from equally credible sources—team A and team B, but the opinion of team A may permit a halving in construction cost compared to the finding of team B. Team B has identified a fault that may have slipped in a magnitude 7 earthquake within 10 km of the power plant in the past few millennia, and team A has interpreted this same fault to have not moved in 5 Myr. At this point, the planning team has a clear dilemma—to include B's higher risk estimate and accept the higher cost, or to examine the credibility of B's calculations in the hope that they can be refuted.

Since expert opinions are involved, and the planning team may not have the expertise to undertake the refutation of team B and it must go to another credible alternative—expert team C for advice, in order to dismiss, or to reconcile, opposing conclusions. The four credibility criteria discussed earlier in this chapter might be invoked to avoid introducing the opinions of team C, but clearly it would be unwise for the planning committee to go to team A to refute the conclusions of team B, because team A believes its conclusions to be correct.

However, there may exist a moral justification for the planning team to both accept the findings of team A and team A's adverse judgment of the findings of team B. Let us suppose that both the planning group and team A share information freely between each other as part of the same funding organization. We could call them the white-flag group. Team B, in contrast, may be part of a not-for-profit research group that carries black flags or no flags. In such a situation, the white-flag planning team may feel compelled to accept the white-flag team A's judgment.

Although, this hypothetical case might be considered a form of corruption, the white-flag participants may have concluded that black-flag carrying team B has a hidden agenda. For example, the planning committee may suspect that the members of team B may have been influenced by the members of the local population at risk from a possible nuclear accident and thereby may have weighted their uncertainties in interpretation toward higher safety margins. Team B may be suspected of being funded by an alternative energy supplier keen to replace nuclear power with renewable technologies. Team B may be a member of an organization known to be against nuclear power. Implicit in these considerations are the planning team's suspicions that the black-flag team may themselves be morally corrupt. A similar suspicion attended Richard Perle's characterization of the motives of seismologists during discussions of the Nuclear Test Ban Treaty in 1986 (Richards, 2015).

Alternatively, members of team A may feel their own credibility is threatened if team B is right and they may be shown to have missed important clues (Darwin's inherited shame factor). They also may feel justified in not releasing the extent of their tests, or in delaying the publication of their site investigation tests to either the white-flag or the black-flag community, in order to undertake further tests to discredit team B's findings.

Various tools are available to the planning committee once it has determined that team B's conclusions are both undesirable and possibly untrustworthy. They are as simple as they are effective—to demolish the credibility of team B's arguments, to highlight minor but inconsequential errors in their scientific assertions, and to use rhetoric to assure the public that "the needful has been done" by their trusted white-flag team, or even to banish members of team B suspected of owning a black flag.

Fortunately, scientists are distanced from such political issues since they have established impersonal methods for scientific discussion. Scientific tests can be undertaken of team A's and team B's hypotheses that once complete, will potentially support decisive conclusions acceptable to both teams, and by the engineering and scientific community at large. If the site is considered unsuitable, a new one must be chosen. If the findings of seismic safety are considered unnecessarily pessimistic, the chosen site will be adopted.

The Planned Jaitapur Nuclear Power Plant in India

The preceding is a conjectural discussion examining aspects of decisions related to the planned Jaitapur nuclear power plant south of Mumbai on the west coast of India, and readers should be aware that it is possible that these suppositions may be biased by the closeness of the author to some of the issues discussed. The Jaitapur nuclear power plant is planned to be the largest in the world (Jaitapur, 2013). The detailed geological site investigation has, at the time of this writing, yet to be made public, and the plans concerning the power plant have yet to be finalized. In late 2012, my coauthor, Professor Vinod K. Gaur, and I published an article suggesting that earthquakes similar to those that occurred approximately 110 km north of the site in 1967 (Mw ≈ 6.3 Koyna) and approximately 400 km to the north east of the site in 1993 (Mw ≈ 6.2 Latur) could not be excluded from occurring closer to Jaitapur, were a capable fault identified near the proposed site (Bilham and Gaur, 2011). This was followed by an article by (Rastogi, 2011), a seismologist advising the Nuclear Power Corporation of India (NPCIL) (Nagaich, 2012), assuring readers that there were numerous errors in our article and "that all apprehensions raised have been considered by the geologists and seismologists involved in site investigations." Specifically, this article describes a 50-km-long inferred fault, 10 km distant from the planned power plant considered inactive, i.e., not *capable* of slip in a future earthquake. A published rejoinder (Gaur and Bilham, 2012) suggested that were this a *capable* fault, its offshore projection would pass within 5 km of the site, and its dimensions would permit a local maximum credible earthquake, Mw > 6.5, and hence thorough testing of its absence of activity in the past several thousands of years would appear to be vital. As of 1 August 2014 it is uncertain that any of these tests have been undertaken.

Shortly after the first of these discussions on the seismicity of Jaitapur was published, I was asked whether I would be willing to answer questions from members of the Indian Press on the 2011 article (Raina, 2012). I did so in January 2012 on my way to give an invited plenary talk in Ahmadabad at a conference on earthquakes in central India. The press who attended this talk highlighted the difficulty of characterizing seismicity near Jaitapur given its short history of earthquakes, and the occurrence of an M > 6 earthquake 110 km to the north in the preceding decades (Bordoloi, 2012). But the headlines used by the media were designed to arrest the attention of readers. These ranged from the factual to the sensational, and required NPCIL to defend their reports concerning seismic safety provided to them by their consultants. The official response (NPCIL, 2013; The Hindu 12 Aug 2012) reasserts the care with which investigations have been undertaken and the concurrence of seismic safety among the three seismologists who have contributed their expertise (also see postscript).

IS M_{MAX} APPROPRIATE FOR PUBLIC CONSUMPTION?

The maximum credible earthquake (e.g., Cornell, 1968; Griffin, 1996; Wyss, 2015a,b) that can occur in a region (termed M_{max}) is an estimated upper earthquake magnitude limit for considerations of future seismicity. For hypothetical future earthquakes, M_{max} can be calculated from the product of earthquake rupture area times the amount of slip that can be sustained by a rupture of this size, given certain information about the strength of the fault that ruptures. The maximum credible earthquake on a 35-km-long fault is of the order of Mw = 6.5, but if the fault is 80 km long, the earthquake could be as large as Mw = 7.0 (one of the uncertainties in the Jaitapur discussion above). In relatively few places has M_{max} been confirmed by observation, since earthquakes often rupture just a part of a fault or plate boundary. Two recent earthquakes, however, exemplified M_{max} for the plate boundary segments on which they occurred: the 2004 Indonesia/Andaman Mw = 9.2 earthquake and the 2011 Tohoku Mw = 9.1 earthquake.

For these two regions, seismologists had failed to quantify M_{max} on the plate boundary. Prior to the occurrence of the 2004 rupture, the previous largest historically known earthquake in the region (Mw = 7.9) occurred in 1881 in the Nicobar Islands. In the Tohoku region, an Mw = 8.4 was considered probable although a previous earthquake in AD 869 may had been larger. In 2004, the death toll was approximately 230,000 and in 2011 it was approximately 20,000. In both cases, the accompanying tsunami was unexpectedly damaging. Would knowledge of M_{max} have resulted in authorities undertaking a substantially different risk-mitigation strategy, or does it take the actual occurrence of M_{max}, to confirm that mitigation measures are warranted? Alternatively, is it possible that these two examples have been sufficient to alert authorities in analogous regions that risk-mitigation strategies for worst-case earthquake scenarios have utility? To what degree is a seismologist morally responsible for assessing M_{max}?

These are interesting questions. Answers to the first must surely be speculative, and although a tsunami warning system has now been initiated in the Indian Ocean as a result of the 2004 disaster, it is by no means certain that it will be functioning before the next mega-earthquake possibly 1000 years hence. Answers to the second might respond favorably to the notion that society has been provided with two warnings in a decade and should be vigilant to a future unexpectedly large earthquake. In neither case was the timing of these two great earthquakes an issue that could be contemplated in a predictive sense, but M_{max} is a measure of energy release that, had it been widely available, could have influenced the height of tsunami defenses in Japan, or the location of emergency generators whose post-tsunami failure led to a meltdown in some of Fukushima's reactors, or could have mandated the availability of a siren-based tsunami warning system on the coasts of India and Sri Lanka.

In that seismologists are responsible for the study of earthquakes, and that most seismologists are funded by the public, a case can be made that they have a moral obligation to inform the public of the largest anticipated earthquake in a region (Kelman, 2015). Wyss et al. (2012) and Wyss (2015a,b) go further and assert that the moral responsibility of the seismologist extends to provide a quantitative evaluation of the societal consequences of the occurrence of this M_{max}. The question arises whether M_{max} has been underestimated in other seismic zones, or whether the seismic community has unwittingly withheld this information from the public at risk elsewhere in the world. A related question is whether the public should be informed of M_{max}. For example, a civil engineer might argue that knowledge of M_{max} in a region has no practical value whatsoever to an individual, its numerical utility being just one of the many inputs used by building authorities responsible for seismic building codes, or for engineers involved in the specific design of critical facilities in a region.

M_{max} in the New Madrid Seismic Zone

For the past 40 years, M_{max} in the central United States has been available to the public as a result of sequence of powerful earthquakes in 1811 and 1812 near the New Madrid Seismic Zone (NMSZ) region of Louisiana and Missouri. In the largest of these earthquakes, the Mississippi River briefly flowed upstream, sand was vented from the ground, masonry buildings near the epicenter collapsed, and people felt the earthquake at great distances. Although deformation rates measured by Global Positioning System are low, small earthquakes are common in the region (Stein, 2010). Since these earthquakes occurred before the development of seismometers, estimates of the magnitudes of the earthquakes must be based on eyewitness accounts, all of which are subject to interpretative errors. Early estimates of the magnitude of the earthquakes suggested magnitudes of Mb $\leq$ 7.4 (Nuttli, 1973), but a 1996 study assigned maximum magnitudes of Mw = 8.0 (Johnston, 1996). This resulted in a general perception that for the purposes of earthquake-resistant construction code, M_{max} should be considered 8.0, a value that demanded considerable investment in civic infrastructure (Stein, 2010).

There is now unanimous agreement that the 1996 magnitudes had been unjustifiably inflated (NEPEC_NMSZ, 2011). The most obvious reason for this is that there is no known fault large enough to sustain an Mw = 8 earthquake near New Madrid, given a reasonable physical frictional strength and inferred slip of the fault system that is believed to have caused the earthquake. In 2009 (Fact Sheet 2009–3017), the USGS described the earthquakes as "several of magnitude from 7–8 in the past 4500 years." In the most recent study currently available, and with the stamp of approval from the USGS review process, the original nineteenth-century accounts were reevaluated and the magnitude of the first and last strongest earthquakes in the sequence were determined to be 6.7 < Mw < 6.9 and 7.1 < Mw < 7.3, respectively (Hough and Page, 2011). In 2011, due to the wide range of proposed magnitudes, and to the startling (for some) realization that the New Madrid earthquakes may have barely exceeded

Mw = 7, NEPEC formed a special subcommittee to evaluate the credibility of all published evaluations (NEPEC_NMSZ, 2011).

The subcommittee found itself unable to recommend a single correct answer and instead argued "that measures of uncertainty such as standard deviations are large, about 0.5 magnitude units." Yet, despite some of the standard deviations being significantly less than 0.5 magnitude units, the committee undertook a consensus approach effectively averaging disparate conclusions, in lieu of identifying flaws in reasoning in contending studies, an approach that might have led to a more decisive conclusion. In defending this approach, the committee accepted a range of magnitudes (averaging Mw = 7.5) with the caveat that an improved magnitude estimate would attend the availability of new instrumental data from a large earthquake in a similar region.

Unlike the case of M_{max} for the planned Jaitapur power plant, the New Madrid M_{max} dispute concerns earthquakes that many members of the public knew had already occurred. The deliberations of the select committee were indecisive about M_{max}, but the resulting uncertainties (based on a logic tree assessment that included numerous other factors) resulted in estimates of anticipated ground motions that differed by a factor of two (for 0.2% probability of exceedance of peak ground acceleration in the next 50 years).

The select subcommittee specifically defined its moral boundaries in the report as follows: "Our panel has concentrated on the scientific issue of hazard assessment; we did not evaluate engineering or policy issues in our review. Those questions are best left for professionals and politicians who are qualified and/or positioned to make informed societal-risk decisions." In its first sentence, the select committee clarifies its absence of bias and in its second, it passes the burden of its uncertain but expert judgment, onward to those responsible for the safety of people. The report of the select subcommittee was evaluated by NEPEC and forwarded verbatim to the director of the USGS with the comment that it presented "a balanced summary of what is and what is not known about the seismic hazard in the NMSZ" (Tullis, 2011).

This evaluation of M_{max} and its consequences for the NMSZ NEPEC succeeded in clarifying diverse and contradictory information already available to the public and to engineers in the region. That the concluding values for M_{max} for future earthquakes differed little from the median value considered by the subcommittee for the 1811/1812 earthquakes was possibly a coincidence. One can imagine, however, a different scenario in which a single evaluation of M_{max} is available, with no alternative scenarios for NEPEC to consider. In these circumstances, we may speculate that NEPEC would undertake an approach similar to that which they adopted in the Brady-Spence prediction. They would presumably attempt to understand the physics and tectonics associated with the estimation of M_{max} from published materials and, in the absence of clarity, they would then quiz the seismologist or team of seismologists concerned directly, by requesting interviews or invited presentations. The first approach distances itself from details of the method, but the second approach dissects the data, the physics and the interpretation at a much more fundamental level. The members of the committee would no doubt weigh the consequences of their judgment with the moral burden of knowing that their collective verdict would subsequently form the basis of government policy.

M_{max} for the Himalaya

The Himalaya are a seismic belt where knowledge of M_{max} is important for the design of a building code. Earthquakes here affect the populations of six nations: Pakistan, India, Nepal, Bhutan, Bangladesh, and southern Tibet. The total population within reach of damaging shaking in Himalayan earthquakes exceeds 50 million (Bilham et al., 2001). The largest earthquake observed recently in the Himalaya was the Mw = 8.6 Assam earthquake of 1950, although it is probable that the 1505 central Himalayan earthquake (Ambraseys and Jackson, 2003) may have released four times more energy with Mw ≈ 9.0, given that slip locally exceeded 20 m (Kumar et al., 2010). If the entire Himalaya slipped >20 m, the amount observed in trench excavations of the 1505 earthquake (Lavé et al., 2005; Kumar et al., 2010) in a single Kashmir-to-Assam earthquake, its magnitude would be Mw ≈ 9.3 and rupture would have a duration of tens of minutes. This unprecedented magnitude represents M_{max} for northern India. However, our current limited understanding of the behavior of the Himalaya based on the past millennium of earthquakes is that the Himalaya slip in shorter segments with correspondingly smaller magnitudes. Few seismologists in the region have considered an Mw = 9.3 possible, and even the notion of an Mw = 9 earthquake in the Himalaya is considered by some improbable (Srivastava et al., 2013).

A numerical estimate of M_{max} for the Kashmir region, the westernmost segment of the Himalaya was discussed at a scientific meeting in December 2011 (Bilham et al., 2011) and subsequently published (Schiffman et al., 2013). This 8-year study by an Indo/Pak/US team considered 33 scenario earthquakes with magnitudes varying from a minimum of Mw = 7.4 (partial rupture) to a maximum of Mw = 9.0 (rupture of the entire Kashmir/Kishtwar region). The highest magnitude estimate (M_{max}) assumed 23 m of slip of the region between the 2005 Kashmir Mw = 7.6 and the 1905 Kangra Mw = 7.8 earthquakes, which the study concluded had a low probability, but could conceivably occur at >1800-year intervals.

The presentation of this information would normally have gone unnoticed, especially since it was the penultimate 15-min talk of a 5-day meeting of multiple concurrent sessions where the results of more than 20,000 different contributions had been presented. Barely 20 scientists were present, much lower than in many of the previous talks, and conference organizers were already clearing up the room. However, it was selected by the press office of the American Geophysical Union (AGU, 2011) as worthy of a press release and within days numerous newspapers had headlined the news warning of a Mw = 9 earthquake in Kashmir. Accounts in Indian newspapers, and those in Kashmir, included none of the discussion of uncertainties or the methods used, but gave much space to consequent damage, with memories of Japan 2011 and the Andaman islands in 2004 prominent in the reporting. The November 2013 published report (Schiffman et al., 2013) that provided the numerical details of the 2011 discussion in San Francisco passed unnoticed by the press.

In December 2011, the immediate response of Kashmir authorities had been to reassure the public that there was no cause for alarm, no earthquake having been predicted, or even forecast. Subsequently, in the weeks following the news, civic authorities in Kashmir organized various school earthquake safety programs and responded to questions from concerned citizens. In the spring of 2012, a scientific meeting in India, led by concerned seismologists based in Hyderabad, independently evaluated the science underlying the AGU presentation, discredited the observations and concluded that the finding could be dismissed as baseless sensationalism (see postscript). In October 2012, the National Disaster Management Agency and the Ministry of Earth Sciences, government of India, organized a workshop to discuss earthquake disaster preparedness in the Kashmir valley. Since then funding for new seismic and geodetic studies have been approved in Kashmir to supplement, and to presumably test, the sparse data on which the initial findings had been based.

At a fundamental level, M_{max} information is inappropriate for assimilation by local disaster agencies. The information constitutes neither prediction nor forecast since it comes with no estimate of time (with the exception of an unstated and inferred estimate of minimum recurrence interval based on geodetic convergence rates). In the case of publicizing M_{max} for Kashmir M_{max}, the fate of the information and its capacity to mislead and distress was sealed once the news media had captured the story. The people and the civic authorities in Kashmir learned of this information via sensational newspaper reporting, a delivery method that is clearly inappropriate for a measured response, since the authorities were effectively having to absorb the information, and to formulate an appropriate response to the information simultaneously.

A conclusion to this line of reasoning is that since M_{max} can be of no direct utility to disaster management authorities in a region, and is likely to result in local panic in the hands of journalists, this information should be restricted to discussions among scientists and engineers tasked to develop safety measures in the construction industry. However, had these procedures been adhered to by proactively denying journalistic access to the information, research into earthquake hazards in the Kashmir Valley may have languished, and may not have received the infusion of research funding and the enthusiasm for disaster management it now enjoys. The role of the journalist in this chain of events is perceived to have been to elevate a passive scientific conclusion to a confrontational pedestal. To subdue its adverse effects on public morale, those responsible for public safety were compelled to characterize the journalistic announcement and its scientific precursor, as scaremongering. Following the logic of Sol and Turan (2004), one would could conclude that the release of M_{max} to the public would fall into the same morally unacceptable class of behavior as earthquake prediction, whereas Wyss (2015a,b) and Wyss et al. (2012) argue that withholding M_{max} and its consequences from the public cannot morally be defended.

DISCUSSION

Three examples are provided where the evaluations of earthquake information by seismologists have utility to local populations, but where the intervention of journalists attempting to deliver this information directly to the public has resulted in adverse reactions by civil authorities mandated to gather the same information, and to serve the same local populations. Journalistic commentaries in two cases in India have led to a division of scientific opinions, in the absence of which contentious conclusions would have been resolved through the traditional dialogue of discussion, testing, refutation, and eventual concordance. A difference in scientific opinion is confusing to a public provided with no guidance as to the credibility of each opinion. Subsequently journalism played a role in advancing biased conclusions, increasingly divergent from the scientific uncertainties that initiated the discussion.

For the planned Jaitapur nuclear power plant to proceed, its planners require unanimity in the assessment of future seismicity. Specifically its engineers require numerical estimates of shaking intensity before they can design the plant, and their chosen design determines the cost. In the absence of expert opinion among themselves, the planners must apply measures of credibility to its expert opinions, either based on personal feelings or from estimates of the credibility of its consultants and their opponents elicited from independent authorities. As the complexity of questioning increases, the moral allegiances of those involved become more complex. The issues are as yet unresolved.

In the case of the journalistic reporting of M_{max} to the public in Kashmir, responsible authorities took the correct policy of denying its links to earthquake imminence or forecast, but some scientists went further. The credibility of the research and the correctness of its conclusions were questioned by other scientists without supporting data, and advisors to the government of India deemed it necessary to withhold from one of the researchers the privilege of undertaking further collaborative field research on Indian earthquakes (see postscript).

In the case histories discussed in India, international teams considered the data and communicated their findings through normal scientific channels. Subsequently, in both cases, local and foreign researchers have been deterred from further investigations, and future collaborative investigations have been discouraged. In retrospect, it is evident that the intercession of journalists derailed the scientific process, but it is also evident that the withholding of scientific findings would have required a level of secrecy not normally associated with science. In the case of Jaitapur, no article on its seismicity need have been written. In the case of Kashmir, the discovery of the unusual width of its potential rupture zone could have been buried in the conclusions of reports to funding agencies. Most scientists would consider both actions orthogonal to their mandate for reporting research of interest to the public.

At present, there is no simple intermediate mechanism for the delivery of sensitive information to the public. In the United States, sensitive seismic findings are refereed by committees such as NEPEC, who are tasked to deliberate on the validity of earthquake predictions and earthquake forecasts, and to provide recommendations to the director of the USGS, who is responsible for timely warnings of geological disasters. Faced with divergent published conclusions for M$_{max}$ in the NMSZ, NEPEC formed a subcommittee to consider this and other related but contentious issues. The subcommittee, armed with invited opinions from dozens of noncommittee members, chose not to identify a single unique correct solution but to weigh the credibility of all solutions and to examine the potential consequences of these extremes. This was a legitimate stance since the mandate of the committee issued from NEPEC was not to decide, but to comment—"NEPEC is exercising its responsibility to advise the USGS Director on issues bearing on earthquake forecasting by convening a panel of independent experts to comment on the level of hazard posed by future large earthquakes in the NMSZ." The written report was reviewed by NEPEC and passed on to the director of the USGS without further judgment.

In summary, the director of the USGS was (and is) mandated to provide timely warnings of potential geological disasters to the public, the members of NEPEC were mandated to advise and recommend findings to the director, their select subcommittee was mandated to comment on the diversity of findings to the committee, and the subcommittee solicited further opinions to broaden the database upon which their commentary was based. At no point in this process are there recognizable moral forks in the path from opinion to product. The team approach in effect used largely external estimates of credibility in accepting the findings of authors with diverging opinions. The net result was a creditable analysis of community-agreed statistical uncertainty, but it is not clear that an engineer about to build a hospital in the NMSZ would not have preferred a more aggressive investigation of the fundamental reasons why some opinions could not be rejected.

CONCLUSIONS

The seismologist interested in future earthquakes contributes a small but crucial link in a safety chain, which if undertaken while adhering to the tenets of scientific investigation and reporting, requires no moral judgments or ethical considerations. The preceding discussion, however, illustrates that some scientific results require the scientist to withhold information likely to cause public anxiety. Bad news about M$_{max}$, either whispered by seismologists to each other or proclaimed through the megaphone of journalists, follow paths whose outcomes must both be considered undesirable, and hence require moral judgments. Alternative procedures currently available in the United States are cumbersome and ineffective alternatives for the release of sensitive seismic information; although, they make no moral demands on participants. The weakness of the NEPEC model is that in the only case where a decisive evaluation of M_{max} would have been of great benefit to local seismic hazard studies, the process fell short of probing below the surface of some of the published claims of authors. It is not clear how the public in a populated region previously considered seismically quiet would receive a NEPEC confirmation, and a recommendation for public release, of a new discovery for an unexpectedly large value for M_{max}, even were it accompanied by a vanishingly small probability for its imminence.

Postscript

On May 19, 2012, the Indian Government barred the author's entry to India en route to Bhutan (Bagla, 2012). No explanation was given and he was deported to the United States on the same plane on which he had arrived. A few weeks after the publication of Bagla's article, the Indian Home Office attributed deportation to the inappropriate use of the tourist visa that Ahmadabad conference authorities had approved for his January 2012 plenary talk (Economic Times 2013).

REFERENCES

AGU, 2011. Press Release. https://www.youtube.com/watch?v=r9oKqp_CL68.

Ambraseys, N., Jackson, D., 2003. A note on early earthquakes in northern India and southern Tibet. Curr. Sci. 84, 570–582.

Bagla, P., 2012. India barred entry to U.S. author of seismic review. Science. 338, 1275. http://www.sciencemag.org/content/338/6112/1275.

Bilham, R., Gaur, V.K., 2011. Historical and future seismicity near Jaitapur, India. Curr. Sci. 101 (10), 1275–1281.

Bilham, R., Gaur, V.K., Molnar, P., 2001. Himalayan seismic hazard. Science 293, 1442–1444.

Bilham, R., Szeliga, W., Singh Bali, B., Khan, A., Wahab, A., Khan, F., Qazi, S., 2011. Velocity Field in the NW Himalayan Syntaxis: Implications for Future Seismicity. American Geophysical Union Meeting, San Francisco. Abstracts T54B-06.

Boehm, C., 2012. Moral Origins: The Evolution of Virtue, Altruism and Shame. p. 381. Basic Books, NY. ISBN:978-0-465-02919-8.

Bordoloi, S.K., Earthquake Risk to Jaitapur Nuclear Power Plant, YouTube. https://www.youtube.com/watch?v=dHGwLNxedeo.

Close, F.E., 1992. Too Hot to Handle: The Race for Cold Fusion. Penguin, London. ISBN:0-14-015926-6.

Cornell, C.A., 1968. Engineering seismic risk analysis. Bull. Seism. Soc. Am. 58, 1583–1606.

Darwin, C., 1871. The Descent of Man, 2 vols, John Murray, London.

Dolce, M., di Bucci, D., 2015. Risk management: roles and responsibilities in the decision-making process. In: Wyss, M., Peppoloni, S. (Eds.), Geoethics: Ethical Challenges and Case Studies in Earth Science. Elsevier, Waltham, Massachusetts, pp. 211–221.

Economic Times, January 2, 2013. India Bans Journalistic Activities on Tourist Visa. http://articles. economictimes.indiatimes.com/2013-01-02/news/36111430_1_tourist-visa-indian-missions-foreigners.

Fogg, B.J., Soohoo, C., Danielson, D.R., Marable, L., Stanford, J., Tauber, E.R., 2003. How do users evaluate the credibility of Web sites? A study with over 2,500 participants, DUX'03. In: Proceedings of the 2003 Conference on Designing for User Experienceshttp://dx.doi.org/10.1145/997078.997097.

Galadini, F., Falcucci, E., Galli, P., Giaccio, B., Gori, S., Messina, P., Moro, M., Saroli, M., Scardia, G., Sposato, A., 2012. Time intervals to assess active and capable faults for engineering practices in Italy. Eng. Geol. http://dx.doi.org/10.1016/j.enggeo.2012.03.012.

Gaur, V.K., Bilham, R., 2012. Discussion of seismicity near Jaitapur. Curr. Sci. 103 (11), 1273–1278.

Gaur, V.K., 2015. Geoethics: tenets and praxis. In: Wyss, M., Peppoloni, S. (Eds.), Geoethics: Ethical Challenges and Case Studies in Earth Science. Elsevier, Waltham, Massachusetts, pp. 141–160.

Griffin, R.H., 1996. Engineering and Design, Earthquake Design and Evaluation for Civil Works Projects. Regulation ER 1110-2-1806. U.S. Army Corps of Engineers, Washington, DC 20314-1000. p. 26.

Hauser, M.D., 2006. Moral Minds. Harper Collins. p. 489.

Hough, S.E., 2010. Predicting the Unpredictable: The Tumultuous Science of Earthquake Prediction. 272 pp.. Princeton University Press. ISBN:9780691138169.

Hough, S.E., Page, M., 2011. Towards a consistent model for strain accrual and release for the New Madrid, central U.S., Seismic Zone. J. Geophys. Res. 116, B03311.

IAEA (International Atomic Energy Agency), 2010. Seismic Hazards in Site Evaluation for Nuclear Installations, Chapter 8: Potential for Fault Displacement at the Site. Specific Safety Guide No. SSG-9, ISBN:978-92-0-102910-2, STI/PUB/1448. .

Jaitapur Nuclear Power Project, 2013. http://en.wikipedia.org/wiki/Jaitapur_Nuclear_Power_Project (accessed March 2013).

Jayaraman, J.S., 2012. Geologists Differ on Seismic Risk to Jaitapur Nuclear Power Plant. http://twocircles.net/2012dec11/geologists_differ_seismic_risk_jaitapur_nuclear_plant.html (last accessed March 2012).

Johnston, A.C., 1996. Seismic moment assessment of earthquakes in stable continental regions–III. New Madrid 1811–1812, Charleston 1886 and Lisbon 1755. Geophys. J. Int. 126, 314–344.

Jones, R.V., 1978. Most Secret War: British Scientific Intelligence 1939–1945. Hamish Hamilton, London. ISBN:0-241-89746-7.

Jordan, T.H., Chen, Y.-T., Gasparini, P., Madariaga, R., Main, I., Marzocchi, W., Papadopoulos, G., Sobolev, G., Yamaoka, K., Zschau, J., 2011. Operational earthquake forecasting: state of knowledge and guidelines for utilization. Ann. Geophys. 54 (4), 316–391. http://dx.doi.org/10.4401/ag-5350.

Kelman, I., 2015. Ethics of disaster research. In: Wyss, M., Peppoloni, S. (Eds.), Geoethics: Ethical Challenges and Case Studies in Earth Science. Elsevier, Waltham, Massachusetts, pp. 37–47.

Kumar, S., Wesnousky, S.G., Jayangondaperumal, R., Nakata, T., Kumahara, Y., Singh, V., 2010. Paleoseismological evidence of surface faulting along the northeastern Himalayan front, India: timing, size, and spatial extent of great earthquakes. J. Geophys. Res. 115, B12422. http://dx.doi.org/10.1029/2009JB006789.

Lavé, J., Yule, D., Sapkota, S., Basenta, K., Madden, C., Attal, M., Pandey, M.R., 2005. Evidence for a great medieval earthquake (1100 A.D.) in the central Himalayas, Nepal. Science 307, 1302–1305.

Meghraoui, M., Atakan, K., 2014. The contribution of paleoseismology to earthquake hazard evaluations. In Wyss, M., (Eds.), Earthquake Hazard, Risk and Disasters. Elsevier, Waltham, Massachusetts, pp. 237–271.

Nagaich, N., January 17, 2012. Seismicity Considerations for Jaitapur. http://www.npcil.nic.in/pdf/Seismicity_Considerations_for_Jaitapur_NPP.pdf.

NPCIL, 2013. Misconceptions and Facts about Jaitapur Nuclear Project. http://www.npcil.nic.in/main/misconceptions_combine_final.pdf.

NEPEC_NMSZ, 2011. Report of the Independent Expert Panel on New Madrid Seismic Zone Earthquake Hazards. http://earthquake.usgs.gov/aboutus/nepec/reports/NEPEC_NMSZ_expert_panel_report.pdf.

Nuttli, O.W., 1973. The mississippi valley earthquakes of 1811 and 1812: intensities, ground motion and magnitudes. Bull. Seism. Soc. Am. 63, 227–248.

Raina, K., 2012. US Geologist Speaks of Earthquake Risk in Jaitapur. (last accessed March 2013) http://www.greenpeace.org/india/en/Press/US-Geologist-speaks-of-earthquake-risk-in-Jaitapur/.

Rastogi, B.K., 2011. Seismicity near Jaitapur, India. Curr. Sci. 103, 130–131.

Richards, P.G., 2015. When scientific evidence is not welcome. In: Wyss, M., Peppoloni, S. (Eds.), Geoethics: Ethical Challenges and Case Studies in Earth Science. Elsevier, Waltham, Massachusetts, pp. 109–118.

Schiffman, C., Bali, B.S., Szeliga, W., Bilham, R., 2013. Seismic slip deficit in the Kashmir Himalaya from GPS observations. Geophys. Res. Lett. 40, 5642–5645. http://dx.doi.org/10.1002/2013GL057700.

Schwarz, J., Morris, M.R., May 7–12, 2011. Augmenting Web Pages and Search Results to Support Credibility Assessment. Vancouver, BC, Canada.

Sol, A., Turan, H., 2004. The ethics of earthquake prediction. Sci. Eng. Ethics 10, 655–666.

Srivastava, H.N., Bansal, B.K., Verma, M., 2013. Largest earthquake in Himalaya: an appraisal. J. Geol. Soc. India 82, 15–22.

Stein, S., 2010. Disaster Deferred: How New Science Is Changing Our View of Earthquake Hazards in the Midwest. Columbia University Press. 296 pp.

Tullis, T., 2011. Letter from NEPEC to McNutt. http://earthquake.usgs.gov/aboutus/nepec/reports/NEPEC_LettertoMcNutt4-18-11.pdf.

Wyss, M., Nekrasova, N., Kossobokov, V., 2012. Errors in expected human losses due to incorrect seismic hazard estimates, Nat. Hazards 62, 927–935. http://dx.doi.org/10.1007/s11069-012-0125-5.

Wyss, M., 2015a. Do probabilistic seismic hazard maps address the need of the population? In: Wyss, M., Peppoloni, S. (Eds.), Geoethics: Ethical Challenges and Case Studies in Earth Science. Elsevier, Waltham, Massachusetts, pp. 239–249.

Wyss, M., 2015b. Shortcuts in seismic hazard assessments for nuclear power plants are not acceptable. In: Wyss, M., Peppoloni, S. (Eds.), Geoethics: Ethical Challenges and Case Studies in Earth Science. Elsevier, Waltham, Massachusetts, pp. 169–174.

Chapter 12

Geoethics: Tenets and Praxis

Two Examples from India

Vinod K. Gaur

CSIR, Centre for Mathematical Modelling and Computer Simulation (C-MMACS), Bangalore, India; Indian Institute of Astrophysics, Bangalore, India

Abstract

Two case histories of evaluating seismic safety factors at the sites of major Indian power projects are discussed: one, a 260-m high dam in the central Himalayas, and the other, a 10-GW nuclear power plant on the west coast of India. A review of the assumptions made in the calculation of ground acceleration figures for their engineering design, reveals that in both cases advisors associated with these projects favored interpretations of lower seismic hazards than the numerically larger hazards assessed by independent scientists. The resulting lower safety factors in the case of the Tehri Dam became the basis for administrative decisions, gravely compromising future public well-being. I argue that the evaluation process in each case exemplifies an infringement of Immanuel Kant's concept of a "categorical imperative" applied to socially evolved "rights issues" using the axiom of noncontradiction.

Keywords: Categorical Imperative; Ethical principles; Jaitapur nuclear plant site; Rigour in scientific analysis; Tehri dam.

PERSPECTIVE

Two major Indian power projects, one completed and the other in progress, called for quantitative estimates of potentially damaging future shaking to be evaluated and passed to decision makers tasked with engineering design and public safety. I examine the moral judgments associated with these evaluations

Geoethics. http://dx.doi.org/10.1016/B978-0-12-799935-7.00012-5
141

in the context of a tenable ethical principle developed from Immanuel Kant's concept of a "categorical imperative." This principle has the force of an unconditional judgment authenticated by the test: "Act only according to that maxim whereby you can, at the same time, will that it should become a universal law."

If this is adopted as a rationally acceptable canon (even if unprovable logically), one can derive from it, using the principle of noncontradiction,[1] Kant's second postulate: "Act in such a way that you treat humanity, whether in your own person or in the person of any other, never merely as a means to an end, but always at the same time as an end." I base my tenets of geoethics on these principles, taken as axioms, and use the principle of noncontradiction to distil certain eigenconditions to be applied to ethical judgments on socially evolved "rights issues," For example, if we consider that it is a citizen's inalienable right to dwell in a safe environment, then those who have freely chosen to develop expertise in the evaluation of environmental safety, and offered to provide this knowledge to society in return for a gainful employment, have a concomitant obligation in performance of which the best available reasoning, knowledge, and capabilities must be deployed. An extended discussion on the context of ethical judgments appears in the appendix. I now review discussions concerning the Tehri Dam and the proposed site for the Jaitapur nuclear plants in the perspective of the above geoethical principles.

TEHRI

The Tehri Dam is located at 30.5° N, 78.5° E in the Central Himalayan fold and thrust belt created by the bulldozer action of the Tibetan plate that advances southward over India at ~2 m per century. This view of Himalayan tectonics, consistent with numerous investigations in the region and now widely accepted, was proposed by Argand (1924) in a prescient visualization of continental deformation. Later, during their Swiss expedition, Heim and Gansser (1939) mapped the coherent deformation structures of this region, further elaborated by Valdiya (1980). These were characterized by three subparallel sets of large thrust sheets that have advanced southward over India, as a result of the continued convergence of the Indian Subcontinent with the rest of Eurasia. The underlying plate tectonic mechanism for Himalayan compression had already been formulated when the first design for the proposed dam site was completed in 1972, although evidence for ongoing deformation in the Himalayas was as yet available only indirectly in the form of two centuries of major historical earthquakes that pointed to critical levels of stress in the Himalayas and their intermittent release. Guided by this brief historical seismicity of the area, the design was based on the assumption that a maximum credible earthquake (MCE) at the site would be $M_w = 7.2$ and the resulting peak ground acceleration (PGA) equal to 0.25g.

This 260.5-m-high dam, the tallest in India, was completed in 2006 at a cost of approximately $1000 M. Its 2.6 cubic km reservoir floods a narrow valley 30 km long, which approaches 1 km in width just north of the dam. Controlled release of reservoir water currently generates 1 GW (proposed 2.4 GW) of electrical power. The project was the subject of three decades of debate due to concerns over the wisdom of constructing a major earth fill structure (approximately an equidimensional[2] triangular prism 575 m wide across the valley at its crest, with a 1.1-km footprint along the valley at its base) in the Himalayas, which has suffered three devastating earthquakes in the first half of the twentieth century. During the two decades of discussions starting around the mid-1980s, the MCE magnitude was raised from $M_w = 7.2$ to 8.5, the latter corresponding to a 100 × 200 km rupture area suffering 10 m of slip, slightly larger than the $M_w = 8.4$ Bihar–Nepal earthquake of 1934.

In August 1986, when I was asked by the Department of Environment, Government of India, to quantify the estimated PGA anticipated during the lifetime of the Tehri Dam, the mechanism of plate tectonics had been firmly established requiring that the Indian plate slides beneath Tibet through intermittent slips on a plane of detachment (the décollement), manifested as earthquakes. However, the recurrence interval for damaging earthquakes and their maximum magnitudes and the convergence rate across the Himalayas were poorly known at the time (Khattri et al., 1983; Baranowski et al., 1984; Gaur et al., 1985). A published (Sinvhal et al., 1973) creep rate of ≈9 mm/year at the frontal Nahan thrust suggested a minimum convergence rate, but we now know Himalayan convergence to be twice this number,[3] and that no interseismic creep occurs on the décollement south of the Greater Himalayas. However, it was recognized that the Tehri Dam lay ≈15 km above the Himalayan décollement (Ni and Barazangi, 1984) and since this represented the closest distance to the thrust fault the rupture of which would shake the dam, these data and published equations (Boore and Joyner, 1982) relating moment magnitude and distance to shaking intensity yielded a PGA at Tehri of 0.56g for $M_w = 8.0$. This exceeded the design PGA for the proposed dam (0.25g) by a factor of two. Accordingly, I suggested that the dam design be revised.

My report (quoted in THDC (1990) document as the National Geophysical Research Institute report) was opened for discussion at a meeting in 1986, presided over by the Chairman of the Central Water Commission. It was met with ridicule by those scientists of the Department of Earthquake Engineering of the University of Roorkee (DEQ-R) who were consultants to the Tehri Hydro Development Corporation (THDC) constituted to construct the dam.

They questioned my basic premise that plate tectonics was responsible for the ongoing deformation in the Himalayas, and therefore considered it preposterous that the Tehri region having no recorded history of a great earthquake, would in the future, experience one stronger than $M_w = 7.2$, their original figure for the MCE as a design basis for the dam. Finally, they proposed that even if a PGA of 0.56g were to shake the dam, the appropriate acceleration used by engineers responsible for dam safety was not PGA, but (PGA/2), the effective peak ground acceleration (EPGA) as defined by Bolt and Abramson (1982).

The EPGA is a frequency-dependent parameter that is significantly less than the PGA in the presence of high frequencies (2–10 Hz) (Newmark and Hall, 1982) but is arguably inappropriate for the longer periods (>2 s) associated with a $M_w = 8$ rupture with several meters of slip and shaking duration exceeding a minute (e.g., Iyengar, 1993). By a majority vote, however, the committee adopted an MCE of $M_w = 7.2$ and the notion of EPGA as a scientifically established principle. These recommendations were then distilled into an endorsement of the originally proposed (DEQ report 1983) value of 0.25g as the design basis for the Tehri Dam.

The Government response to this finding appears in a transcription of the Parliamentary Proceedings of 29 August, 1991, as follows: (http://parliamentofindia.nic.in/ls/lsdeb/ls10/ses1/1729089101.htm).

B.C. Khanduri (Gharwal Province): *"In August 1986 the Expert Committee said that the project should be abandoned, although a considerable amount of money, around Rs 232 crore, had been spent on it. But a very interesting thing happened thereafter. In November, a Russian delegation, headed by Mr. Gorbachev came, and aid of Rs 2000 crore was managed for this Tehri Dam. However, the Russians said that they wanted a technical clearance, which [at that time] did not exist. Therefore, a Technical Committee [to further investigate the adequacy of the dam design] was ordered. It consisted of people who were not seismologists. It is very interesting to note that this Technical Committee, which sat after the earlier Technical Committee had recommended abandonment of the project, met for just one day, i.e. on 16 October, 1986, and gave a recommendation that the construction of a dam was safe. Based on this the Russians gave [approved] the aid of Rs 2000 crore."*

In 1990, following numerous public protests over, among other things, the October 1986 committee's selection of an MCE of 7.2, the Government appointed a five-member Group of Experts under the Chairmanship of the Director General of the Geological Survey of India (GSI) to reevaluate the seismic design accelerations at the dam site. The committee accepted an MCE of $M_w \leq 8.5$ and proceeded to estimate the PGA at Tehri using a seismic source proposed by Brune (1976). But in so doing, one member of the committee adopted a low value for Q (an attenuation characteristic) tentatively used by Brune as an illustrated example, a procedure which led to an estimated PGA of 0.5g, (and hence an EPGA of 0.25g). When consulted, Brune responded that the Q = 50 value in his illustrative example was inappropriate for the Himalayas, where Q might locally exceed 500, resulting in significantly larger accelerations (Brune, 1993). He also faulted the arbitrary use of EPGA as a design acceleration and pointed to the availability of new data that suggested potentially higher PGA (Campbell, 1989).

Brune is quoted verbatim in the transcriptions of the August 1991 parliamentary discussion: *"There is no question in my mind that Tehri dam should be subjected to the most rigorous state-of-the-art design analysis and should be designed for peak ground acceleration of about 1g."*

The lively parliamentary debate that accompanied this revelation is reproduced below (complete with dotted lines from the parliamentary report cited above):

Shri Kamal Nath: *The dissenting note of Dr. Gaur, with the observations of Prof. Brune, was referred to the High Level Committee, which gave a supplementary report in July, 1990 reiterating its earlier opinion. Prof. Gaur refused to sign the supplementary report. The supplementary report of the High Level Committee was considered by the Government in August, 1990 and it was decided to refer the matter to another expert Prof. Jai Krishna who opined in September, 1990 that the proposed dam section for the Tehri Project is safe from the point of view of seismicity...* (Interruptions).

Shri M. B. Wasnik (Buldana): *What is the background of this expert?* (Interruptions).

Shri Kamal Nath: *Prof. Jai Krishna was one of the consultants to the Tehri Dam Authority. It was a single man Committee, which was appointed...* (Interruptions).

Shri B. C. Khanduri: *He is not a seismologist. He is only an earthquake engineer...* (Interruptions).

Shri M. B. Wasnik (Buldana): *How has a person connected with the Authority been employed?...* (Interruptions).

Shri Kamal Nath: *Prof. Jai Krishna, a one man Committee, in September 1990 opined that the proposed dam section for the Tehri Project is safe from the point of view of seismicity. The Ministry of Mines had, however, been receiving representations from various quarters expressing doubts and misgivings regarding seismic safety of the dam. Therefore, the Director General, GSI was asked to constitute an Expert Group for a critical reappraisal of the safety aspects of Tehri Dam taking into account the misgivings raised. This Expert Group gave its report in July, 1991 observing that the design of the dam has been subjected subsequently to a* [simulated] *peak ground acceleration of 0.5g and found satisfactory. The Department of Mines, this month in August, 1991, observed as follows. "Based on the report of the Director General, Geological Survey of India, the Department of Mines accepts the recommendation of Prof. Jai Krishna that the proposed dam section for the Tehri Project is safe from the point of view of seismicity of the region."*

However, while giving its approval, the Department of Mines had placed two caveats: experimental determinations of the local Q value and depth of the décollement using a seismic array, and the evaluation of a computer simulation of the dam's response to the 1968 Gazli accelerogram (one of the few observed records at that time of vertical and horizontal accelerations exceeding 1g) "to be designed and carried out in coordination between Geological Survey of India, National Geophysical Research Institute (NGRI) and Tehri Hydro Development Corporation." Despite the first of these caveats, no experimental determinations of Q value have yet been published.[4]

In response to various representations on the circumstances leading to earlier approvals, the Government, in 1996, appointed another high-level committee consisting of three seismologists (Kailash Khattri, Ramesh Chandra, and Vinod Gaur) and two earthquake engineers (Narayan Iyenger and Naveen Nigam). In the spirit of transparency, the committee invited the Tehri Dam consultants to freely participate in the discussions at all its meetings. The proceedings began with the task of evaluating the most credible ground motion scenario for the Tehri Dam site, consistent with an improved understanding of Himalayan convergence rates and subsurface geometry. These included the implications of recently constrained geological displacements along the Main Himalayan Thrust (Mugnier et al., 1994), and geodetic constraints of Himalayan convergence (Paul et al., 1995). The dam consultants informed the committee that the dam design had been tested by their Russian consultants, for 1.3g PGA ground motion using the observed accelerations measured in the 1976 Gazli $M_s = 7.0$ earthquake as a template. About 8 s into this 14-s simulated earthquake the dam showed evidence for vibratory deformation reaching a maximum at ≈ 12 s. The increasing displacements of the dam toward the end of the simulation was of concern, because the strong shaking duration for a $M_w \approx 8$ earthquake with 150-km rupture may approach 50 s, and for the 600-km-long rupture considered by Khattri, it may exceed 200 s. One of the members of the 1996 committee judged that a longer simulation was unwarranted and stated so in a written report. The remaining four members of the 1996 committee, however, dissented, and in a separate report recommended that the inadequacies of previous studies be addressed by undertaking (1) a three-dimensional (3-D) nonlinear simulation of the dam for long-duration, long-period shaking appropriate for an MCE of $M_w \geq 8$, and (2) a simulated dam failure analysis to evaluate the consequences of catastrophic downstream flood following a hypothetical breach of the dam. The two reports were submitted to the Government on 18 February, 1998, and the National Committee on Seismic Design Parameters favored the minority judgment, which permitted dam construction to proceed (Ramachandran, 2001).

Subsequently, the dam consultants tested the response of the dam to an artificially extended Gazli accelerogram by applying it thrice (end-to-end) successively to extend the shaking duration from 14 to 42 s (Ramchandran, 2001), corresponding to the rupture duration anticipated in a $M_w = 8$ earthquake. The dam survived this simulation, providing a measure of confidence in its resilience to a great earthquake. However, the ratio of vertical to horizontal motion and frequencies of vibrations in this artificial "triple event" are unlike those expected in a Himalayan rupture. Since acceleration in the first and last 4 s of the record is greatly diminished, its triplication results in three periods of high acceleration separated by two 7-s periods of comparatively modest shaking. It would be appropriate to subject the dam to the accelerations recorded in real $M_w > 8.5$ earthquakes, although such data are only available from submarine earthquakes and not from those immediately above the slipping décollement. In the absence of a real record, however, Sengupta (2010) subjected the dam to a computer two-dimensional (2-D) model using the 23-s-long Michoacan $M_w = 7.5$ earthquake record linearly scaled in amplitude to simulate an $M_w = 8$

earthquake. She expressed concern over her finding that the maximum compaction was ≈1 m, and downstream translation of the crest of the dam attained ≈7.5 m, sufficient to compromise its safety. However, the equidimensional form of the Tehri Dam is poorly modeled by 2-D simulations, and no 3-D simulation of the dam responding to a great décollement rupture beneath the dam has been undertaken.

Despite the absence of the mandated experimental measurements of Q or a full 3-D simulation of dam performance in response to an $M_w = 8.5$ earthquake beneath the dam, and an evaluation of the consequences of a simulated dam breach, construction of the dam proceeded and was completed in 2007. The completed dam, therefore, falls short of the recommended compliance with International Standards (ICOLD-57, Bulletin 46) that require that (1) dam simulations should demonstrate the absence of significant damage when subjected to the Operating Basis Earthquake and (2) catastrophic failure and the uncontrolled release of water should not occur when subjected to the MCE. A study of the effects of a dam breach has repeatedly been considered unnecessary since none of the extant simulations indicated that structural failure was possible.

With the exception of the 1996 committee, members of the several expert committees appointed to evaluate dam safety repeatedly elected to accept solutions to future scenarios that favored the completion of the dam, declining to undertake the quantitative evaluation of worst-case scenarios to which the dam might be subjected in its lifetime, and raising serious questions as to whether they fulfilled their ethical responsibilities to the public at risk from dam failure. In some cases, arguments of limited validity (EPGA vs PGA) for reducing anticipated hazards were invoked to yield accelerations conformable to existing designs. The completed design passes the test of those simulations to which it has been subjected, but no fully realistic simulation has yet been undertaken. Thus, there is a real concern as to whether at several points in the past two decades expert committee members, singly or collaboratively, may be considered to have betrayed their ethical obligations to the public at risk from catastrophic damage in a future earthquake. In several cases, decisions to approve dam completion required experimental determination of geophysical quantities characterizing the Himalayas. The significance of the desired requisite data seems to have been either ignored or not understood by committee members. The perceived insistence by individual scientists and institutions, to adhere to concepts and numbers borrowed from the works of others, even after these concepts and numbers had been refuted and clarified by their original authors, is particularly worrisome.

JAITAPUR

Jaitapur (16° 35′ N, 73° 20′ E), on the west coast of India, about 250 km south of Mumbai, is the chosen site for India's largest nuclear power plant with a capacity of 9900 MW, also the largest in the world. The site lies between two creeks on a ~25-m-high laterite-covered marine terrace on the Konkan coast of the Western Ghats,[5] ostensibly a suitable location for a water-cooled power plant with apparently little risk from coastal tsunami, and with no known earthquakes in recent history. Nuclear power plants can be designed to withstand strong ground shaking provided their design engineers are furnished with quantified estimates of shaking intensity and duration (estimated to ~84-percentile confidence for critical structures). Therefore, those responsible for plant safety are faced with the difficult task of estimating the possible magnitude and location of a future earthquake where none has been recorded in the past few hundred years.

Recent seismic quiescence unfortunately is a poor indicator of potential seismicity, and circumstantial evidence does point to recent tectonism. The proposed power plant is located on the footwall of a fault that has offset a recently emerged and deformed marine terrace (Powar, 1993). The age of terrace emergence and faulting is unknown. Widdowson (1997) recognized three genetically distinct paleosurfaces and interpreted these as signifying a period of uplift and extensive erosion during the lower to mid-Tertiary. Powar (1993), noting the occurrence of Neogene lignites along the Konkan coast, at altitudes of 40–50 m, as well as the strong concentration of N–S lineaments in the region, suggested continuing uplift of the region through the Neogene, particularly during the intense fluvial erosion event 50,000 years before present.

Significant geological features in the vicinity of the Jaitapur site call attention to their existence for reckoning the safety factor to be incorporated in the design of the plant structure(s), notably the 35-km-long NNW trending escarpment 10 km south of Jaitapur that forms a 25-m-high step in the marine terrace, and is clearly expressed in SRTM (Shuttle Radar Topographic Mission) digital elevation data, with its eastern side dropped down relative to the west. This fault scarp is arguably the largest to have been mapped in southern India but no paleoseismic investigations have been undertaken across it to determine when it last slipped. Submarine faults have been mapped by Gujar et al. (1986) near its northward offshore expression, and a 200-km-long gravity trough over the continental shelf (Majumdar et al., 2011) is apparently collinear with it offshore. No earthquakes have been recorded within 45 km of Jaitapur, although an M = 6.5 earthquake in 1967 shook Koyna, nearly 100 km NNE. This moderate but potentially damaging earthquake ruptured a buried fault in the West Indian shield, with no previous history of earthquakes or surface expression. Its occurrence surprised many, as did the 1993 Latur earthquake of a similar size in central India about 300 km ENE of Koyna. A cursory glance at the past half century's epicentral map of M ≥ 4.0 earthquakes in the Indian shield reveals the widespread distribution of small earthquakes indicative of a pervasive stress field that poise the faults of the Indian Shield in a state of self-organized criticality. The faults are diffuse, of limited length, and of uncertain continuity and can be triggered by small variations in the ambient tectonic stress.

It was against this perspective that Bilham and Gaur (2011) examined the seismic potential of the proposed Jaitapur site. Publically available reports and unpublished materials cited by the Nuclear Power Corporation (NPCL) focus on geotechnical issues and the findings of local seismic networks, without referring to the broader context of the tectonic setting of Jaitapur. Based on recent historical quiescence of the area and the absence of any small earthquakes recorded by these local seismic networks, a Koyna-type earthquake occurring within a radius of 30–40 km of Jaitapur was considered impossible (Rastogi, 2012). Consultants to the NPCL concluded that the prominent fault scarp and other minor faults in the region near the planned nuclear site have been inactive since Miocene times (5–23 MyBP).

No detailed investigation of the 25-m-high scarp of the Vijaydurg fault near Jaitapur has been published, nor has an examination of the date of the emergence of the marine terrace, the timing of its westward tilt, and plausible causes of the observed flexure and faulting been included in site investigations. Two decades ago, Powar (1993) offered a quantitative estimate for recent marine terrace emergence (50,000 years before present). Although this could have formed a starting point for applying newly developed age dating investigations, the concept of its emergence as an active tectonic process was not considered of interest in geotechnical investigations. No estimate of the MCE on the Vijaydurg fault was considered, although its subaerial length and inferred depth permits earthquakes in the range $6 < M_w < 7$. Assuming a 50 KyBP age for the marine terrace would require about 25 earthquakes at recurrence intervals of 2000 years to account for the present height of the escarpment (Gaur and Bilham, 2012). Such earthquakes would not be recorded in the written history of India, which at these latitudes extends back in detail to less than 400 years.

Although such considerations could easily be tested by transparent independent paleoseismic investigations of the surface fault, or from an offshore search for turbidites as would be triggered by large earthquakes, the official policy (NPCL Press Release of January 17, 2012) of the NPCL is that all concerns have been accounted for by experts.[6] In 2012, the Government of India took the unusual step of barring one of the authors from entering India (Bagla, 2012). At the time, the reasons for this government intervention were not disclosed. Although the Indian Home Office subsequently attributed this to the inappropriate use of a tourist visa that had been approved for a talk the author in question had given earlier in the year, it is difficult to escape the suspicion that deterring continued scientific and public discussion of seismic hazards in India may have played a part.

The discounting of the possible occurrence of a nearby damaging earthquake is clearly imprudent, for should one occur during the planned lifetime of the power plant, it would be embarrassing for government consultants to have advised engineers of its impossibility. Future embarrassment may or may not occur, but the immediate ethical cost to India resulting from the NPCL of India's resistance to open scientific debate is that the public trust in the Government of India's decision-making process is shaken.

Currently, the engineering design of the power plant is claimed by its French designers AREVA to withstand an $M_w = 8$ earthquake, but the proximity of this design earthquake to the epicenter is not specified. Should the Vijaydurg fault slip, the nuclear plant would lie less than 10 km from the rupture, on the hanging wall of a $6.5 < M_w < 7$ earthquake. The cost of reinforcing the nuclear plant to resist damage from a nearby rupture is presumably high, but the costs of geological studies that would demonstrate whether the Vijaydurg fault is a currently inactive Micocene feature would be far less.

THE CRITIQUE

Both the cases described above relate to future risks to engineering structures, and concern two fundamental ethical issues. The first is the trust that society places on autonomous institutions and individuals to render well-researched safety design factors warranted by the scale of the perceived risk. The second is the concomitant obligation of those who have willingly agreed to perform the implied tasks, to do so by taking into account all relevant knowledge that is globally available, and engaging in such scientific investigations that are considered necessary to reduce uncertainties in the reckoning of critical design factors.

Despite these two engineering projects being separated by two decades, the responses of the government show similarities: an initial incomplete assessment of hazard analysis and an apparent reluctance thereafter to embrace new tectonic findings and emerging improved investigative methods. The importance of plate tectonics was denied as a working hypothesis in the early stages of hazard discussions for the Tehri Dam, and uncertainties in intraplate seismicity are being resisted in the assessments of seismic hazard for Jaitapur.

Common to each of these investigative processes is the perceived existence of a possible conflict of interest. Do government-funded scientists and engineers have a duty to provide answers that minimize impediments to government plans, or do they have a duty to provide answers that minimize public risk? Clearly, these investigations and their findings must satisfy both equally (and by definition, it is in the government's interest that they should); however, given the government's treatment of those who have questioned the thoroughness of current investigations at Jaitapur, one might conclude that political pressures play an important role in influencing internal reports.

It is to be regretted that an unfortunate consequence of government rules on "information security" applicable to internal reports by scientific establishments in India is to effectively mask incomplete data and their analyses from critical examination. This absence of transparency means that many Indian scientists forfeit rewarding opportunities to participate in the generation of rigorously tested knowledge and shared data with international scientists. To cite four well-known examples: Global Positioning System (GPS) measurements with millimeter precision are officially secret in India, yet GPS navigating devices with up-to-date road maps are available in Indian cars. Maps of many parts of India are restricted, yet Google imagery with much higher precision is publically available worldwide. Data from worldwide seismic stations are routinely accessed by Indian seismologists to characterize seismicity in India, yet India provides no reciprocal availability of high-quality data to the worldwide community. Gravity measurements are officially secret in India, yet are routinely measured by gravity satellites.

THE EPILOGUE

The two cases described above are not isolated instances but typify the prevailing mental culture in many government scientific establishments, even in Academies. For example, India's Planning Commission has, for over a decade, circulated highly optimistic figures of the country's annual utilizable fresh water balance equal to 1123 cubic km, despite repeated cautions (e.g., Narasimhan, 2008) that the implied 40% figure for evapotranspiration is grossly unrealistic, considering its measured values of over 60% obtained by scientists in countries of similar climates. This factor when correctly applied to India's water budget and sensibly accounted for the minimum ecological flow, reduces the total utilizable water to 654 cubic km (Narasimhan and Gaur, 2010a,b), perilously close to the current utilization of 634 cubic km, the widespread deprivations notwithstanding. However, the issue has been allowed to remain contentious, creating no brighter light than incommensurate debate, instead of making an experimental test to quantify the prevailing evapotranspiration rates in different regions of the country. A pernicious effect of such miscalculations is to shift the emphasis of sociotechnological interventions from conservation of a scarce resource to engineering of supply systems for a nonexistent one.

Another example relates to the balance of costs and benefits that may accrue from the introduction of genetically modified (GM) seeds in India. In 2010, the Ministry of Environment asked the Science Academies of the country to prepare an independent investigative report concerning the potential impacts on health and the environment of a genetically engineered, insect-resistant, aubergine. The resulting report was devoid of objective or independent research and instead included verbatim passages from an earlier report assuring the safety of the modified aubergine, written by a known GM proponent (Shetty, 2010). The Minister, following protests about this apparent plagiarism rejected the Academies' report. The lament highlighted by *Nature* in September 2010, "Where is the rigour that is expected from such distinguished bodies" exposes the basic deficit in India's science culture that proliferates in the absence of the intrinsic values of Science mandating an ineluctable openness to independent scrutiny. The "rigor" in approach to and analysis of issues cannot blossom in an ambience of paranoiac sensitivity to data and information and parochial treatment of technological solutions borrowed piecemeal and shorn of the integrating values of their origin. Therefore, when government decisions on matters related to science and development technologies appear to be more a result of authoritarian judgment than the result of a reasoned analysis, it comes not as a surprise, but as a disappointment at missed opportunities. Indeed, the Fallacies of the Expert[6] pose a challenge to developing societies today where the lives of ordinary citizens are increasingly being invaded by the materials and systems, not organically evolved from their milieu but planted in their midst without the values of the cultures that produced them. Thus, ethical considerations of public safety issues entailed in high-technology interventions, which become increasingly more exacting with the complexity of a given technological solution, would remain unrealized in countries like India until such time when the value of life and rights is organically internalized in its social fabric and governance.

It is obvious, and perhaps not unexpected, that some topics in the affairs of government are politically sensitive. In India, their scientifically resolvable concerns are obfuscated by focusing on sweeping challenging scientific issues under the carpet and ensuring immunity to criticism of their approach and analysis under the garb of data sensitivity. This would explain the continuing practice of classifying data even where, as pointed above, such restrictions have little practical value other than discouraging scientific thoroughness and reasoned debate. An all-pervasive hierarchy helps support this potentially degenerate situation by securing substantial compliance of a scientific position by the authority of office rather than by scientific judgment. In this ambience of positive reinforcement, cautionary voices of a few scientists who have the courage to point out errors appear as pointless meddling even to well-meaning politicians with little patience for the philosophical values of scientific skepticism. Clearly, the probability that scientific recommendations made to the government by expert committees would be sound will depend on the proportion of critically thinking scientists in the country. India's poor showing in Science and in scholarly endeavors generally, as periodically endorsed by several international ratings, is, therefore, the basic cultural deficit that afflicts Indian society of which ethical failings of experts is only a piece. Meanwhile, the widespread reliance by the government on criteria that bear little correlation with intellectual abilities, in the choice of experts, encourage the proliferation of rhetorical and authoritarian values both among scientists and the citizenry, further reducing the potential for sound judgments.

APPENDIX: THE CONTEXT OF GEOETHICS

Bertrand Russell (1961) identified two quintessential human traits as drivers of progress: the urge to *understand* the world, and the urge to *reform* it (*to a better state*). The Greeks recognized this intuitively and looked for canonical principles that would authenticate the processes entailed in pursuing these urges. Accordingly, they searched for *ideals of perfection*, which they believed to underlie all reality, the ideal way to *reason* (**logic**), the ideal *form* (**aesthetics**), the ideal *social organization* (**politics**), and the ideal *conduct* (**ethics**). For a long time, it was assumed that logical bases could be found as *self evident truths* to define the *criteria* whereby the "rightness" of these concepts could be judged, and as *analytical procedures* devised to construct a *sure* pathway. While this has proved elusive, much has been learned to refine the "reasoning process" and the analysis of "forms," of both objects and ideas, that have greatly advanced our knowledge of the world. The last two ideals of social organization and of perfect conduct, however, have defied an analytical treatment. Social organizations indeed share quite a few universal traits rooted in basic human urges and psychological instincts, but as historically testified, they are dominated by the belief systems of the respective communities and the specificities of their natural environment.

Matters of ethics, however, have continued to be regarded by many as flowing from a self-evident principle of reason. Spinoza (1632–77), whom Russell (1961) called the noblest and the most lovable of the great philosophers, valiantly endeavored to deduce these principles a-la-Euclid (*Ethics*; Spinoza, 2000), only to founder on the grounds of its extralogical origins. Yet, Kant (1724–1804), one of the most analytical philosophers of the eighteenth century, maintained that ethical *principles* could be distilled as *categorical imperatives* from certain conscience-inspired judgments, notably "'treat people as an end, and never as a means or instrument to achieve an end." For, it would otherwise violate the inalienable principle of autonomy. These academic questions of ethics, however, suffered a phase change around the middle of the last century in consequence of (1) our improved understanding of the workings of the dynamic planet we live off and (2) our growing ability to intervene in the progression of natural processes both in the physical and biological realms to outfit our increasingly designed world.

The 1970s saw the birth of the standard model of the earth, which alone, of its siblings in the solar system, evolved the unique machinery of plate tectonics to transfer its internal heat to the surface and thereby organize itself into an interactive system of rock, water, and air. This perspective greatly illuminated our understanding of how the earth works as well as that of its constantly evolving critical states, which are chronicled in the relics of its past ages, particularly its more frequent paroxysms that often grievously unsettle our normal lives. The new understanding highlighted the fundamental principles of conservation, of both mass and energy, and the importance therefore, of working out the paces at which various components of the earth system, process the exchange of energy and materials between them. For, these process rates driven by equilibrium conditions determine the chemistry of the atmosphere and the dynamic state of the environment forever poised to change with any alteration in the state of any of the earth's subsystems. The comfortable state of the environment that has been largely responsible for the flourishing of our vaunted civilization, and now necessary for its continued survival, appeared late in geological evolution, only about 3 million years ago.

The new understanding also catalyzed the development of rational approaches to designing our habitats from land use planning to engineering structures that would minimize risk to which local communities may be exposed. These emerging possibilities, concomitantly with those predicated by the newfound understanding of the workings of life systems at the cellular and molecular level and of our growing capability to manipulate materials by novel approaches to molecular synthesis, inevitably provoked the extant, staid concept of ethics around the middle of the last century. The result was its differentiation into a number of derivatives to deal with ethical questions specific to a professional or disciplinary field: bioethics, geoethics, business ethics, and so forth. Since ethical questions in each of its differentiates require considerable knowledge and understanding of specific details, the flourishing approach has been to treat the various issues as *situation specific* without adherence to any unifying order. Resolution of discipline-centered ethical questions, unmindful of collateral issues relating to social, economic, and political implications, is thus often reduced to being *prescriptive* and therefore of limited applicability in a globalized world. This brings us to the desirability of exploring geoethical issues critically, with a view to examining whether some *rational criteria* could be found or constructed by the torch of a higher principle that would be difficult to refute, even if it is not logically derivable, or to accept an unbridgeable arbitrariness, forcing one to adopt a situation specific approach.

At the dawn of explicit philosophical enquiry, it was recognized that rational approaches to reasoning could be derived from some natural rules of inference based on commonsense, such as a proposition, defined as a statement or an event that can either be true or false, cannot simultaneously hold along with its negation (p and *not* $p \equiv$ False). This simple formulation is indeed consistent for any situation whose existence in a given domain is incompatible with that of its annihilator. Such too is the case with our socially evolved principle of fundamental "rights": the right to life with dignity and security from fear of natural and man-made threats, entailing for its exercise, the right to authentic and accurate information to the extent that available knowledge and technology of the day are capable of offering.

Fortunately, there are citizens who have autonomously chosen to professionally engage in activities that would secure these fruits for society and, in making that choice, implicitly accepted the obligation to provide for their chosen domain, the best available desired knowledge products that would serve the exercise of the corresponding "rights." This obligation is implicitly covenanted with and paid for by society, in return for diligently analyzed risk scenarios under a suite of plausible hazard conditions, and qualified by their respective probabilities, to illuminate an optimal, informed choice. Failure to discharge this obligation for want of diligence, expertise, or integrity, thus constitutes the annihilator of that right. In addition, its simultaneous existence in society thereby becomes untenable. Conversely, therefore a rational tenet of geoethic as indeed of any branch of applied ethics would be to require through the formulation of public policy and practice, that individuals and institutions opting to provide the complement of rights issues must be provisioned with the requisite physical and intellectual resources to perform their *inalienable* obligations. In particular, one may consider the following list as constituting the basic minimum of eigenconditions that must be weighed in ethical judgments on earth issues.

1. That the composite earth system possesses natural rhythms and paroxysms in its variously pacing realms, arising from an unceasing exchange of a near finite stock of materials and energy to maintain a dynamic state of equilibrium. Perturbations of these multiply connected exchange process rates, whether by intrinsically forced transitions or those induced by human engineering, have the effect of forcing the entire earth system to move to a new equilibrium that often proves grievous to society: extreme weather events in the wake of an increasingly carbonized atmosphere, transformed river regimes downstream of dams, and urban flooding with loss of natural drainage, to cite a few. Modern understanding of the behavior of coupled dynamical systems predicates that such perturbations, if they exceed some critical levels, may lead to irreversible changes.

2. That diversity of landforms and associated ecological systems is critical for the stability of the interconnected earth system, particularly for its performance of a host of invisible ecosystem services that we are so abjectly dependent upon. These services are sustained by a synergistic network of material and energy flow. Once disrupted, they take millennia to be optimally reorganized through generations—long transitional states that often prove hazardous to human communities. Human intervention to alter any component of the earth system critical for the sustenance of subsisting human communities, therefore, requires utmost care in its engineering design to ensure (1) that natural regimes of biogeological processes are minimally disrupted and (2) that the new foisted structures are robust enough not to become a source of cascading hazard in the event of their failure.

3. That right to life with dignity and security, compatible with the human heritage of knowledge and technology, is an *inalienable* "right" of citizens on this planet, and this condition entails an equally inalienable obligation on the part of individuals and institutions who have autonomously chosen to engage in the requisite scholarly and professional activities to make the exercise of these rights possible.

4. Specifically, these obligations consist in (1) identifying the sensitivities of the earth system to the stability of a habitat and social milieu chosen for engineering interventions, or infusion of new materials and systems, based on a transparent analysis of available knowledge supported, wherever necessary, by new generated evocative data/information and (2) openly placing the entire proceedings entailed in the conclusion in the public domain. The last is the sacred value of science committed to a perpetual possibility of eliminating errors and fallacies in a given set of propositions and reduction of their uncertainties through independent analysis made possible by open availability of details.

NOTES

1. The law of noncontradiction is one of the few fundamental principles that have the form of self-evident truths. For, it asserts that a proposition that is a declaratory sentence whose meaning can be either true or false and its opposite cannot at the same instant be true. Or, symbolically, *p* and *not p* is logically false. The use of this principle in investigating questions of ethical behavior can be traced as far back as the fourth century BC in Socrates' *elenctic method*, although traditionally attributed to *Aristotle's Metaphysics*. It is also found in ancient Indian logic as a meta-rule in the *Shrauta Sutras* of Panini's grammar.

2. The completed Tehri Dam crosses a 574-m-wide triangular canyon across the Bhagirathi River in Central Himalayas, with its crest rising to 260 m above the deepest point of the foundation. The fundamental period of vibration of this tall structure, if approximated by a 2-D model would be given by the classical shear beam equation, $T_f = (2.61 \ H/V_{shear})$, predicting a value of 2.1 s for the central section of the Tehri Dam. The natural periods used by the designers are similar (THDC report), suggesting that they followed this very approach. However, for this approximation of plane strain to be valid, the minimum aspect ratio L/H must be greater than 6, thrice the aspect ratio of the Tehri Dam. Makadisi et al. (1982) in their classic paper show that the three-dimensional effects caused by the stiffened structure of a small-aspect-ratio dam would reduce its fundamental period by a factor of 1.8. Furthermore, it not only would produce larger displacements and accelerations (Gazetas, 1987) in the body of the dam, but would have the possibility of inducing longitudinal vibrations. Intriguingly, this assumption was not referred to by the dam consultants at DEQ-R, despite considerable literature and computational results already created by the late 1980s. When this apparent design lapse was pointed out, they claimed that the 2-D analysis was in fact more conservative compared with its 3-D counterpart (e.g., Chandrasekaran of DEQ-R quoted by Ramachandran (2001)).

3. At this time, the tectonic setting of the dam was known imperfectly. In particular, it was not recognized that the 2400-km-long Himalayas converges with southern Tibet at rates of 16–18 mm/year (Bilham et al., 1997). The seismically active plate boundary consists of a 100-km-wide thrust fault, a décollement, that emerges at the surface 60 km SW of the dam, and underlies the dam site attaining a depth of ≈20 km, some 35 km SW of the locking line where the accumulation of tectonic strain is accompanied by modest microseismicity. To the north of this seismically active location spanning the Greater Himalayas, the plate slips aseismically beneath Tibet. The accumulating strain energy NE of the dam is released in earthquakes that rupture southward, either in the form of moderate and major earthquakes that propagate toward the surface (e.g., the 1991 Uttarkashi $M_w = 6.8$ earthquake) or in the form of great earthquakes that permit the rocks of the Himalayas to slide southward over the Indian plate. The last time that a great earthquake shook Tehri was on 6 June, 1505, and is recorded as slip of the surface frontal faults by 15–20 m (Ambraseys and Jackson, 2003; Kumar et al., 2006). Presumably, in 1505, the Tehri Dam site was displaced SE by more than 15 m. The rupture parameters of this historical earthquake suggest that the moment magnitude of the 1505 earthquake, if it occurred as a single rupture, may have approached $M_w = 9.0$ (600 km × 100 km × 20 m), with a renewal time based on the local convergence rate of more than 1200 years. The largest earthquake considered in the initial design of the Tehri Dam was $M_w = 7.2$. During two decades of discussions starting in 1980, this was subsequently raised to $M_w = 8.5$, corresponding to a 100 × 200 km rupture with 10 m of slip, slightly larger than the $M_w = 8.4$ Bihar–Nepal earthquake of 1934.

4. [Sh. Kamal Nath] The Ministry of Mines had, however, been receiving representations from various quarters expressing doubts and misgivings regarding seismic safety of the dam. Therefore, the Director General, GSI was asked to constitute an Expert Group for a critical reappraisal of the

safety aspects of Tehri Dam taking into account the misgivings raised. This Expert Group gave its report in July, 1991 observing that the design of the dam has been subjected subsequently to a PGA of 0.5g and found satisfactory. The Department of Mines, this month in August, 1991, observed as follows.

Based on the report of the Director General, GSI, the Department of Mines accepts the recommendation of Prof. Jai Krishna that the proposed dam section for the Tehri Project is safe from the point of view of seismicity of the region. Based on these reports, the Department of Mines also considers that the following scientific studies need to be undertaken. It means that the Department of Mines while clearing it has made two observations. In addition, whether we call them as observations or as conditions, which is a separate thing. The Department of Mines has made the following observations.

(i) Timebound micro-seismic investigation (with the help of digital seismometers) in the Tehri area within about a year's time in order to have better estimates of the critical seismic parameters including determination of the depth of the plan of the detachments under-thrusting and estimation of Q-Value.

Therefore, Q-value determines Prof. Brune's formula. It is a very important ingredient of Prof. Brune's formula.

If the Q-value is wrong, the whole of Brune's formula goes astray. The Department of Mines has asked for a proper determination of Q-value.

(ii) The design of the Dam to be tested for actual accelerogram. That means the PGA, which has been estimated at 0.5g.

For the purpose of its clearance, certain presumptions were made.

Those presumptions have to be checked out by these two conditions of the Department of Mines.

(iii) The design of the Dam to be tested for actual accelerogram of the Gazli-the Gazli earthquake aspect is the worst-case scenario aspect, it is the worst scene and for PGA higher than 0.5g just to test the stability of the design.

The studies should be designed and carried out in coordination between Geological Survey of India, National Geophysical Research Institute and Tehri Hydro Development Corporation. If the results of the above studies necessitate any modification of the design, the same could be taken care of by the concerned authorities.

The Department of Power has agreed to fund these studies.

For the time being, the Project authorities propose to continue with the present design on the plea that it would be able to withstand the seismic forces.

As far rehabilitation is concerned, an observation was made and I would like to say something on that. I will say some points for the benefit of the House. The total submergence is 112 villages plus Tehri town. The hon. Members who hail from Tehri. I suppose, will have to look for another abode.

The Power Generation is 2000 MW at Tehri and 400 MW at Koteshwar.

This will be the installed capacity, but the firm power will be 487 MW. This is a Hydel Project. So, it will be used for peaking. The Irrigation potential as envisaged by the Project authorities is 2.7 lakh ha of additional land and Stabilization of irrigation where irrigation is not adequate or to sustain it. 6.04 lakh ha. 3000 cusec of drinking water has to come to Delhi. The Project cost in 1990 cost is Rs. 3500 crores. This is on the prices prevailing in 1990. There is an environmental cost, which, for the time being, has not been taken into consideration. As far as life of the Dam is concerned, there is a presumption being taken on Siltation and Assumed sediment rate by the Project authorities is 14.5 ha m per 100 km^2 per year. The CAG has observed that cost-benefit ratio has reduced from 1:11.7 to 1:1.349 in 1986. I do not know if any cost benefit ratio has been done after 1986. Uptillnow, an expenditure of about Rs. 600 crores has been incurred; Rs. 450 crores has been incurred on the power components and Rs. 150 crores roughly has been incurred on the irrigation component. There is a provision in the current plan. This Project not yet has the P.I.B. clearance. It is not correct to say that my Department is sitting back and not monitoring it.

I will just read out the last para of the conditions of the clearance, which was given by my Ministry while approving this project. The hon. Member has said that we have not given clearance, which is not correct. We have given a conditional clearance. Our condition was very categorical and it says: "The completion of studies, formulation of action plans and their implementation will be scheduled in such a way that their execution is *paripassu*, that means at the same time, with the engineering works failing which the engineering works would be brought to halt without any extraneous considerations. These conditions will be enforced, among others, under the provisions of the Environment (Protection) Act, 1986." Now, the status of these conditions is as follows:

MR. CHAIRMAN: Will you take more time?

SHRI KAMAL NATH: Sir, I was just trying to be as elaborate as possible for the benefit of Members. It is up to you. If you allow me more time, I will take, if you do not, I will not.

MR. CHAIRMAN: If the House agrees, we will extend the time by few minutes so that we can have a complete reply.

SEVERAL HON. MEMBERS: Yes, Sir.

[Translation]

SHRI RAJVEER SINGH (Aonla): Sir, let this be completed. This is a very important matter.

[English]

SHRI KAMAL NATH: Sir, the items which were a part of the conditional clearance, I will very briefly dwell on those as to what happened and what is their current status. Those are: the safety aspects and the design of the dam. The design of the dam had to be approved by the Ministry of Mines. The Ministry of Mines has given clearance on the design of the dam, subject to those conditions. The question is when those conditions are fulfilled, whether they will require modifications in the design or not. The Tehri Dam authorities have stated that any modifications required in the design of the Dam arising out of those two studies to be carried out will be made. So after the High Level Committee gave its Report in April, 1991 with a dissent note of Prof. Gaur, subsequently the Department of Mines have conveyed that the design is acceptable with these two conditions.

5. The Western Ghats form a ~1600-km-long elevated coastal tract of average height 1200 m. They mark the faulted western margin of the Indian shield carved ~90 million years ago when Madagascar rifted away from India creating the Arabian Sea. Subsequently, the northern half of the Ghats was covered with Cretaceous outpourings of the Deccan flood basalts, represented at many sites and altitudes by their flat-topped lateritic relicts of chemical weathering that are relatively resistant to erosion. The Ghats in the region consist of a ~50-km-wide coastal belt of narrow, flat-topped terraces ranging in altitude from 3 to 750 m (Powar, 1993) and the upland Sahyadris to the east, across the spectacular Western Ghats Scarp forming a water divide. The Sanskrit word Ghat means steps leading down to water.

6. The Greeks, in their relentless bid to maintain the purity of logical arguments, identified two broad classes of fallacies: *formal* and *informal*. The former can be easily detected by identifying the form of the argument, e.g., the various syllogisms of deductive logic. However, to grapple with the tentativeness of inductive propositions, they turned their attention *to informal fallacies whose* legitimacy could not be decided by inspecting their form but only from the contents of the argument. Aristotle himself identified 13 but logicians have qualified over a dozen more. One of these is called the "Fallacy of the Expert," which arises because the premises of the argument are not logically relevant to the conclusion, but are made psychologically persuasive by the use of rhetoric.

REFERENCES

Anon, 1983. Seismic Parameters for Tehri Dam. Department of Earthquake Engineering, University of Roorkee (DEQ-R) EQ-83-4.

Anon, 1990. Aseismic Design of Tehri Dam. Tehri Hydro Development Corporation (THDC) Ltd., Roorkee.

Argand, E., 1924. La tectonique de l' Asie. Proc. 13th Int. Geol. Cong. 7, 171–372.

Ambrasseys, N., Jackson, D., 2003. A note on early earthquakes in northern India and southern Tibet. Curr. Sci. 84, 571–582.

Anuradha, S., 2010. Estimation of permanent displacements of the Tehri dam in the Himalayas due to future strong earthquakes. Sadhana, Ind. Acad. Sci. 5 (Part 3), 373–392.

Bagla, P., 2012. India barred entry to U.S. author of seismic review. Science 338, 1275.

Baranowski, J., Armbruster, J., Seeber, L., Molnar, P., 1984. Focal depths and fault plane solutions of earthquakes and active tectonics of the Himalayas. J. Geophys. Res. 89, 6918–6928.

Bilham, R., Larson, K., Freymueller, J., 1997. GPS measurements of present day convergence across the Nepal Himalayas. Nature 386 (6620), 61–64.

Bilham, R., Gaur, V.K., 2011. Historical and future seismicity near Jaitapur, India. Curr. Sci. 101 (110), 1275–1281.

Bolt, B.A., Abrahamson, N.A., 1982. New attenuation relations for peak and expected accelerations from strong ground motion. Bull. Seismol. Soc. Am. 72, 2307–2321.

Brune, J., 1976. The physics of earthquake strong motion. In: "Seismic Risk and Engineering Decisions" Lomnitz, C., Rosenblueth, E. (Eds.), Elsevier Publication, pp. 141–177.

Brune, J.N., 1993. The seismic hazard at Tehri dam. Tectonophysics 218, 281–286.

Boore, D.M., Joyner, W.B., December 1982. The empirical prediction of ground motion. Bull. Seismol. Soc. Am. 72, S43–S60.

Campbell, K.W., 1989. Emerical Prediction of Near Source Ground Motion for the Diablo Canyon Power Plant Site, San Luis Abispo County, California. Dep. Interior, U S Geol. Survey. Open file report, pp. 89–484.

Gaur, V.K., Chander, R., Sarkar, I., Khattri, K.N., Sinvhal, H., 1985. Seismicity and the state of stress from investigations of local earthquakes in the Kumaon Himalayas. Tectonophysics 118, 243–251.

Gaur, V.K., Bilham, R., 2012. Discussion of seismicity near Jaitapur. Curr. Sci. 103 (11), 1273–1278.

Gazetas, G., 1987. Seismic response of earth dams: some recent developments. Soil Dyn. Earthquake Eng. 6 (1), 1–48.

Gujar, A.R., Rajamanickam, G.V., Ramana, M.V., 1986. Indian J. Mar. Sci. 15, 241–245.

Gazetas, G., 1987. Seismic response of earth dams: some recent developments. Soil Dyn. Earthquake Eng. 6 (1), 2–47.

Heim, A., Gansser, A., 1939. Geological Observations of the Swiss Expedition. Publ: Zurich: Gebruder Fretz.

Iyengar, R.N., 1993. Dynamic analysis and seismic specification of Tehri dam. In: Gaur, V.K. (Ed.), Earthquake Hazard and Large Dams in the Himalayas. Indian National Trust for Art and Cultural Heritage, New Delhi, pp. 137–152. Publ.

Khattri, K.N., Tyagi, A.K., 1983. Seismicity patterns in the Himalayan plate boundary and the identification of areas of high seismic potential. Tectonophysics 96, 19–29.

Kumar, S., Wesnousky, S.G., Rockwell, T.K., Briggs, R.W., Thakur, V.C., Jayangondaperumal, R., 2006. Paleoseismic evidence of great surface rupture earthquakes along the Indian Himalayas. J. Geophys. Res. 111, B03304. http://dx.doi.org/10.1029/2004JB003309.

Makadisi, F.L., Kagawa, T., Seed, H.B., 1982. Seismic response of earth dams in triangular canyons. J. Geotech. Engrg. Div., ASCE, 108, CT10, 1328–1337.

Majumdar, T.J., Bhattachaerya, R., 2011. A comparative evaluation of the gravity signatures over a part of the western Indian offshore for lithospheric studies. Indian J. Geo-Mar. Sci. 40 (4), 491–496, 25.

Mugnier, J.L., Huyghe, P., Chalaron, E., Mascle, G., 1994. Recent movements along the main boundary thrust of the Himalayas: normal faulting in an overcritical thrust wedge? Tectonophysics 238, 199–215.

Narasimhan, T., 2008. A note on India's water budget and evapo-transpiration. J. Earth Syst. Sci., Indian Acad. Sci. 117 (3), 237–240.

Narasimhan, T., Gaur, V.K., 2010a. A framework for India's water policy. Econ. Polit. Weekly XLV (30).

Narasimhan, T., Gaur, V.K., 2010b. A Framework for India's Water Policy. Report No. R 4–09 of the National Institute of Advanced Studies, Bangalore.

Ni, J., Barazangi, M., 1984. Seismotectonics of the Himalayan collision zone: geometry of the underthrusting Indian plate beneath the Himalayas. J. Geophys. Res. 89.

Newmark, N.M., Hall, W.J., 1982. Earthquake Spectra and Design, Engineering Monographs on Earthquake Criteria, Structural Design, and Strong Motion Records. Earthquake Engineering Research Institute, University of California, Berkeley, CA.

Paul, J., Blume, F., Jade, S., Kumar, V., Swathi, P.S., Ananda, M.B., Gaur, V.K., Burgmann, R., Bilham, R., Namboodri, A., Mencin, D., 1995. Microstrain stability of India. Proc. Indian Acad. Sci. (Earth Planet. Sci.) 104 (1), 131–146, 1995146.

Powar, K.B., 1993. Geomorphological evolution of Konkan coastal belt and adjoining Sahyadri uplands with reference to quaternary uplift. Curr. Sci. 64, 793–796.

Russel, B., 1961. History of Western Philosophy. Publ. George Allen and Unwin Ltd.

Ramachandran, R., May 12–25, 2001. The Tehri turnaround. Frontline 18 (10).

Rastogi, B.K., 2012. Seismicity near Jaitapur. Curr. Sci. 103 (2), 130–131.

Shetty, P., September 29, 2010. Plagiarism plagues India's genetically modified crops. Nature 467, 504–505. http://dx.doi.org/10.1038/news.2010.503.

Sinvhal, H., Agrawal, P.N., King, G.C.P., Gaur, V.K., 1973. Interpretation of measured movement at a Himalayan (Nahan) thrust. Geophys. J. R. Astron. Soc. 34 (2), 203–210. ISSN 0016-8009.

Spinoza, 2000. Ethics, Edited and Translated by GHR Parkinson. Oxford University Press.

Valdiya, K.S., 1980. Geology of Kumaon Lesser Himalayas. Publ. Wadia Institute of Himalayan Geology.

Widdowson, M., 1997. Tertiary palaeosurfaces of the SW Deccan, Western India: implications for passive margin uplift. In: Palaeosurfaces: Recognition, Reconstruction and Palaeoenvironmental interpretation. Geol. Soci Spec. Publ. 120:221–248.

Chapter 13

Corporate Money Trumps Science

William Gawthrop (Retired)
President of Green Mountain Geophysics, Boulder, Colorado, USA

Chapter Outline

Abstract

A nuclear power plant was designed and built using questionable understanding of the earthquake hazards at that site. Although there have been scientific reasons to question the design, the politics behind some of the design decisions seem to more reflect attempted cost savings as opposed to safety based on science.

Keywords: Nuclear power plants; Seismic hazard; Seismic risk.

INTRODUCTION

Being a junior in college at Cal Poly in San Luis Obispo, I was quite naive about the politics of a nuclear power plant. I believed that scientific research was unbiased and not prone to political wills. My experience working near the Diablo Canyon nuclear power plant effectively killed those beliefs.

My love of earthquakes started in 1957, during the Daly City earthquake. Watching what I had previously thought to be a stable earth suddenly shook violently. My world would never again be the same. I started excitedly reading everything I could find about earthquakes and other earth processes. When in high school, I designed an instrument to measure the creep along a fault. Fortuitously, I lived in Menlo Park, home of the National Center for Earthquake Research (NCER), part of the U.S. Geological Survey (USGS). When I entered to introduce myself one day in 1968, I met Bob Burford who was in the process of developing a similar instrument for installation along the creeping section

Geoethics. http://dx.doi.org/10.1016/B978-0-12-799935-7.00013-7

161

of the San Andreas Fault. We pooled our efforts and I became the first volunteer scientist for the NCER in Menlo Park. When it came time to install the first instrument, Bob became uncomfortable about working with a volunteer, so he arranged to have me hired as a paid employee of the government. I was in heaven. We installed alignment arrays to determine the precise locations of the fault in about a dozen places before installing the creep meters.

The summer after I graduated from high school in 1969, another researcher at the USGS, Robert Page, needed workers to go to Alaska to install a seismograph network in the area of the 1964 Prince William Sound earthquake. He had heard about the crazy kid willing to work for nothing just to learn more about earthquakes, and decided to put me under his wing. We installed permanent, telemetered seismographs by land, sea, and helicopter, a truly amazing job for a bright-eyed kid just out of high school.

I also assisted in building portable smoked paper seismograph recorders for short-term studies, mostly of aftershocks. I even led some aftershock studies using these recorders until getting into trouble one day. After an earthquake in Southern California, we put out these recorders in several areas near the rupture zone. One instrument was placed behind a campground in a county park. Upon leaving in our government truck after the installation, a bystander asked what we government guys had been doing up there. Being a good civil servant, I responded that we were setting up seismographs to record earthquakes. His follow-up question was "why were we setting up seismographs today when the earthquake was yesterday." My response was that we were setting up for the big one tomorrow. He looked a little stunned when I drove off. Upon returning a few hours later, all the campers were gone, probably to some distant location out of state. My superiors decided I was not the proper person for disseminating information to the public.

Where the Trouble Began

When it came time to do a senior project in 1974, a requirement of all physics students at Cal Poly, the USGS was kind enough to allow me to borrow four of these smoked paper seismograph recorders for a couple of weeks to do a short-term study of the seismicity around San Luis Obispo. At the time, there were no permanent seismographs in the vicinity, so I figured this would be an interesting area to check out. I remembered experiencing two earthquakes my freshman year in 1969 from the offshore region, so I knew the area was seismically active. My goal was to place the instruments in a large enough array, so that in conjunction with permanent stations nearby, they could adequately locate earthquakes around an area about half the size of the county or about 1600 square miles. Three of the inland sites were easy to establish, but I required a site for the fourth instrument as far to the west as possible. This is where the trouble began.

PG&E, the local power company in Northern California, was building a nuclear power plant at Diablo Canyon, which was near the coast west of San

Luis Obispo. It was the ideal location for my fourth instrument. I merely had to obtain permission to put my portable seismograph on their property for my 3-week project. I went to the local office, which had only office clerks to handle the power company's billings. My request was followed by numerous blank stares. They at least agreed to contact the main office in San Francisco about my request. A few hours later, I received a telephone call from a senior civil engineer of PG&E. He wanted to fly down to meet me the following day to talk. I was surprised that they would send such a senior person for such a simple request of letting me set up my seismograph.

The day began with meeting him at the San Luis Obispo airport and going for coffee. He explained that my request was unusual and that they had already done extensive studies to show that there was no significant earthquake activity that would affect their power plant and that there was of course no need for me to do another study. I explained my motivation was for scientific and personal interest only and that I merely wanted a location to place my portable instrument that would give me a good distribution for recording earthquakes. He asked me if I would like a tour of the power plant, which I of course enthusiastically agreed. We drove out for an extensive tour of the power plant, even going into the future core for the nuclear reactors. He went into detail about how solidly every part of the power station had be constructed and tested. While this was extremely interesting for a young physics major, I did not see the relevance to finding a place to put my portable seismograph.

At the end of the tour, he said PG&E would graciously grant my request. My only requirement was to give a report of my results to the engineer who had hosted me at the plant and perhaps a few others at their San Francisco headquarters before talking to anyone else. It seemed like a simple request and I was always happy to talk about earthquakes with anyone willing to listen. We went to an area above the plant where the pools were to later store the nuclear waste. I picked out a site with easy access, but with little traffic noise. It proved to be an excellent site, completing the array for my study.

Over the next 20 days, my network recorded 14 earthquakes large enough to be accurately located. Eight of the earthquakes were within 30 km of San Luis Obispo, with four located right along the coast. After analyzing these earthquakes, I wrote up a report (Gawthrop, 1973) and returned the smoked paper seismographs to Menlo Park. Then as promised, I went up to San Francisco to talk to my PG&E contact about what I had found. After about 5 min in his office, he said we should move to another room where some additional people would also be interested in hearing my report. I was led to an auditorium where I was shocked to see about 50 people waiting. This was my first public speaking situation, so my nerves were probably on display at the lectern. After my short presentation, people in the audience started asking questions, not at all about the research, or what I had learned from it, but about who was I to do such a study and why did I think I was qualified to do it. From the multitude of questions, I felt that this was an ambush and all they were really interested in was disqualifying anything I had found. I left feeling astonished and dismayed about my experience, and wondering what their motivation was for treating me so poorly.

The Larger Study

Friends at the USGS encouraged me to continue this study, by going to UC Berkeley and Caltech in Pasadena where the two major networks of seismographs had been recording since the early 1930s. My plan was to look at older records of earthquakes in my study area and by combining the information from both networks getting a better picture of the seismicity of my study area. I learned that in the early days of these networks, Perry Byerly of Berkeley and Beno Gutenberg of Caltech, who ran these two networks did not like or respect each other. They had established a line that coincidentally went east–west through my study area, where the one network was not to study earthquakes on the other side of this line. Because of this, neither network put seismographs near this region. This line later became known as the Byerly–Gutenberg discontinuity because it resulted in severely reduced coverage of the earthquake activity. Looking at California seismicity maps, it falsely appeared that this was an area of fewer earthquakes. This artifact of misplaced politics became a controversy related to the Diablo Canyon nuclear power plant. My analysis (Gawthrop, 1975) revealed that the area did in fact have as much activity as other areas in coastal California, and much of that activity was concentrated along the coast.

At about the same time, I talked with Holly Wagner and Dave McCulloch of the USGS marine science center who were following up on a Shell Oil report (Hoskins and Griffiths, 1971) that had mapped a significant fault just offshore of central California. Their studies indicated that this large fault, which they had named the Hosgri using a combination of the beginnings of its discoverers names, showed recent activity by the fresh sediments at the surface having been cut by fault rupture and different geology from one side to the other (Page et al., 1979). This study, along with the earthquakes that I had reanalyzed, combined to make a strong argument that the Hosgri Fault was currently active and capable of producing significant earthquakes. This fault was located 3–5 km offshore of the Diablo Canyon plant. The Nuclear Regulatory Commission (NRC) at that time had a regulation that no nuclear power plant was to be built within 5 miles (8 km) of an active earthquake fault. Since the power plant construction was mostly completed at the time, I assume they felt they had no other option than to change the regulation, as that is what they ended up doing.

The NRC required PG&E to show the plant could withstand forces subjected on it by the largest credible earthquake (see Bilham, 2015) from the Hosgri Fault. PG&E spent a significant effort, trying to prove the Hosgri Fault was much less significant than it appeared to some seismologists (Evernden, 1975; Gawthrop, 1975, 1978). Some of these reports seemed to use questionable science to achieve their findings (Gawthrop, 1981; Hanks, 1979, 1981). Much of this work centered on pushing the rupture of a magnitude 7.3 earthquake in 1927 from any possibility of being along or near the Hosgri Fault (Hanks, 1979). From underestimating the extent of the Hosgri Fault (Page et al., 1979; Wells and Coppersmith, 1994) and ignoring the likelihood that the 1927 earthquake could have occurred on this fault, the NRC settled on the largest credible earthquake on the Hosgri Fault was a magnitude 6.5 from a purely strike-slip rupture. Their new engineering analysis for the power plant was based on this largest credible earthquake. Later, after the completion of the power plant and after more scientific studies, the NRC did require that the maximum credible earthquake be increased to magnitude 7.5.

The earthquake that occurred in the region on November 4, 1927 with an estimated magnitude of 7.5 has been subject to much scrutiny. This was the second largest earthquake Perry Byerly from UC Berkeley had investigated, and after being severely chastised by local business people after the 1925 Santa Barbara earthquake, he was reticent to say the quake was near the coast (Byerly, personal communication, 1974). His purpose of studying the quake was to analyze the seismic wave velocities to his Berkeley seismograph. The distance from Berkeley was important but not the azimuth, so he told me that he purposely pushed his estimate of the earthquake location away from the shore where he was confident it had been.

I used P wave arrivals from seismographs around the world in an attempt to relocate the epicenter. In 1927, timing was poor on many of the Earth's seismographs, so my location was subject to large potential errors. The center of the error zone of my location was near the Hosgri Fault near Point Sal.

Looking at records from the Santa Barbara seismograph, it was clear that a large number of aftershocks had occurred in the few days following the 1927 main shock with S-P time intervals of between 11 and 17 s. Because the S waves (secondary) travel slower than the P waves (primary), the delay with which they arrive at a station allows an accurate estimate of the distance of the source from the recording point.

No aftershocks appeared to have S-P intervals greater than 17 s. This clearly constrained the limits of the rupture zone to be from about 88 km at the nearest to Santa Barbara to about 140 km at the far end. Using this range of distances and assuming the earthquake was located on the Hosgri Fault, the rupture zone was from near Purisima Point close to Lompoc north to the approximate area of Diablo Canyon (Gawthrop, 1975, 1978). The possibility of this magnitude earthquake having occurred on the Hosgri Fault blew holes through PG&E's claim of the maximum credible earthquake being only magnitude 6.5.

Two seismologists from Caltech had written the seemingly flawed analysis for PG&E that Diablo Canyon was in an area of much lower seismicity, which seemingly was due to a lack of understanding of the Byerly–Gutenberg discontinuity. This report was part of the scientific studies section of PG&E's original report to the NRC, justifying Diablo Canyon as a good site to locate a nuclear power plant. A personal communication from the then editor of the

Bulletin of the Seismological Society of America where I had published my analysis of the 1927 earthquake explained that at least one of the anonymous reviewers was very aggressive in their critic of my studies and, according to the editor, was told to mellow his language (Thomas McEvilly, personal communication, 1981).

In a more recent Caltech study of seismograms in the Berkeley area during the earthquake, Helmberger et al. (1992) showed that the epicenter had to be further away from Berkeley than my poorly constrained location had been, comparing this earthquake to two earthquakes from the offshore Santa Lucia Bank in 1969. They concluded that even though the study did not have any constraint in the southeast direction perpendicular to the general shoreline, that the earthquake must have occurred well offshore, possibly in the region of the Santa Lucia Bank, over 100 km from Diablo Canyon. Based on the constraints of the rupture zone by the aftershock information, the ruptured fault had to come very close to the shoreline. Helmberger's analysis proves the epicenter was further south than my location, but his epicenter location is outside of the rupture zone defined by the largest S-P time interval of the aftershocks observed at the Santa Barbara seismograph (Gawthrop, 1975, 1978, 1981). So it is highly unlikely that the earthquake occurred further to the west than 140 km from Santa Barbara, defined by the largest S-P time intervals of the aftershocks, east of the Santa Lucia Bank.

Evernden, who studied the effects of attenuation on Modified Mercalli intensities in California (Evernden, 1973, 1975), in a personal communication, argued that the fault must have ruptured during the 1927 earthquake very near to the Lompoc area (about 80 km southeast of the Diablo Canyon site) where Modified Mercalli intensity IX was observed, and likely trended northwesterly, nearly parallel to the coast. Wells and Coppersmith (1994) developed equations that predict the length of rupture from this magnitude earthquake to be 60–100 km in length, depending on the depth assumed for the fault rupture. This would extend the rupture to near the Diablo Canyon power plant if the Hosgri Fault was the source.

Intensities as high as Modified Mercalli IX are rarely observed far from earthquake ruptures in California. So if the earthquake originated on a fault in the Santa Lucia Bank region, the fault rupture must have extended to the east to the coast probably near Lompoc. No fault has been mapped that extends this way, so it is more likely that the epicenter was near the coast with a rupture having occurred along the Hosgri Fault or a nearly parallel unmapped fault with an epicenter just offshore of Lompoc and extending to near Diablo Canyon.

Even if the 1927 earthquake did not occur on the Hosgri Fault, there is definitely a strong possibility that earthquakes of similar magnitude could occur there or on neighboring unmapped faults. Other earthquakes along this region of central California (i.e., 1983 Coalinga earthquake and the 2003 San Simeon earthquake) generally show oblique thrust faulting, which produces more vertical forces than does a purely strike-slip fault. The conclusion from this is that an earthquake larger and with more vertical accelerations than the 6.5 strike-slip maximum design earthquake used for the design of the nuclear power plant at Diablo Canyon is quite possible.

The Fallout from These Studies

In 1975, shortly after producing my analysis of the earthquake activity in the central California near-shore region, I received a call from my contact at PG&E. He first inquired about my studies and then asked how my recent visit with my parents in Santa Cruz had gone. This was puzzling, as I had never told him about going to Santa Cruz. A few weeks later, I received another call from him and was asked about my trip with my friends the previous week to Yosemite, which included apparent knowledge of my itinerary there, again something which I had not told him about. I became very concerned that I was being watched. While I cannot be certain of being watched by PG&E, this had occurred shortly after I heard of the mysterious death of Karen Silkwood, a whistle blower at Kerr-McGee's plutonium processing plant in Oklahoma. For a period of time, I became paranoid that I was going to be targeted in a similar manner.

The cost of PG&E's power plant increased from a projected $500 million in 1972 to about $5 billion dollars to complete the plant, probably due to the delays and redesign they had to complete because of the earthquake hazard. Even though my research was motivated by purely scientific purposes and backed up and checked by researchers at the USGS, I felt like I was being blamed for PG&E's problems of siting the power plant in an earthquake zone. For about a year after these calls, I nervously looked over my car thoroughly prior to driving, looking for any sign of tampering.

Later that year, I went off to graduate school in Boulder, Colorado. During my time there, I received several accusations, mostly from people involved in the earlier studies of the earthquake hazards to the plant that my work was biased and that I had purposely hidden the data that might have allowed others to disprove my results. This was absolutely not true, but it definitely tainted my opinion about biases of supposedly independent scientists. Too much money and politics seemed to be involved in the geophysical sciences. This had a major influence in changing my lifelong dream of being a research scientist. While I remained involved in geophysics after this, I never completed my PhD and steered clear of the political side of geophysics.

CONCLUSIONS

The Diablo nuclear power plant is currently operating and producing power and profits to PG&E. Based on similar studies on what is known about recurrence intervals of earthquakes along secondary faults in California from the Working Group on California Earthquake Probabilities (WGCEP, 2003), the power plant will likely live out its life without major earthquake damage, but the small possibility of an earthquake exceeding its maximum design is real.

REFERENCES

Bilham, R., 2015. Mmax: ethics of the maximum credible earthquake In: Wyss, M., Peppoloni, S. (Eds.), Geoethics: Ethical Challenges and Case Studies in Earth Science. Elsevier, Waltham, Massachusetts, pp. 119–140.

Evernden, J., 1973. Interpretation of seismic intensity data. Bull. Seismol. Soc. Am. 63 (2), 399–422.

Evernden, J., 1975. Seismic intensities, "size" of earthquakes and related parameters. Bull. Seismol. Soc. Am. 65 (5), 1287–1313.

Gawthrop, W., Preliminary Report on a Short-term Seismic Study of the San Luis Obispo Region, unpublished.

Gawthrop, W., 1975. Seismicity of the Central California Coastal Region. USGS Open File Report 75-134.

Gawthrop, W., 1978. The 1927 Lompoc, California earthquake. Bull. Seismol. Soc. Am. 68 (6), 1705–1716.

Gawthrop, W., 1981. Comments on "The 1927 Lompoc, California earthquake (November 4, 1927; M=7.3) and its aftershocks" by Thomaas C. Hanks. Bull. Seismol. Soc. Am. 71 (2), 557–560.

Hanks, T., 1979. The Lompoc, California California earthquake (November 4, 1927; M=7.3) and its aftershocks. Bull. Seismol. Soc. Am. 69 (2), 451–462.

Hanks, T., 1981. Reply to W. Gawthrop's "Comments on 'The Lompoc, California California earthquake (November 4, 1927; M=7.3) and its aftershocks'". Bull. Seismol. Soc. Am. 71 (2), 561–565.

Helmberger, D., Somerville, P., Garnero, E., 1992. The location and source parameters of the Lompoc, California, earthquake of 4 November, 1927. Bull. Seismol. Soc. Am. 82 (4), 1678–1709.

Hoskins, E., Griffiths, J., 1971. Hydrocarbon Potential of Northern and Central California Offshore: Region 2. AAPG. Special Volume, A128, pp. 212–228.

Page, B., Wagner, H., McCulloch, D., September 1979. Tectonic interpretation of a geologic section of the continental margin off San Luis Obispo, the Southern Coast Ranges, and the San Joaquin Valley, California: cross-section summary. Geol. Soc. Am. Bull. 90, 808–812.

Wells, D., Coppersmith, K., 1994. New empirical relationships among magnitude, rupture length, rupture width, rupture area, and surface displacement. Bull. Seismol. Soc. Am. 84, 974–1002.

Working Group on California Earthquake Probabilities, 2003. Earthquake Probabilities in the San Francisco Bay Region: 2002–2031. USGS Open File Report 03-214.

Chapter 14

Shortcuts in Seismic Hazard Assessments for Nuclear Power Plants are Not Acceptable

Max Wyss
International Centre for Earth Simulation, Geneva, Switzerland

Chapter Outline

Abstract

Nuclear power plants represent a great risk to the population, when they are located near faults capable of strong earthquakes and along coasts prone to tsunamis. Out of a desire to maximize profits, it can happen that estimates of seismic hazard are minimized, to reduce construction costs. This chapter recounts an incident in California resulting from a flawed assessment of seismic hazard at the design stage of an NPP that lead to a costly subsequent retrofit, the financial burden of which was resisted by the power plant operator.

Keywords: Nuclear power plant; Safety of the population; Seismic hazard.

INTRODUCTION

The importance of the safety of nuclear power plants (NPPs) was dramatically moved to center stage by the catastrophic sequence of events initiated by the tsunami attack on the Dai Ichi facility on March 11, 2011. For some time, television coverage followed the efforts of Japanese Government and corporate teams to contain and partially clean up the contamination. Now, 2 years later, sporadic reports show how ineffective robots are sent to clean up the area still inaccessible to humans, only to be turned into junk by the radiation. Why the protective seawall was only 6 m high, when there exists a list of 28 tsunamis in Japan higher than this and 22 of them reach 10 m and more (ITDB/WRL, 2005) remains a mystery.

Geoethics. http://dx.doi.org/10.1016/B978-0-12-799935-7.00014-9

169

The 2011 Tohoku M9 earthquake and tsunami have triggered much criticism of the standard method of seismic hazard assessment (e.g., Stein et al., 2012). As part of the rethinking of this problem, aspects of the history of safety considerations at the Diablo Canyon, California, NPP were summarized by Chapman et al. (2014). These authors report that the Pacific Gas & Electric Company (PG&E) is now collaborating with the US Geological Survey (USGS) and the Nuclear Regulatory Commission in developing methods and knowledge concerning seismic hazard assessment in California. This collaboration is laudable, giving hope that an event like that at the Dai Ichi facility may not happen in the United States. However, the history of the design and construction of the Diablo Canyon NPP is characterized by an astonishing lack of concern for the safety of the population, which is worth reviewing.[1]

Gawthrop (2015) relates his experience with PG&E in 1969, when he stumbled on two discoveries in close proximity to each other: The Hosgri fault zone and the planned NPP. Gawthrop had no other interest but to define active faults, wherever they may be found, by recording microearthquakes, but according to Gawthrop PG&E did not seem to share this interest.

Chapman et al. (2014) report that "The design ground motion spectrum for this (Hosgri) fault zone was anchored to a peak acceleration of 0.75 times the acceleration of gravity (g)." This statement is not correct. The design in the 1960s was for a peak ground acceleration (PGA) of 0.4 g. Only in the 1980s, after seismologists of the USGS had intervened (Coulter, 1975), was the plant retrofitted to withstand a PGA of 0.75 g. Subsequent to this intervention, PG&E applied for an increase in electricity rates to pay for the additional costs of $600 million. The consumers of California formed a consumers defense group, opposing the proposed rate hike. They argued that the company had caused these extra costs by their incorrect original design and therefore the company, not the consumer, had to pay for repairing the error. This chapter recounts the battle about this price hike. I have chosen a narrative format to conform to the chapter by Gawthrop (2015).

THE ASSIGNMENT

In the 1980s, I was Associate Professor for seismology at the University of Colorado, Boulder, and researcher at the Cooperative Institute for Research in Environmental Science. Besides studying earthquakes and teaching, I consulted on occasion for international engineering companies, estimating the seismic

---

1. The Associated Press (AP) published an article on 25 August 2014 entitled: "Expert calls for Diablo Canyon shutdown" (http://www.ktvu.com/news/news/ap-exclusive-expert-calls-diablo-canyon-shutdown/ng8Tj/). The devastation wrought by the Napa Valley, M6, earthquake of 24 August 2014 renewed interest in earthquake safety of the Diablo Canyon nuclear power plant. The AP reported that Michael Peck, who for five years was Diablo Canyon's lead on-site inspector, argued in a 42-page report filed in 2013 that the Nuclear Regulatory Commission is not applying the safety rules it set out for the plant's operation. AP further explains that Peck stated in his report that continuing to run the reactors "challenges the presumption of nuclear safety." Peck's conclusion that no one knows whether the facility's key equipment can withstand strong shaking, as related by the AP, is alarming.

hazard at locations where high dams for reservoirs were planned. This can be a delicate business, when the area is seismically active. On the one hand, companies and the population are interested in keeping the building costs low; on the other hand, the structure has to be built strong enough to resist possible ground shaking due to nearby earthquakes. For the design, a reliable estimate of PGA is necessary (Wyss, 2015). In order to estimate the PGA, a critical parameter is the magnitude M_{max} of the maximum credible earthquake (MCE), of which faults in the vicinity may be capable (Bilham, 2015). The second critical value is the distance of the project site from this fault because the amplitude of seismic waves decrease as a function of distance as they travel away from the epicenter.

One day, I received a phone call from a law firm in San Francisco, asking me if I would be willing to review a report that claimed the Diablo Canyon NPP had been built with all due care, accounting for the seismic hazard adequately, even conservatively, according to the state-of-the-art knowledge at the time (1966). The lawyers suspected that this was not true. I agreed to consider the problem and was visited the following week by two smartly dressed young men. They scanned my office and me with hard, unsmiling eyes. It turned out that the NPP in question had to be retrofitted to be able to withstand approximately double the accelerations it had been designed for. Consequently, the power company had applied for permission to raise their rates, such that they could recoup the $600 millions that retrofitting would cost. However, a consumers defense group had been formed that argued the company had caused the extra costs by their negligence and should not be allowed to make the consumers pay for their error. The two lawyers worked for a San Francisco law firm that represented the consumers defense group pro bono. Hesitantly, because of my relative youth, the two scouts entrusted me to prepare a case against a famous professor of seismology, who had written a report absolving the power company of all negligence.

After carefully reviewing the report that claimed the company could not possibly have acted in a different way at the time of designing the plant, I was asked to visit the law firm in San Francisco to help prepare the people's case. On the 26th floor of one of the most elegant sky scrapers, the Bay Bridge was prominently visible through the wall of glass. The assembled law team relaxed in their chairs, when the voice emanating from the telephone amplifier signaled agreement to join the team. We had not been able to find an earthquake engineer in the United States, willing to testify against the power company. Luckily, a world expert in Mexico City was not similarly hesitant.

A COMPENDIUM OF ERRORS

I next presented the law team with a brief list of fatal flaws in the seismology professor's expert report.

1. His report contained the claim that in 1966 nobody would ever had dreamed that accelerations due to earthquakes as high as 0.75 g were possible, as now suddenly demanded by seismologists of the USGS as the value for which the power plant had to be retrofitted. I recommended that we would stress

the fact that C. F. Richter's famous book, *Elementary Seismology* published in 1958, contained a careful and detailed analysis of several reports on objects being thrown up into the air due to ground shaking in earthquakes, including the famous account of tossed tombstones from nineteenth century India (Oldham, 1899). The acceleration by which objects are attracted to our planet by the force of gravity is termed 1 g. Therefore, anyone familiar with elementary seismology would have known in 1966 and before, that 1 g accelerations can happen in earthquakes. Today, there exist many instrumental measurements of accelerations two to three times larger than 1 g.

2. Next I proposed that we would point out in the hearings that the self-same company was denied a request to build an NPP at Bodega Head, northern California, in 1963/1964 because six experts thought it possible that accelerations in the range of 0.5–1 g could happen at the site selected (Neumann, 1963; Saint Amand, 1963; Maxwell and Farba, 1963; USC&GS, 1964; Newmark, 1963; Eaton, 1964). The contradiction between the above testimony contained in the archives of the power company and the seismology professor's claim could not have been more glaring.

3. The third strong issue concerned the model for the MCE that was constructed by the California engineering firm that came up with the value of 0.4 g for the acceleration to which the power plant was designed. The very first analysis of the seismic hazard commissioned for the power plant in question was only a few pages long and did not include a specific study of the seismicity and tectonics of the area. It was a general discussion of seismicity in California. The two experts from Caltech (California Institute of Technology) concluded correctly that anywhere in western Southern California an M6 3/4 earthquake could not be ruled out, even in places where no surface trace of an active fault was detectable (Benioff and Smith, 1967).

After this expert report, the engineering firm entrusted with calculating the acceleration to be expected had to accept M6 3/4 as the MCE magnitude and had to position it beneath the project site. Such a hypothetical earthquake source, if shallow, would have generated relatively high accelerations. Therefore, the expert engineer pushed the hypothetical earthquake to a depth of 19.3 km (12 miles). This way, a relatively large distance was created from the deep source to the plant at the surface. As a consequence, the hypothetical MCE was said to generate only a 0.2 g acceleration, which the engineer "conservatively" doubled as safety margin to 0.4 g as the value for the design. The reasons why an informed expert would not have made this mistake are given in the next two paragraphs.

In 1966, a USGS and a Caltech team (including myself) had recorded 100 s of aftershocks of the Parkfield earthquake on the San Andreas fault with accurately calculated depths mostly between 4 and 10 km (Eaton et al., 1970; Steward et al., 1967). Given this at the time well-publicized observation, the deep model assumed by the expert engineer was unrealistic, decreasing expected accelerations to an unacceptably unsafe level.

4. The claim that a magnitude M6 3/4 earthquake in central California would result in 0.2 g accelerations only, was brazenly bold, considering that in the same year (1966), the M6.4 Parkfield earthquake at a distance of about 100 km from the project site generated accelerations as high as 0.6 g total horizontal acceleration (Cloud and Perez, 1967).

I presented the lawyers with a long list of additional concerns about statements in the expert professor's and the expert engineer's reports (Wyss, 1988), however, we all thought that the aforementioned errors were sufficient to show that the design value of 0.4 g was not derived based on sound evidence This showed in our opinion that the need for retrofitting the plant was the company's fault. Therefore, it was not correct to make the Californian consumers pay for the mistake.

THE RESOLUTION

After the tense meeting on the 26th floor, I was taken to an exceedingly charming small hotel, where I could relax in a room stuffed full with embroidered pillows. The owner of the hotel was a key supporter of the struggle against the proposed unfair raise of rates.

The case was never heard before the California Board responsible for authorizing the rate hikes for electricity. This was much to my regret because I was well prepared to thoroughly refute the incorrect claims of the expert professor. The consumers defense group accepted an out-of-court settlement of dividing the cost 50/50. The company was allowed to raise rates enough to recoup $300 million. My interpretation of this result is that the lawyers on our side assumed that not all members of the Board would have been fully convinced of our arguments, perhaps by reasons other than logic.

CONCLUSIONS

The incorrect statements that I discovered in expert's reports that had the effect of minimizing the perceived seismic hazard and thus also minimizing the safety of the population of California are not unique. Gaur (2015) and Bilham (2015) report that a possibly active fault near the location of a planned NPP in India has not been trenched to determine what magnitude earthquakes this fault may be capable of. This type of shortcut endangers the public. I fear that the cases reported in this book represent only the tip of the iceberg of similar cases worldwide, which raise serious questions about research integrity as summarized by Mayer (2015). Similar concerns appear to arise when large mining interests are at stake (e.g., Das, 2015; Hostettler, 2015). However, the report of the current collaboration of PG&E with the USGS and the Nuclear Regulatory Commission to work toward better understanding and assessing the seismic hazard, and the quality presentations on the topic at meetings of the American Geophysical Union, show that progress in greater collaboration with parties focused on profits can be achieved. However, this requires that someone stands up and insists that the seismic hazard is estimated based on facts.

REFERENCES

Benioff, H., Smith, S., 1967. Seismic Evaluation of the Diablo Canyon Site. Pasadena, Exhibit 4709.

Bilham, R., 2015. Mmax: ethics of the maximum credible earthquake. In: Wyss, M., Peppoloni, S. (Eds.), Geoethics: Ethical Challenges and Case Studies in Earth Science. Elsevier, Waltham, Massachusetts, pp. 119–140.

Chapman, N.A., Berryman, K., Villamor, P., Epstein, W., Cluff, L., Kawamura, H., 2014. Active faults and nuclear power plants. EOS 95 (4), 33–40.

Cloud, W.K., Perez, V., 1967. Accelerograms – parkfield earthquake. Bull. Seismol. Soc. Am. 57, 1179–1192.

Coulter, H. W., 1975. USGS review of geologic and seismologic data relevant to the Diablo Canyon site, units 1 and 2, January 28.

Das, M., 2015. Ethics and mining – moving beyond the evident: a case study of manganese mining from Keonjhar District, India. In: Wyss, M., Peppoloni, S. (Eds.), Geoethics: Ethical Challenges and Case Studies in Earth Science. Elsevier, Waltham, Massachusetts, pp. 393–407.

Eaton, J.P., 1964. Geologic and Seismic Investigations of a Proposed Nuclear Power Plant on Bodega Head, Sonoma County, California. Part II – Seismic Hazards Evaluation. Menlo Park.

Eaton, J., et al., 1970. Aftershocks of the 1966 Parkfield-Cholame, California, earthquake: a detailed study. Bull. Seismol. Soc. Am. 60, 1151–1197.

Gaur, V., 2015. Geoethics: tenets and praxis. In: Wyss, M., Peppoloni, S. (Eds.), Geoethics: Ethical Challenges and Case Studies in Earth Science. Elsevier, Waltham, Massachusetts, pp. 141–160.

Gawthrop, B., 2015. Corporate money trumps science. In: Wyss, M., Peppoloni, S. (Eds.), Geoethics: Ethical Challenges and Case Studies in Earth Science. Elsevier, Waltham, Massachusetts, pp. 161–168.

Hostettler, D., 2015. Mining in indigenous regions: the case of Tampakan, Philippines. In: Wyss, M., Peppoloni, S. (Eds.), Geoethics: Ethical Challenges and Case Studies in Earth Science. Elsevier, Waltham, Massachusetts, pp. 371–380.

ITDB/WRL, 2005. Integrated Tsunami Database for the World Ocean. TSLI SD. Russian Academy of Sciences, Novosibirsk.

Maxwell, B.W., Farba, N.S., 1963. A Review of the Seismic Factors Pertaining to the Bodega Bay Atomic Power Unit No 1, Technical Operations Branch.

Mayer, T., 2015. Research integrity: the bedrock of the geosciences. In: Wyss, M., Peppoloni, S. (Eds.), Geoethics: Ethical Challenges and Case Studies in Earth Science. Elsevier, Waltham, Massachusetts, pp. 71–81.

Neumann, F., 1963. Earthquakes in the Bodega Bay Area. San Francisco.

Newmark, N.M., 1963. Structural Design Considerations and Safety against Earthquakes, Bodega Bay Tomic Park Unit No 1 San Francisco.

Oldham, R.D., 1899. Report on the Great Earthquake of 12th June 1897, vol. 29, Mem. Geol. Surv. India, Calcutta. p. 1379.

Richter, C.F., 1958. Elementary Seismology. W. H. Freeman and Company, San Francisco. p. 768.

Saint Amand, P., 1963. Geologic and Seismologic Study of Bodega Head. Northern California, Association to Preserve Bodega Head and Harbor, Bodega Head.

Stein, S., Geller, R.J., Liu, M., 2012. Why earthquake hazard maps often fail and what to do about it. Tectonophysics 562, 1–25.

Steward, S.W., et al., 1967. Seismic Refraction Survey and Aftershock Locations in the Vicinity of the Parkfield-Cholame Earthquake of June 27, 1966. EOS Trans AGU.

U.S. Coast and Geodetic Survey, Seismicity and Tsunami Report, Bodega Head, California, October, 1964.

Wyss, M., 1988. The Inadequacy of PG&Es Seismic Hazard Evaluation for the Diablo Canyon Nuclear Power Plant. Boulder, Colorado, p. 55.

Wyss, M., 2015. Do probabilistic seismic hazard maps address the need of the population? In: Wyss, M., Peppoloni, S. (Eds.), Genethics: Ethical Challenges and Case Studies in Earth Science. Waltham, Massachusetts, pp. 239–249.

Chapter 15

Geoethical and Social Aspects of Warning for Low-Frequency and Large-Impact Events like Tsunamis

Stefano Tinti, Alberto Armigliato, Gianluca Pagnoni and Filippo Zaniboni
Settore di Geofisica, DIFA, Università di Bologna, Bologna, Italy

Chapter Outline

Abstract

Real-time warning of large-impact events like tsunamis, that may attack the nearest coast soon after the generation, implies that there is no time to collect enough data before tsunami arrival to provide accurate estimates of the wave height and tsunami duration, which may result in issuing inaccurate alarm messages to the population. Assessments of tsunami hazard and analyses of tsunami risk, that are at the basis of long-term prevention policies, can be affected by large uncertainties as well, since they suffer from systematic scarcity of real observations because large tsunamis are very infrequent events. This paper discusses the role of geoscientists in providing predictions and the related uncertainties. It is stressed that through academic education geoscientists are formed more to improve their understanding of processes and the quantification of uncertainties, but are often unprepared to communicate their results in a way useful for society. Filling this gap is crucial for improving the way geosciences and society handle natural hazards and devise effective means of protection.

Geoethics. http://dx.doi.org/10.1016/B978-0-12-799935-7.00015-0

Keywords: Assessment of large-impact disasters; Assessment of low-frequency disasters; Communicating uncertainties; Geoscientists responsibility; Managing large uncertainties in launching alerts; Tsunamis.

INTRODUCTION

Geoscientists deal often with hazardous processes like earthquakes, volcanic eruptions, tsunamis, and hurricanes, and their research is aimed not only to a better understanding of the physical processes, but also to provide assessment of the spatial and temporal evolution of a given individual event (i.e., to provide short-term prediction) and of the expected evolution of a group of events (i.e., to provide statistical estimates referred to a given return period, and a given geographical area).

One of the main issues of any scientific method is how to cope with measurement errors, a topic that in case of forecast of ongoing or of future events translates into how to deal with forecast uncertainties. In general, the more the data available and processed to make a prediction, the more accurate the prediction is expected to be if the scientific approach is sound, and the smaller the associated uncertainties are. However, there are several important cases where assessment is to be made with insufficient data or insufficient time for processing, which leads to large uncertainties.

Two examples can be taken from the science of tsunamis, since they are rare events that may have destructive power and very large impact. One example is the case of warning for a tsunami generated by a near-coast earthquake. Warning has to be launched before the tsunami hits the coast, a few minutes after its generation. This may imply that data collected in such a short time are not yet enough for an accurate evaluation, also because the implemented monitoring system (if any) could be inadequate (for instance one reason of inadequacy could be that implementing a dense instrumental network could be judged too expensive for rare events). The second case is the long-term prevention from tsunami strikes. Tsunami infrequency may imply that the historical record for a given piece of coast is too short to capture a statistically sufficient number of large tsunamis, which entails that tsunami hazard has to be estimated by means of speculated worst-case scenarios, and their consequences are evaluated accordingly and usually are associated with large uncertainty bands.

In case of large uncertainties, the main issues for geoscientists are how to communicate the information (prediction and uncertainties) to stakeholders and citizens and how to build and implement together responsive procedures that should be adequate. A discussion of how large uncertainties can affect the decision making process, especially within the civil protection systems, in cases of disasters management is given by Dolce and Di Bucci (2015).

MEASUREMENT ERRORS AND PREDICTION UNCERTAINTIES

Any branch of natural hazards sciences has to deal with measurements of physical quantities when monitoring dynamic processes. The fact that errors are unavoidable and that the results of a measurement process may differ from one run to the other, even when the experiments are conducted with the highest care, is long recognized in experimental sciences. Measurement error theory was founded to build a conceptual framework and provide guidelines on how to treat random and systematic errors in case of repeated experiments, with the recommendation that the result of a scientific measurement procedure cannot be only the measured value of the quantity, say X, but also an associated band, say ΔX, that in a way indicates how accurate the measurement is, so that the result is expressed as $X \pm \Delta X$.

Predictions are also a common practice in many natural hazards sciences, though in some cases, like for earthquakes, a substantial unpredictability is explicitly advocated by many scientists. Making a prediction or a forecast (these terms will be used as synonym in this paper) means to provide the value, say Y, that a physical quantity will assume in a specified future time and in a given place. Usually the predicted value is the result of computations based on a theoretical model that may provide also a band ΔY, so that the prediction is given as $Y \pm \Delta Y$.

In this paper we will assume that (1) a real process can be described by one or more variables that assume defined or "true" values, (2) these variables can be subjected to measurement procedures with result $X \pm \Delta X$ (this should be taken as a scalar expression in case of a single variable and as a vector expression of N components in case of N observables), (3) these variables can be the object of a time-evolution model resulting in predictions $Y \pm \Delta Y$. The measurement band ΔX will be denoted as a measurement error, while the prediction band ΔY will be called prediction uncertainty, or simply uncertainty.

In view of the above, we can define a measurement interval $I_M = [X - \Delta X, X + \Delta X]$ and similarly a prediction interval $I_P = [Y - \Delta Y, Y + \Delta Y]$ where the respective measurement and predicted values X and Y are the midpoints. Indeed, the last constraint is not needed and we can simply introduce a measurement interval $I_M = [X_1, X_2]$ and a prediction interval $I_P = [Y_1, Y_2]$ by specifying the lower and upper ends of such intervals and implying that X belongs to the interval I_M and that Y belongs to the interval I_P.

Since we are dealing with geosciences, we will not consider here that the observation of variables is a process that has the potential to perturb the variable itself as underlined by quantum mechanics and the uncertainty principle. For the sake of simplicity, we will further assume that measurement procedures have no bias and therefore the "true" value of a variable falls always within the measurement interval I_M.

Moreover, since we consider only a conceptual frame, we will avoid all possible complications, and discard probability distributions within the intervals I_M

and I_P introduced above. To be more precise, it is sufficient for our discussion to assume that there is a prediction process providing three values $Y_1 < Y < Y_2$ and a measurement procedure providing three further values $X_1 < X < X_2$ and we assume also that the "true" value of the variable equals X. In this view, the prediction uncertainties and measurement errors can be defined a posteriori as the respective semidifferences, $\Delta Y = (Y_2 - Y_1)/2$ and $\Delta X = (X_2 - X_1)/2$.

ACCURATE, SUCCESSFUL, USEFUL PREDICTIONS

With the premises of the previous section, it is straightforward to introduce a criterion to evaluate if a prediction is correct. First, one can introduce the prediction error or residual R as the discrepancy between the predicted and the measured value, i.e., $R = Y - X$. Second, one can see if Y falls within I_M or not. If the value of Y happens to be comprised between the two endpoints X_1 and X_2, then the prediction can be said to be correct or accurate. In the special case of symmetric intervals (i.e., X being the midpoint of I_M), the prediction is accurate if the absolute value of the prediction residual $|R|$ is less than ΔX or equivalently if $|R|/\Delta X < 1$. The ratio of the residual absolute value over the measurement error ΔX can be taken as a measure of the discrepancy. Moreover, one can say that a prediction is very accurate if the ratio is much smaller than 1, accurate if it is smaller than 1, inaccurate or wrong if it is larger than 1 and very inaccurate or very wrong if it is much larger than 1.

Taking into account the prediction interval, one can introduce the concept of successfulness and define that a prediction is successful if I_P contains the true value X and it is unsuccessful if X falls outside such interval. In case of symmetric intervals, it is easy to see that a prediction happens to be successful if the residual $|R|$ is less than the uncertainty ΔY or if the ratio $|R|/\Delta Y < 1$. Therefore, by definition, the accuracy depends on the measurement error, while the successfulness depends on the prediction uncertainty.

Under many circumstances, one finds that ΔY is larger or much larger than ΔX. In this case it can occur that $|R| < \Delta X < \Delta Y$, or in other words that the prediction is accurate and successful, or that $\Delta X < |R| < \Delta Y$, meaning that the prediction is inaccurate but successful, or finally that $\Delta X < \Delta Y < |R|$, implying that the prediction is both inaccurate and unsuccessful. These different cases can be visualized in Figure 1, panels (a)–(c) respectively.

In a modern complex society, there is a strong need to know how natural potential hazardous processes evolve in time, and subjects elaborating predictions are normally different from subjects that can benefit from such predictions. If we distinguish the categories of prediction providers or forecasters and prediction recipients or end users (like coastal communities), one can see that forecasters tend to be satisfied by successful predictions ($Y_1 < X < Y_2$), while end users tend to privilege the aspect of prediction accuracy ($X_1 < Y < X_2$).

This can be elaborated further and slightly modified by introducing the concept of a useful prediction. Usually ΔX is very small since for many observables

PREDICTION

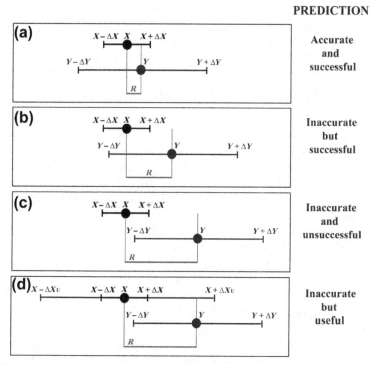

(a)	Accurate and successful
(b)	Inaccurate but successful
(c)	Inaccurate and unsuccessful
(d)	Inaccurate but useful

FIGURE 1 Graphical representation of the "true" value of a physical quantity and its uncertainty (black line) compared with the prediction and its band (blue line) in different cases of predictions: accurate (a), successful (b), inaccurate and unsuccessful (c), and inaccurate, but useful (d). The residual R (difference between real and predicted value) is marked in red.

the monitoring networks are based on high-quality sensors devised to minimize the measurement errors. In view of the above considerations, this implies that it is hard for a forecaster to provide an accurate prediction. It is also true, however, that the final user can be satisfied to receive a prediction that differs from the real one by an amount that is less than a threshold, say ΔX_U targeted to his/her own needs (usually it is such that $\Delta X_U > \Delta X$). Let us use an example for clarification. If we suppose that a weather forecaster likes to predict the maximum air temperature at the ground for the next day in a given place, where a meteorological station with temperature sensor is installed, it is reasonable to suppose that the sensor can measure the local temperature with an associated error ΔX about 0.1 °C or less. If the end user of the forecast is the local community, nobody will complain if the predicted maximum temperature Y differs from the measured value by some degree decimals. Instead, there would be some complains if the difference is larger than $\Delta X_U = 1$ °C. The selection of ΔX_U is subjective, since by definition it depends upon the need of the final user, but conceptually it is relevant to define it, since this allows one to state that a prediction is useful if the predicted value Y falls within the prediction interval $I_U = [X - \Delta X_U, X + \Delta X_U]$, as depicted in Figure 1, panel (d). And in terms of user's satisfaction, what really matters is not to get an accurate prediction, but rather a useful prediction, since it can be used to make decisions and take appropriate actions.

SOURCES OF UNCERTAINTIES IN TSUNAMI FORECAST SCIENCE

Tsunami science is surprisingly new, compared to many other disciplines. Actually, mention of tsunamis is already made by Aristotle (see his explanation of the tsunamigenesis by an earthquake due to the effect of contrary winds in Meteorology, Book 2, Part 8). Moreover, descriptions of inundation and damage caused by tsunamis can be found in historical documents of ancient Greece and Rome (see e.g., the numerous catalogs of historical tsunamis in the Mediterranean region and references therein: Galanopoulos, 1960; Antonopoulos, 1979; Papadopoulos and Chalkis, 1984; Guidoboni et al., 1994; Tinti and Maramai, 1996; Soloviev et al., 2000; Ambraseys, 2002; Tinti et al., 2004; Papadopoulos et al., 2007). Tsunami science development suffered from a scarcity of data due to the infrequency of such events and the lack or poorness of specific monitoring instrumental networks and of data collection through in situ surveying and field work. It was only after the catastrophic Pacific-Ocean tsunami sequence in the 1950s and in the early 1960s in the past century (including the November 5, 1952 Severo-Kurilsk tsunami, the May 22, 1960 Chilean tsunami, the Good Friday 1964 Alaska tsunami) that (1) post-tsunami surveys to gather data on damage and on tsunami run-up heights became a systematic activity undertaken by tsunami scientists; (2) tide-gauge networks tailored to the tsunami community needs started to be devised and implemented, at least in the Pacific Ocean; and (3) tsunami simulation models started to be used as a means to reconstruct past tsunamis and to assess the possible impact of future events.

Nowadays, tsunami science is in full development as regards conceptual analysis, methods, instrumental monitoring networks and operational tools. This is also due to the fact that the catastrophic events of December 26, 2004 in the Indian Ocean and of March 11, 2011 in Japan dramatically further raised the awareness of scientists, authorities, and the public that tsunamis are a major threat and that specific tools are required to protect coastal population and property. Nonetheless, even today the density of networks recording tsunami waves offshore and along the coast is much less than needed, and much less than the density of the present-day networks installed for weather observations or for monitoring of seismic waves, just to mention two companion disciplines (meteorology and seismology) that similarly to tsunami science deal with multiscale processes with potential for local as well as for regional disasters.

The insufficient coverage of tsunami observation data (sea level, and even more current velocity) is the main cause of errors in tsunami science. Indeed, there are no direct observations of the generation of a tsunami and very few observations of tsunami propagation offshore, and most of tsunami records are provided by coastal tide-gauges usually installed within harbors where they can be easily maintained and protected from vandalism. Typically, a local tsunami covering a propagation of some hundreds of kilometers can be recorded by about ten instruments or even fewer, while a transoceanic tsunami in the Pacific, that is the most densely instrumented ocean, can be covered by tens to hundreds of instruments. In this respect, see the analyses of the records of the December 26, 2004 tsunami by Rabinovich et al. (2006), and Rabinovich and Thomson (2007), and of the March 11, 2011 Tohoku tsunami by Bressan and Tinti (2012). However, since the observational networks are far from being equally distributed, there are seas where local tsunamis can pass unrecorded and large tsunamis can be seen only on a few tens of records.

Because of the scarcity of experimental waveforms the role of modeling is more important for tsunami science than for other disciplines. Computations can fill the gap left by observations and provide data even in those areas that are instrumentally uncovered. To better clarify the major sources of error in tsunami science, let us take a tsunami of tectonic origin. An earthquake generates a tsunami since it abruptly displaces the bottom of the sea, and the water mass above the seismic deformation area is bound to follow this sudden movement. This phase of the process is theoretically dominated by the vertical motion of the water that follows the bottom displacements and in doing so reaches a condition of hydrostatic disequilibrium accumulating potential energy for gravity waves. When the building up of the potential energy ends, the sea surface ends up being deformed with appearance of crests (troughs) above the areas of sea bottom uplift (subsidence), and the tsunami starts traveling across the ocean as a long period gravity wave. During this propagation phase, the horizontal velocities prevail in magnitude over the vertical velocity. All the tsunami generation process, in general, is not measured by any instruments and must be inferred. Hydrodynamic theory and laboratory experiments suggest that the amplitude of the initial crest–trough system is an increasing function of the sea bottom vertical displacement, which in turn increases with the magnitude of the earthquake and depends on a number of factors such as the position of the fault, the coseismic on-fault slip distribution, the elastic properties of the Earth's crust, etc.

Seismologists have developed theories and models over the years to infer the hypocenter location, earthquake size (seismic moment, magnitude, etc), and focal mechanism from processing seismic waveforms. Therefore, seismologists using their own data (seismic records) and their own models, provide input data to tsunami science to infer the initial sea surface displacement.

This in turn can be used to compute tsunami propagation by means of hydrodynamic numerical models suitable for long period gravity waves. The propagation of tsunamis is governed strongly by the bathymetry and also by bottom friction that may have a key role under certain circumstances. Even restricting to the influence of the sole water depth, it is clear that bathymetric data, and in case of land flooding, also topographic data form a set of experimental data needed to compute tsunami propagation. Usually even these data, provided either in form of a set of depth (altitude) isolines or in terms of regular grids, are the result of complicated raw-data processing from marine, aerophotogrammetric or satellite observations.

Tsunami scientists know that the largest sources of uncertainties for any estimation regarding tsunamis are the ones mentioned above, i.e., the earthquake data as well as the topobathymetric data. Tsunami predictions are made by running models where the above data are used as input for the models and where the set of the observables to predict is a subset of the model output (see e.g., Tinti et al. (2012), for the effects of the uncertainty of the coseismic slip distribution). Using the notation of the previous section, one can state that the prediction uncertainty ΔY depends on the uncertainties of the input parameters of the model, and a sensitivity analysis can provide a ranking identifying the parameters that have most influence on the magnitude of ΔY.

TSUNAMI WARNINGS

Tsunami Warning Systems (TWSs) started to operate in the Pacific Ocean in the form of national warning systems (in Japan since 1941, in USA since 1949, in Russia since 1958) and as an international warning center known as Pacific Tsunami Warning Center located in Hawaii, USA, established in 1968 (Imamura and Abe, 2009). After the 2004 Indian Ocean tsunami the international systems expanded to cover also the Indian Ocean, the Caribbean, and the North-east Atlantic and the Mediterranean, under the coordination of the UNESCO-IOC, and many national systems were also established.

The main purpose of a TWS is to assess an ongoing tsunami and to launch adequate alerts to the population in case of an impending threat (Joseph, 2011). The population needs to receive information on what is going on, and in case of danger to receive instructions or suggestions about the most appropriate response to protect life and property from the tsunami attack. The organization of such a response is a matter involving the civil protection of a nation or of a local community that has the legal mandate to take care of the emergency management.

Typically a TWS gets signals from seismic stations, makes estimates of hypocenter location and of magnitude in a very short time, and based on these seismic data runs a tsunami model. This can be done in real-time or a prerun simulation can be selected from an archived database containing a large number of precomputed tsunami scenarios. The selection process can be more complex, and imply a weighted combination of scenarios or a scenario ensemble, but for the purpose of this paper this is immaterial. What matters here is that the TWS produces a simulation scenario depending on the seismic estimates made. In a later stage, after some offshore or coastal instruments have recorded the tsunami, also these readings can be used in combination with the seismic data to optimize the choice of the scenario, that is to refine the prediction.

Let us suppose that for a given community living in a village with a harbor where a tide-gauge is installed it is relevant to know the maximum sea surface elevation produced by the tsunami in the basin. In our notation this observable is Y, while the maximum elevation measured by the tide-gauge is X. Usually tide-gauge readings are given with a measurement error ΔX of 1 cm, but in order to make decisions about the community response to the tsunami attack even an error of 30 cm can be tolerable, that is the amplitude of the interval for a useful prediction ΔX_U can be taken as large as 30 cm. Usually this value can be defined by the local authorities, such as the port authorities or by the local civil protection managers. We stress that the size of ΔX_U can change in dependence of the tsunami size. If the expected tsunami is small, a tolerance of 30 cm can be required, but if the tsunami is large, even a larger tolerance can be accepted. For example, if the local authorities estimate that a tsunami beyond 5 m is disastrous for the community and the immediate general evacuation is the only adequate response, and if the real tsunami happens to exceed this critical threshold value (i.e., $X > 5$ m), then every prediction $Y > 5$ m is a useful prediction, which means that the useful prediction interval is defined as $[X_{U,1}, \infty]$ with $X_{U,1} = 5$ m. In principle there is no upper limit to the residual $R = Y - X$.

The TWS may decide to provide Y or a succession of values Y_i ($i = 1, 2, \ldots n$) representing different predictions that are updated as more seismic and tsunami data become available and released at the successive time T_i. In normal situations, predictions tend to converge as time passes and more data constrain the solution of the assessment procedures. The TWS should also specify the prediction uncertainty ΔY. However, this value is usually not provided in the operational practice. After the tsunami has ended, the goodness of the prediction can be evaluated against the useful interval, which means that one can verify if the predictions Y_i fall within the interval $[X - \Delta X_U, X + \Delta X_U]$, or in the case mentioned before of a lower threshold, if Y_i are larger than the threshold.

TSUNAMI HAZARD

This paper mainly focuses on predictions associated with tsunami warning. Nonetheless, for the sake of completeness, we also touch here on the topic of long-term predictions that are inherent in tsunami hazard estimates. The question scientists are called to answer is typically related to the expected characteristics of tsunami occurrences in a given place within a given interval of time. To give an example in line with the one given in the previous section, one can estimate the maximum sea elevation expected in the next 10 years in a harbor where a tide-gauge is installed. Usually the interval of prediction is larger and for some applications it can be conventionally taken to be on the order of hundreds or even thousands of years.

Though the question above seems not to be too different from asking to estimate the maximum sea level in the same harbor produced by an ongoing tsunami, indeed the way to provide an answer is in principle quite different, since probability plays a fundamental role in case of hazard estimation. Indeed tsunami scientists regard the tsunami occurrence as a stochastic event. If, for simplicity, one restricts the attention to tsunamis of seismic origin, one has to first evaluate the probability of occurrence of the tsunamigenic earthquake, and then, given the earthquake occurrence, the maximum sea level induced by the tsunami in the target harbor. Since in general the sources of earthquakes that can generate tsunamis affecting the harbor are very many, the probability computations become quite complex, because one has to deal with the occurrence of several concomitant stochastic processes (see e.g., the hybrid probabilistic-deterministic analysis applied to Italian coasts by Tinti (1991) and by Tinti et al. (2005), and to the Moroccan coasts by El Alami and Tinti (1991); the probabilistic tsunami hazard analysis, PTHA, applied to the U.S. Pacific Northwest coastline by Geist and Parsons (2006); the study on the Makran zone in the Indian Ocean by Heidarzadeh and Kijko (2011)). The major problem for scientists, however, resides in the definition of the features of the sources relevant for tsunami generation: more precisely, the identification of the seismogenic areas, the return period of earthquakes of a given magnitude, and given focal mechanism. This is a typical field of investigation for seismologists, who for a long time have devoted their studies to produce seismic hazard maps on global down to local scales. These are widely used especially to estimate the maximum seismic acceleration, velocity, and displacement expected in a given place that are of interest to build earthquake-resistant structures and facilities.

The result of the probabilistic computations is not a single value Y but a probability distribution $p = p(Y)$. This means that after making a series of assumptions concerning the return period of earthquakes, the characteristics of the stochastic seismic process (for example the assumption of a no-memory Poissonian process), and after running simulation of tsunamis, one gets a probability density curve providing the probability of occurrence of any given sea elevation maximum in the next 10 years in the harbor. The possible elevation values go from $Y_1 = 0$, if no tsunami occurs in the next decade, to an upper limit Y_{max}, that is reached if the largest probable tsunami affects the harbor. Mathematically this implies that $p(Y)$ is defined over the domain $[0, Y_{max}]$. The probability density function describes fully the prediction, but often one prefers to provide only the average, or the mode or the median value of the distribution complemented by pairs of percentiles Q_s (typically the 5th and 95th Q_5 and Q_{95}, or the 1st and 99th Q_1 and Q_{99}) that give a measure of the distribution dispersion. In this way, one can provide one value of prediction Y (for instance the average of $p(Y)$, technically known as expected value) and the associated lower and upper ends of the prediction interval Y_1 and Y_2 (for example the percentiles Q_1 and Q_{99}). Indeed, quite often what really matters is only the upper limit Y_2, since this value is used to take a decision. In this example $Y_2 = Q_{99}$ is the sea elevation produced by a tsunami that will be exceeded only with 0.01 probability in the next 10 years, which can also be rephrased as the sea level that can be expected to be exceeded once every 1000 years.

In case of probability-based long-term predictions like the ones associated with tsunami hazard assessments, it is not straightforward to determine the correctness of the prediction, and often it is even not simple for the general public to interpret the result of the hazard analysis itself. To this purpose, interesting are the considerations made by Wyss (2015) regarding the parallel topic of the seismic hazard assessments. Concerning the evaluation of the goodness of the prediction, one can see two problems. The first one is related to the determination of the real value X. In the example given above regarding a prediction in the next 10-year interval, in principle one should wait until the end of the interval before reading the maximum sea elevation on the 10-year tide-gauge record to determine the value of X. This is impractical and indeed impossible if the time interval is on the order of 10^2–10^4 years (see the analysis on New Zealand coasts by Lane et al. (2013)) as is required in many hazard studies regarding critical facilities and infrastructure.

A second problem relates to the probability concept itself. Let us suppose that one finds that the upper limit Y_2 is exceeded within the 10-year interval, i.e., $X > Y_2$ (consider that it may be not necessary to wait until the end of the interval since a tsunami surpassing the predicted value Y_2 might occur even at the interval beginning). Does it imply that the prediction was inaccurate? Technically not, because if Y_2 is taken to be equal to Q_{99}, this means that the limit is expected to be exceeded in 1% of the cases. So in principle, it is not obvious how to evaluate the accuracy of the tsunami hazard prediction, i.e., how it can be either verified or falsified. In a sense, this situation is optimal for the forecaster, since he/she can claim that his/her predictions are successful, since they cannot be proven to be inaccurate. From the user point of view, however, the upper limit Y_2 is taken as a value that cannot be surpassed since the user will construct buildings, structures, industrial facilities, and devise evacuation plans, depending on this value, and adverse consequences will result if it is exceeded. In our terminology, Y_2 becomes a threshold to falsify the usefulness of the prediction rather than its accuracy: as soon as X is observed to be larger than Y_2, the user judges the prediction as not having matched the criterion of usefulness.

TSUNAMI PREDICTIONS AND TSUNAMI WARNING

Forecasting natural hazards for the society is a process that, schematically, involves three actors: the forecasters who formulate the prediction, the end users (the communities) that in principle should be the beneficiaries of the prediction, and the authorities (for instance national and local civil protection agencies) that are an element in between, since they receive the prediction from the forecasters and pass it, either directly or reelaborated, to the users.

In this simplified scheme, geoscientists are the forecasters: they are experts of the natural processes evolution and of the monitoring systems and can process signals and data, elaborate and apply theories, devise methods of analysis, and run numerical models.

In the previous sections it was stated that predictions can be broadly split into short-term and long-term predictions, and that as regards tsunamis, TWS forecasts are probably the best example of the first category, while tsunami hazard assessment were taken as an example of the second. Since the paper is focused on tsunami warning, we restrict to considerations that are mainly tailored to short-term tsunami predictions though they can also be of value for the long-term. The conceptual scheme is the one sketched in the previous sections where the ideas of accurate, successful, and useful predictions were introduced and where the main sources of uncertainties for tsunami science were highlighted.

One of the main scopes of any one of geosciences is to improve the knowledge of the processes of nature, in order to provide better estimates of their ongoing evolution and future occurrences. Better estimates in this paper mean better predictions. Like all other scientists, even geoscientists, by education and in many cases by personal character, tend to reduce as much as possible the uncertainties in predictions. The ideal situation would be that the prediction interval I_P be so small to be contained in the measurement interval I_M, implying that forecasters issue accurate predictions, but this is an unrealistic possibility. At the present technology level, indeed, it is impossible for any TWSs to forecast the maximum tsunami amplitude in ports and coastal towns with a centimeter precision or the arrival time of a tsunami with a precision on the order of seconds. This achievement will remain probably out of reach for several generations from now, even after a substantial improvement of the monitoring system, of the communication technology enabling real-time transmission of massive data, and of data processing and tsunami modeling. Therefore, this can be put in the category of the long-term objectives of the tsunami forecast science, but a strategy to cope with today's limitations has to be considered.

Since a geoscientist cannot produce accurate forecasts, the main goal for him/her is to produce useful forecasts. This is a crucial point. Indeed, the stress on the need to produce useful forecasts comes not only from the above mentioned impossibility of elaborating accurate predictions, but it is also more intimately and philosophically related to the nature of predictions in geosciences itself. Predictions released by geoscientists related to natural hazards are addressing the society. In this context, when issuing predictions, geoscientists provide a service that is embedded and founded in scientific knowledge, but remains a service. Under this point of view, it is consequentially logic that the service has to be satisfactory for the users, and therefore the main aim of predictors should be the production of useful predictions, which is fully achieved when the prediction interval I_P is contained within the interval I_U, based on the needs of the users.

The last condition highlights that there must be an open communication channel between geoscientists and users that involves procedures and even the design of the TWSs: the users should communicate to the forecasters their needs, and the forecasters should implement systems able to provide useful predictions. In case of TWSs, it would be fundamental to know the highest tolerable uncertainties for the users (ΔX_U) regarding the maximum tsunami sea level, the maximum tsunami currents, and the tsunami arrival time. Rough estimates for the authors of this paper could be in the order of 20–30 cm for the first, of 20–30 cm/s for the second and of 1 min for the third of such variables. However, unfortunately no effort has been put so far to produce these estimates and to put them at the basis of the operational prediction systems for tsunamis, and at the moment the question is open.

One further way to address the problem of the user's satisfaction is to consider why the user needs the prediction. In case of an impending tsunami, the users need the forecast since they have to respond to the potential hazard by taking or not taking actions. These actions can be usually distinguished in a few categories. In the Pacific, the Pacific TWS considers three levels of alert, called information, watch, and warning from the least to the most severe, for bulletins distributed over the ocean-wide pacific region, and four levels of alerts (information, advisory, watch, and warning) for local messages in the Hawaii region (http://ptwc.weather.gov/ptwc/about_messages.php). The body coordinating the activity for the TWSs in the region covering the North-east Atlantic and the Mediterranean has defined a protocol based on three types of messages with increased level of alerts called respectively information, advisory, and watch (see the NEAMTWS Interim Operational Users Guide approved in 2011 and published as a technical document at the address http://neamtic.ioc-unesco.org/).

For the present discussion, let us suppose that the classes are three: class 1 = do nothing; class 2 = protect harbors and beaches; class 3 = class 2 + evacuation of lowland. In this case, the user expects to receive predictions that are of help to choose the right option out of the possible three. This implies that the quantitative predictions concerning tsunami variables are in turn discretized in classes, with corresponding thresholds. For example, if one takes the sea level as the only relevant parameter, one can fix the value Y_{12} to discriminate between class 1 and class 2 and the value Y_{23} as a discriminant between the other classes, and accordingly fix the intervals $I_1 = [0, Y_{12}]$, $I_2 = [Y_{12}, Y_{23}]$ and $I_3 = [Y_{23}, \infty]$ corresponding to each action (see Figure 2 for a sketch of this discretization). In this frame, the prediction message issued by the TWS may contain the forecast Y for the parameter and the prediction interval I_P, which allows the user to easily verify which interval I_K it belongs to and to take the corresponding action, or may directly suggest to the user the action to take by using coded keywords. This second alternative is the one commonly adopted by the operational TWSs and technically this means that the message is a warning message rather than a prediction message since it gives recommendations rather than predicted values.

In case of warning based on discretized classes, the criterion by means of which the warning can be evaluated should be slightly changed. Instead of the real value X of the observable variable, one can consider the real class C, which is the class within which the real value X falls. If the prediction interval I_P happens to be contained in the real class C, the prediction (warning) is useful (Figure 2—panel (a)), while if it belongs fully to a different class, it is wrong

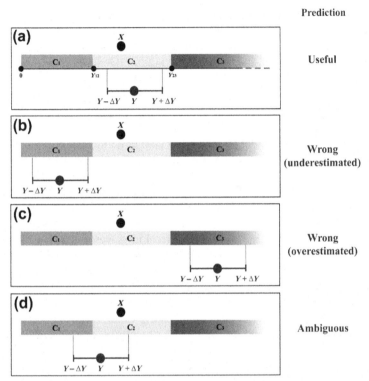

FIGURE 2 Illustration of three discretized classes (C_1=do nothing; C_2=protect infrastructures; C_3=evacuate people) in different cases of prediction (blue lines). It is assumed that the real value of the observable X falls in the class C_2. In (a) the prediction is useful since the entire prediction interval I_U falls in the class C_2. In (b) and in (c), the prediction is wrong since there is no overlap between the prediction interval and the class C_2. In (d) the prediction is ambiguous, since I_U crosses the boundary between two classes, C_1 and C_2.

(Figure 2—panels (b) and (c)). On the other hand, no matter how small the prediction interval I_P may be, however, if it happens to be close to the boundary between two classes, it may inevitably touch both of the classes. In this case the user is in front of an ambiguity (Figure 2—panel (d)). If one selects the less severe action, the risk is an underestimation of the real danger, while in the other case the risk is an overestimation. The precautionary principle would suggest solving the dilemma by taking the more severe response.

THE OPTIMAL TWS

In view of the considerations made in the previous section, one is able to depict a TWS that is optimal for the user as regards the warning precision. An additional crucial element for defining an optimal system is the prediction time. Of course, the TWS should launch messages early enough that the population has time to react. Launching an evacuation to a population already affected by a disastrous flooding has no value. The definition of the prediction time is a critical issue, but this is left to future contributions and not treated here.

One can state that two conditions have to be satisfied by an optimal TWS. First, a number of classes have to be established that correspond to adequate responses to the attacking tsunami. The number and the boundary of the classes are to be specified in dependence of local conditions, since they are certainly site dependent. Indeed the same sea level rise induced by a tsunami could be quite disastrous in some places such as very densely populated lowland, and almost inconsequential in other places where there is high topography or there are reliable natural or artificial tsunami barriers. The second condition is the minimization of classes' mismatches, which can be obtained by minimizing the prediction interval I_P.

To implement the action correctly, the user should be aware of the classes and their boundaries, and when receiving the prediction, he/she should know if the prediction interval is large enough to touch two contiguous classes to be able to make his/her own decision about the action to take.

What should be avoided by the forecaster or the issuer of tsunami warning is to disseminate messages where the action that the user can take is not clear or, which is the same, the messages are too generic. Generic messages contain text where the predictor says that the magnitude of the earthquake is large enough to produce damaging or disastrous local tsunamis, and that it is the responsibility of the local authorities to take proper actions to save population and property. A message of this type poses the entire responsibility of the decision to the Civil Protection Agencies (CPAs) that are the intermediate element between the TWS and the public. Unfortunately, most of the time the CPAs have no independent or additional information, and hence they have no means to take their decision. Hence, they may decide to pass the responsibility of the decision to the population itself by simply disseminating the same or a similar message, or they may decide to launch the alert for the highest-level response, that in general is a full-scale evacuation of the coastal zone.

Overresponding to a danger always has social costs, that in general are only or mostly economic in the short term, and that can be mostly psychological in the long term, influencing the public opinion. A series of repeated overresponses undermines the confidence of the public in the TWS and in the CPA with the consequence that the TWS alerts might tend to be ignored and the effective response to a real disastrous tsunami could be inappropriately less than needed.

Geoscientists, in their role of forecasters, should contribute to building up optimal TWSs. This implies to work jointly with CPAs and users in the specification of the site-dependent action classes, and the minimization of the prediction interval to such extent that most of the times the alert messages give no rise to any action ambiguities. Working together with the society is a must for forecasters of natural hazards, since they are not only geoscientists who speak of science with other scientists through scientific papers in specialized journals. Their work should be oriented by the societal needs, and establishing a full two-way communication bridge between forecasters and the population lies at the basis of the construction of a warning system.

In case where the present TWS is not able to fulfill the requirements for an optimal user-needs-oriented TWS, geoscientists should make any possible efforts to make the recipients of their messages (CPAs and general public) well aware of the limitations of their forecasting tools and of their forecasting results through specific information and educational campaigns. A good practice would be to inform the users of the uncertainties ΔY associated with the forecasts Y, which in principle is not impossible since all forecasts are the result of computations and numerical tsunami simulations. Therefore, appropriate sensitivity studies of the numerical models and algorithms used in the TWS operations could provide relevant information on such uncertainties. Such studies however are not routine and are not included as requirements in the standard operating procedures of any existing TWS to the knowledge of the authors.

CONCLUSIONS

This paper has considered the role of geoscientists in the TWSs. The authors have spent time in defining the uncertainties associated with forecasts since they found that there is a lack in the theory, notwithstanding the abundant literature concerning measurements and prediction errors. The main contribution here is the introduction of the concept of a useful forecast based on uncertainties that are tolerable to the user versus the concept of a successful prediction that is mostly forecaster-oriented.

We have further outlined the main sources of uncertainties that affect the tsunami forecasts, though in doing so we have restricted our analysis to tsunamis induced by earthquakes. Generalization to other sources such as landslides and volcanic eruptions can be made, but are left to further studies. We have mainly considered the uncertainties related to the short-term predictions required by the TWS, though we have also briefly addressed uncertainties for long-term predictions like the one carried out in tsunami hazard evaluation. In this regard, we have mainly stressed the difficulties for the user to verify or falsify the long-term predictions that are based on probabilistic analyses.

In considering the TWS, we have seen that the current practice is to discretize the forecast in classes of responses and we have adapted the analysis of uncertainties of continuous variable provided in the first part of the paper, to discretized variables. We have identified the characteristic requirements for optimal TWS as regards warning. They are two and cover two different aspects of the activity of the geoscientists: one is the definition of the classes, that has to be done in collaboration with the users, and therefore can be successful only if a real and fruitful communication channel is open between the geoscientists and the users (intermediate and final); the other is the minimization of the prediction interval, which is a typical scientific and operational issue since it implies enhancement of the monitoring system (seismic and oceanographic), of the ancillary data sets (e.g., bathymetry and topography) and of the tsunami simulation models.

Geoscientists should do their best to identify the possible faults of the today's TWS and to produce forecasts that help the users to take the proper actions if a tsunami attacks. This paper has identified the target (the building of optimal TWSs) and has identified a viable strategy in case a TWS is found not to be optimal. In this case, the main task is the correct information of the users of the limitations of such a TWS which also includes the specification of the prediction intervals.

ACKNOWLEDGMENTS

This paper contains research that has started under the FP7 project TRIDEC (contract no. 258723) and continues under the FP7 project ASTARTE (contract no. 603839).

REFERENCES

Ambraseys, N.N., 2002. Seismic sea-waves in the Marmara Sea region during the last 20 centuries. J. Seismol. 6, 571–578.

Antonopoulos, J., 1979. Catalogue of tsunamis in the Eastern Mediterranean from antiquity to present times. Ann. Geofis. 32, 113–130.

Bressan, L., Tinti, S., 2012. Detecting the 11 March 2011 Tohoku tsunami arrival on sea-level records in the Pacific Ocean: application and performance of the Tsunami Early Detection Algorithm (TEDA). Nat. Hazards Earth Syst. Sci. 12, 1583–1606.

Dolce, M., Di Bucci, D., 2015. Risk management: roles and responsibilities in the decision-making process. In: Wyss, M., Peppoloni, S. (Eds.), Geoethics: Ethical Challenges and Case Studies in Earth Science. Elsevier, Waltham, Massachusetts, pp. 211–221.

El Alami, S.O., Tinti, S., 1991. A preliminary evaluation of the tsunami hazards in the Moroccan coasts. Sci. Tsunami Hazards 9, 31–38.

Galanopoulos, A.G., 1960. Tsunamis observed on the coasts of Greece from antiquity to present time. Ann. Geofis. 13, 369–386.

Geist, E., Parsons, T., 2006. Probabilistic analysis of tsunami hazards. Nat. Hazards 37, 277–314.

Guidoboni, E., Comastri, A., Traina, G., 1994. Catalogue of Ancient Earthquakes in the Mediter-Ranean Area up to the 10th Century. Pub. Ist Nazion. Geofisica, Rome. p. 504.

Heidarzadeh, M., Kijko, A., 2011. A probabilistic tsunami hazard assessment for the Makran subduction zone at the northwestern Indian Ocean. Nat. Hazards 56, 577–593.

Imamura, F., Abe, I., 2009. History and challenge of tsunami warning systems in Japan. J. Disaster Res. 4, 267–271.

Joseph, A., 2011. Tsunamis: Detection, Monitoring, and Early-Warning Technologies. Academic Press. p. 436.

Lane, E.M., Gillibrand, P.A., Wang, X., Power, W., 2013. A probabilistic tsunami hazard study of the Auckland region, Part II: inundation modelling and hazard assessment. Pure Appl. Geophys. 170, 1635–1646.

Papadopoulos, G.A., Chalkis, B.J., 1984. Tsunamis observed in Greece and the surrounding area from antiquity up to the present times. Mar. Geol. 56, 309–317.

Papadopoulos, G.A., Daskalaki, E., Fokaefs, A., Giraleas, N., 2007. Tsunami hazards in the Eastern Mediterranean: strong earthquakes and tsunamis in the East Hellenic Arc and Trench system. Nat. Hazards Earth Syst. Sci. 7, 57–64.

Rabinovich, A.B., Thomson, R.E., Stephenson, F.E., 2006. The Sumatra tsunami of 26 December 2004 as observed in the North Pacific and North Atlantic Oceans. Surv. Geophys. 27, 647–677.

Rabinovich, A.B., Thomson, R.E., 2007. The 26 December 2004 Sumatra tsunami: analysis of tide gauge data from the World Ocean. Part 1. Indian Ocean and South Africa. Pure Appl. Geophys. 164, 261–308.

Soloviev, S.L., Solovieva, O.N., Go, C.N., Kim, K.S., Shchetnikov, N.A. Tsunamis in the Mediterranean Sea 2000 B.C.–2000 A.D. Kluwer Academic Publishers, Dordrecht, The Netherlands, p. 237.

Tinti, S., 1991. Assessment of tsunami hazard in the Italian seas. Nat. Hazards 4, 267–283.

Tinti, S., Maramai, A., 1996. Catalogue of tsunamis generated in Italy and in Côte d'Azur, France: a step towards a unified catalogue of tsunamis in Europe. Ann. Geofis. 39, 1253–1299.

Tinti, S., Maramai, A., Graziani, L., 2004. The new catalogue of the Italian tsunamis. Nat. Hazards 33, 439–465.

Tinti, S., Armigliato, A., Tonini, R., Maramai, A., Graziani, L., 2005. Assessing the hazard related to tsunamis of tectonic origin: a hybrid statistical-deterministic method applied to southern Italy coasts. ISET J. Earthquake Technol. 42, 189–201.

Tinti, S., Armigliato, A., Zaniboni, F., Pagnoni, G., 2012. Influence of the heterogeneity of the seismic source on the timely detectability of a tsunami: implications for tsunami early warning in the Central Mediterranean. In: Proceedings of the Twenty-second (2012) International Offshore and Polar Engineering Conference, Rhodes, Greece, June 17–22, 2012, Copyright © 2012 by the International Society of Offshore and Polar Engineers (ISOPE). ISBN: 978-1-880653-94-4 (Set); ISSN: 1098-6189 (Set).

Wyss, M., 2015. Do probabilistic seismic hazard maps address the need of the population? In: Wyss, M., Peppoloni, S. (Eds.), Geoethics: Ethical Challenges and Case Studies in Earth Science. Elsevier, Waltham, Massachusetts, pp. 239–249.

Chapter 16

What Foreign Rescue Teams Should Do and Must Not Do in Muslim Countries

Mohamad amin Skandari
Disaster Management Expert, Social Security Organization, Tehran, Iran

Chapter Outline

Abstract
First responders with the best intentions may inadvertently offend the people they are trying to help, if they do not know local taboos and rules of conduct. In the stressful aftermath of a disaster, the affected population may be especially sensitive to what might appear to them as disrespectful behavior. Islam contains a number of very strongly held believes and rules concerning respect for objects and persons related to the Muslim faith. This chapter summarizes some of the rules that a first responder should follow in a Muslim society. Helpers who rush to disaster stricken regions are full of good intentions, and with a little consideration of the main rules of Muslim culture can avoid to unintentionally provoke confrontations.

Keywords: Ethical advices; First responders; Muslim communities.

Geoethics. http://dx.doi.org/10.1016/B978-0-12-799935-7.00016-2
193

INTRODUCTION

When we talk about ethics, it could be considered in different aspects, for example as an academic discipline, in which case its definition is: "The branch of philosophy that deals with the general nature of good and bad and the specific moral obligations of and choices to be made by the individual in his relationship with others" (American Heritage Dictionary)[1]. Or as an applied discipline ethics refers to "standards of conduct, standards that indicate how one should behave based on moral duties and virtues, which themselves are derived from principles of right and wrong" (Josephson Institute on Ethics). As professional discipline, ethics means: "The rules or standards governing conduct especially of the members of a profession" (American Heritage Dictionary)[1]. The most relevant definition of ethics, we are considering in this chapter is the professional one, meaning the rules and standards that first responders should respect.

According to this definition, two different sets of rules and standards will be discussed in this chapter.

- What humanitarian workers and first responders should do and respect generally?
- What humanitarian workers and first responders should do and respect when they are working in an affected Muslim community?

WHAT FIRST RESPONDERS SHOULD DO AND RESPECT GENERALLY

When we arrive as first responders in an affected area, we should do our best to help the suffering people. Our work should be coordinated with other responders and responsible organizations. In addition, if we want to reach the best results; we also have to strongly respect the following items.

Humanity

As a first responder we should bring assistance without discrimination, to prevent and alleviate human suffering wherever it may be found. The purpose is to protect life and health and to ensure respect for the human beings (American Red Cross, 2006)[2].

Impartiality

We should not discriminate because of nationality, race, religious beliefs, class, or political opinions. We must endeavor to relieve the suffering of individuals, being guided solely by their needs, and to give priority to the most urgent cases of distress (American Red Cross, 2006).

---

1. training.fema.gov/EMIWeb/edu/docs/crr/CAT%20-%20Session%204%20-%20Ethics.doc.
2. http://www.redcross.org/about-us/mission.

Neutrality

As a humanitarian worker in an affected area we must not take sides in hostilities or engage at any time in controversies of a political, racial, religious, or ideological nature (American Red Cross, 2006).

Fair Allocation of Resources

When a disaster occurs, victims will be encountered who are in an imbalance situation; there resources are reduced and limited, while their needs and demands are increased. One of the first responders' duties in this situation could be fair allocation of limited resources to people who need them. Many experts believe the most pressing, difficult ethical issues prompted by disaster preparedness and mitigation activity concern the problem of distributive justice (Jennings and Arras, 2008). This task will become exceedingly challenging when we encountered the old disturbing question: Who shall live when not all can live? How shall we choose who lives and who dies? (Jennings and Arras, 2008). Answering this question is very difficult, and if you are in a disaster situation under aftermath conditions, finding a rational response for the vital question who shall receive help when not all can is incredibly hard. It is making decisions under fire.

Making Decisions under Fire

Scientists believe human beings have developed by evolution, including the capability of making logical decisions. However, most of our decision-making is not designed for today. Our brain is programmed to err on the side of safety for ourselves, not for other people, and also to do it quickly—whether it is right or wrong. In emergency situations the body first initiates an adrenaline response and in high adrenaline situations the logical part of the brain switches off (Ian Moore, 2013). For being able to make a correct decision in a disaster situation, one needs to be prepared and have an exercised action plan at the ready for any probable condition. Therefore, to be successful as a first responder, we must be prepared, make plans, and go through exercises before disaster strikes.

Meeting the Special Needs of Vulnerable Populations

In a disaster affected community, especially when we encounter a devastated society, there are some groups of people which are in a dangerous situation and need immediate attention to survive. They usually are called the vulnerable population. One of our duties as first responders is protection and service to those who, in an emergency event and its aftermath, will be especially vulnerable to harm and injustice, especially loss of life, health, or dignity (Jennings and Arras, 2008). Vulnerable populations are pregnant women, children, disabled persons, elderly persons, the sick, the wounded, indigenous populations, environmentally displaced persons, ethnic and religious minorities, and the most disadvantaged members of society, including persons or groups of persons who are victims of racism. These people should receive the benefit from disaster prevention measures tailored to their already existing vulnerability (Eur-Opa, 2011).

WHAT HUMANITARIAN WORKERS AND FIRST RESPONDERS SHOULD DO AND RESPECT, WHEN THEY ARE WORKING IN AN AFFECTED MUSLIM COMMUNITY

When you are working in a foreign community you should respect their rules and standards. Especially when you are working there in a stressful situation after a disaster observing this rule may become vital. In this situation you and your audience (victims who survived) are in a stressful situation. First responders usually are prepared to work under this condition, but your audience consists of ordinary people who may be injured and are suffering. They are stressed, their heart rate will be up and their adrenaline and cortisol levels will be high. Under these conditions, they are likely to be more sensitive than normal and they are prone to make poor decisions. In particular they may react in a dangerous way to seemingly innocent actions of a first responder, if they believe them to be offensive or rude.

Therefore, if you are planning to work as a humanitarian helper or a first responder, while you are not a Muslim, and you are not familiar with their culture and their rules, we strongly encourage you to learn and consider at least the most important points mentioned in the following.

Islam

Islam is a monotheistic and Abrahamic religion articulated by the Qur'an, a book considered by its adherents to be the verbatim word of God (Arabic: الله Allāh) and by the teachings and normative example (called the Sunnah and composed of Hadith) of Muhammad, considered by them to be the last prophet of God. An adherent of Islam is called a muslim (en.wikipedia.org/wiki/Islam).

Most Muslims are of two denominations, Sunni (75–90%), or Shia (10–20%). The religion (Islam) is very important for Muslims, and usually no Muslim has permission to change or reject his/her religion. There are very hard penalties for rejecting Islam in many Muslim countries (even death). If a Muslim consciously and without coercion declares their rejection of Islam and does not change their mind after the time allocated by a judge for researcher-consideration, then the penalty for male apostates is death, and for women life imprisonment. However, this view has been rejected by some modern Muslim scholars.

Muslims are living in many countries around the world. About 13% of Muslims live in Indonesia, the largest Muslim country, 25% in South Asia, 20% in the Middle East, 2% in Central Asia, 4% in the remaining South East Asian countries, and 15% in Sub-Saharan Africa. Sizable communities are also found in China and Russia, and parts of the Caribbean. Converts and immigrant communities are found in almost every part of the world (Wikipedia).

The culture of Muslims is not identical in all countries and in different denominations. Many of them are not too rigid in their religious believes, but some of them are strongly rigid.

Almost all Muslims (Sunni and Shia) strongly respect the following: Islam, Quran, Mosques, and The World of Allah. Therefore it is important to observe several rules and restrictions in interactions with these parts of Muslim religion. Some Muslims respect the rules of Islam even more than their life; they may ask non-Muslim people to behave according their Muslim rules, while in their countries and if foreigners do not respect their rules, they might punish people whom they see as transgressors very hard.

Note

When you are working in a Muslim country, do not try to invite local people to join your religion. It might be very dangerous for them and you. For example, in August 2010, ten aid workers were killed by gunmen in Afghanistan. The Taliban have claimed responsibility for these murders, claiming the aid workers were trying to convert Muslims and alleging that the group had been carrying Bibles translated into the Afghan language Dari. How deep the hostility toward missionaries runs is demonstrated by the following reaction by an educated expert on disaster response: "I'm not sorry about this, they must not abuse the chaos to advertise their believes."

It is clear, your help and your attitude could be the best recommendation for the moral standards in your nation and followed by your religion. The people whom you help will see you and your behavior and they will respect you and your attitudes (religious, national, personal believes), if you are a completely fair first responder.

Mosque:

A Mosque is a place of worship for followers of Islam.
You can recognize a mosque based on its architecture; a common feature in mosques is the minaret, the tall, slender tower that usually is situated at one or two of the corners of the mosque. The other common feature in the mosques is their dome. Domes normally are constructed in the shape of a hemisphere, often placed directly above the main prayer hall.

Under most interpretations of Islam, non-Muslims may be allowed into mosques, as long as they do not sleep or eat there, but some Muslims believe that non-Muslims may not be allowed into mosques under any circumstances. However, several rules, such as not allowing shoes in the prayer hall, are universal.

Note:
We strongly recommend that if you must enter into a mosque (for help or relief), do it accompanied by at least one local responder or resident and respect the advice they may give you.

Quran, Also Qur'an or Koran

This book is the central religious text of Islam, which Muslims believe to be the verbatim word of God (Allah). Muslims regard the Quran as the main miracle of Muhammad, the proof of his prophethood.

The book exists in different sizes, but many of them have a typical shape and words on their cover to recognize them. This book is very important for Muslims and they respect it very much, Muslims are required to be clean (make Wudu, religious Ablution) when handling and reading the Qur'an.

Note

If you see books that might be the Quran, don't push them away or walk on them, even if you are trying to save life; many Muslims prefer to die rather than to see their Quran treated with disrespect. (It's better not to touch the Quran in public if you are not a Muslim).

A foreign expert told me that he took along a copy of the Quran when he was asked to work in a Muslim country. The purpose was to learn something about Islam and he thought showing interest for the Muslim culture would strengthen his relationship with his hosts. However, the brow of an engineer who saw him reading the Quran clouded over and he told the expert that he was reading it all wrong, without devotion. The expert had no idea that there are some requirements to touch the Quran. He discovered that his well intentioned effort probably had the opposite effect of what he had expected: Instead of pleasing his host, he had offended him.

The word "ALLAH"

This word is used mainly by Muslims to refer to God in Islam.

Muslims are also required to be clean (make Wudu, religious Ablution) when touching the word الله on paper or any other objects.

Note

If you encounter an object on which you can see the word الله or Allah, please respect it like we are told by the Quran.

The factors discussed above are crucial and they are common between almost all Muslims. However, there are some additional important elements, which some Muslims care about. The following elements are less important than the aforementioned ones, but should also be considered.

Pictures of Holy Persons (Imams)

Some Muslims in some countries have painted pictures of their Imams (historical holy leaders) in their homes and mosques. Many of these portraits have shiny or green faces. Although they are not common in all Muslim communities, but where the pictures exist; it means they are highly respected by the residents.

Note
You can handle or carry them respectfully, but don't touch the written words on them; they may be words from the Quran.

Prayer Beads

Prayer beads are used by members of various religious traditions such as Hinduism, Buddhism, Roman Catholicism, Orthodox Christianity, Anglicanism, Sikhism and Bahá'í Faith and Islam to count the repetitions of prayers, chants or devotions, such as the rosary of Virgin Mary in Christianity and dhikr (remembrance of God) in Islam.

Note
There are no special requirements for handling them, but be careful don't push them away or walk on them when you are working in a praying place. (Usually there are many prayer beads in mosques)

Turbah

A Turbah is a small piece of soil or clay, often a clay tablet, optionally used by some Shia schools during the daily prayers (Salat).

Note
Although they are not common in all Muslim communities, but where the pictures exist they are highly respected by the residents; particularly when the name of God (الله) or Imams are written on them.

Some other important cultural aspects you may encounter in Muslim communities are:

- *Dress and Hijab*: For women: In general, standards of modesty call for a woman to cover her body, particularly her chest. The Quran calls for women to "draw their head coverings over their chests" and the Prophet Muhammad instructed devout women to cover their bodies except for their face and hands. Most Muslims interpret this to require head coverings for women. Some Muslim women cover the entire body, including the face and/or hands (girls older than 9 years are included in this rule). For men: The minimum amount to be covered is between the navel and the knee.

If you are a woman, please note that many people suggest women traveling to Muslim countries (not all of them) should cover their arms and legs with loose clothing. Many female travelers find it a good idea to cover their hair in Islamic countries to avoid unpleasant attention from local people.

- *Food*: Muslims only eat Halal food, the word "Halal" means permitted or lawful. So, when we are talking about Halal food it means any food that is allowed to be eaten according to Islam. The opposite of Halal is Haram (forbidden). Muslim followers must not consume the following:

  - pork or pork by-products;
  - animals that died without being slaughtered;
  - animals not slaughtered properly or not slaughtered in the name of Allah;
  - blood and blood by-products;
  - carnivorous animals;
  - birds of prey; and
  - alcoholic beverages.

Muslims usually prefer their food to be prepared by a Muslim. Experience shows that some Muslims in devastated areas prefer to go hungry rather than to consume food prepare by non-Muslims. First responders can use the cooperation of local people to provide acceptable food.

CONCLUSION

Although the practice of Islam varies somewhat in different regions and as a function of personal persuasion, there are some fundamental aspects that should be respected by foreigners visiting a Muslim area. The need to be tactful is especially strong in emergency situations where everyone, helpers and victims alike, are subject to heavy stress. First responders without experience and ignorant of strongly held believes of Muslims may create counterproductive situations. Therefore it is strongly advisable that helpers in any capacity, who are planning to work in Muslim regions, familiarize themselves with the rules of respect outlined in this chapter, before their deployment.

REFERENCES

American Red Cross, 2006. New Employee and Volunteer Orientation (Module 1).

Eur-Opa, 2011. Ethical Principles Relating to Disaster Risk Reduction and Contributing to People's Resilience to Disasters. European and Mediterranean Major Hazards Agreement (EUR-OPA).

Jennings, B., Arras, J., 2008. Ethical Guidance for Public Health Emergency Preparedness and Response: Highlighting Ethics and Values in a Vital Public Health Service. Centers for Disease Control and Prevention.

Moore, I., 2013. Decision-Making in a Crisis Understanding the Brain. Accessed at: http://www.securitynewsdesk.com/?s=Decision-making+in+a+crisis++Understanding+the+brain.

Wikipedia, 2014. http://en.wikipedia.org/wiki/Islam.

Section IV

COMMUNICATION WITH THE PUBLIC, OFFICIALS AND THE MEDIA

Chapter 17

Some Comments on the First Degree Sentence of the "L'Aquila Trial"

Marco Mucciarelli
Seismological Research Centre, OGS, Trieste, Italy

Chapter Outline

Abstract

The appeal trial in L'Aquila of the seven scientists and civil protection officers is ongoing, and the first degree sentence could be either confirmed or reversed. It is, however, interesting to discuss some technical details that may have gone unnoticed by the international audience, being hidden in a 900-pages-long legal document without an available translation in English. While it is now clear that it was not a trial of science, the sentence focuses on miscommunication of risk and how this may have contributed to the number of fatalities. However, the sentence misses some important seismological points, such as the role of local seismic amplification (unknown prior to the quake) and to the impossibility of recognizing foreshocks before the main shock has occurred, and includes a worrisome paragraph about the impossibility of reducing seismic risk decreasing the vulnerability.

Keywords: Earthquake risk; L'Aquila earthquake; L'Aquila trial; Risk communication; Site effects.

Two years ago, and three years after the quake that killed more than 300 people in L'Aquila, Italy, a group of seven seismologists, engineers, and civil defense officers were sentenced for manslaughter after a trial that raised international attention. It was the first time that people who were not directly involved in the

Geoethics. http://dx.doi.org/10.1016/B978-0-12-799935-7.00017-4

205

cause of damage (bad design and errors in construction) were considered guilty for the casualties. Other papers have been devoted to discuss the consequences of risk communication (Dolce and Di Bucci, 2015) or the inappropriateness of a trial of science (e.g., Alexander, 2014).

This chapter has a more limited purpose, that is to discuss some technicalities of the sentence that could be otherwise unclear for non-Italian speakers. To fully discuss the motivation of judgment of the Italian High Risk Committee (Commissione Grandi Rischi, CGR) at L'Aquila would take an expert with two degrees, one in geophysics and one in law. To translate properly from Italian to English what is a very complex reasoning would require probably also an interpreter degree, thus this paper is devoted only to five of the most relevant paragraphs of the sentence, a bulky legal document of more than 900 pages.

WHAT SHOULD HAVE BEEN "PREDICTED"

The first comment that has to be made is that the focal point of the sentence is not the predictability of earthquakes but the debate on "the enslavement of science" as the judge calls it.

The judge states "*science* is not subjected to trial for not being able to predict the 06/04/09 earthquake. Already it is said that the current scientific knowledge does not allow a deterministic earthquake prediction (that is) to accurately predict the year, month, day, time, location, magnitude and depth of an earthquake."

Shortly after this statement, however, the judge accuses the defendants of an omission, for failing to properly assess the risk and, in particular, says "the law did not require an answer in terms of certainty on the prediction of the earthquake, but a risk assessment in terms of completeness and adequacy and... there is a big difference between the predictability of an earthquake and the predictability of risk: the earthquake is a natural phenomenon that is not predictable; risk is a potential situation that can be analyzed."

Can we agree with the judge? The following is a generally accepted definition of risk:

$$R = H * V * E$$

i.e., the risk is given by the convolution of three factors: hazard (H), vulnerability (V), and exposure (E).

With definitions also shared in the judgment, the exposure takes into account the number of people and the total value of assets exposed to risk, while the vulnerability defines the possibility that a building suffers damage due to an event and the hazard is the probability that an event occurs in a given period.

In the case of an earthquake, hazard is defined by the probability that a certain level of ground motion (e.g., peak ground acceleration) is exceeded in a fixed period of time. The hazard map at the base of most national seismic codes expresses this probability for a standard time window (475years, see Wyss (2015) and Bilham (2015) for further discussion about the usefulness of such estimates).

If the time window is reduced to a very short period, and one wonders what will happen within one or two weeks, for being able to predict the expected ground motion for a certain earthquake (let alone all the problems of directivity and site effects) one should be able to do just that prediction that the court considers impossible. Alternatively, considering the limited number of seismic sequences preceding a strong earthquake that turn out to be foreshocks of the main event (in Italy, according to the various opinions from 2 out of 100 to 2 out of 1000 earthquake swarms are followed by a main shock) the occurrence of a sequence raises the probability in relative terms, but leaving it still low, on the order of 1/1000. And what actions should one take confronted with such a low chance?

THE ROLE OF SITE EFFECTS

The second point that seems striking to a seismologist is that the long disquisition on the effects of the earthquake lacks a fundamental aspect: consideration of the local variability of the effects of the earthquake. The judge says that he has been informed by experts on peak accelerations and other parameters of the recorded ground motion and concludes that since the collapsed buildings are a small percentage, the earthquake had no abnormal properties. This is only partly true, if we look at where the collapsed buildings were located. All of the victims whose family members started the judicial action were residing in Onna, San Gregorio, and in the Via XX Settembre neighborhood. Anyone who has worked on the area affected by the earthquake of 2009 knows that the damage in Onna is a striking case of site effects. In less than a kilometer, the intensity ranges from of IX MCS degree (Mercalli–Cancani–Sieberg scale) in Onna to VI MCS degree in Monticchio. Onna has been one of the most studied areas during the phase of microzoning for reconstruction, and the effects of site amplification were recognized to be strong in that and other localities, with amplification exceeding a factor 2. Among the many papers published on site effect at L'Aquila see Boncio et al. (2011) and Gallipoli et al. (2011) for Onna and San Gregorio, and Del Monaco et al. (2013) for Via XX Settembre. The extensive geological, geophysical, and geotechnical studies that now explain the strong variability observed in site amplification were not available before the earthquake and it would have been impossible to recognize beforehand that those two neighborhoods were much more dangerous than others.

THE BENEFIT OF HINDSIGHT

A third point mentioned by the judge is that he does not use the benefit of hindsight and does not contest that at the time of the meeting of the CGR the defendants should have known that the earthquake was going to happen. However, there is a problem with the so-called "predictive" shock that occurred the night before the main event. The judge maintains that the CGR "reassured" people and that advice changed their "natural" behavior, which would have been to flee the house after a perceivable shock, fearing that a bigger one might come. Thus, the victims were killed or injured because they decided to stay inside their houses instead of getting out. The point is that the quake of magnitude 3–4 alarming the population could very well not have occurred. It is not this "foreshock" that determined the occurrence of the main shock. At that time, no one could say, if in a few hours the main shock would occur. For example, there had been another earthquake 2 days earlier with a magnitude about 3, also mentioned in the judgment, not immediately followed by a main shock.

The fact that people remained in their houses after the (fore)shock because they had been "reassured" by the experts and authorities is the main thrust of the process. However, if there had been no (fore)shock, the people would have stayed in the doomed houses anyway. The misinformation the experts are accused of having distributed becomes such only if there is a premonitory shock, which becomes such only with the benefit of hindsight.

If there had not been foreshocks, could there not have been a prosecution? This is for the laymen, one of the mysteries generated by the prosecution. A symmetrical counterpart in the prosecution is that there was no investigation of the collapsed buildings with no casualties: the best-known examples are a three-story house in San Gregorio (see Mucciarelli et al., 2011), and the primary school in Goriano Sicoli. The legal system does not seem to be interested in finding out whether there have been illegal practices in the design and implementation, unless there were no casualties.

WHO REASSURED WHO

The fourth point to be discussed is how the sentence seems to take dozens of pages interpreting single words from the official minutes of the meeting of the CGR, scrambling to find a reassurance that is not so evident. Some CGR experts expressed their concern about the hazard and the building vulnerability in L'Aquila and were not at all "reassuring" people.

A worrying new idea introduced by the judgment is that for scientists it seems mandatory to appear on TV or communicate to other media, after the meeting of the CGR. In the opinion of the court, they should have corrected public statements from any other member of CGR if they did not agree with his interviews. One may think that a scientist's job could have been to collect and analyze data, present the results, separated from opinions, to the proper committee, agree on the written statements in the minutes, and then go back to do their work. However, according to the sentence, the scientists should instead have talked to the press and appeared on radio and TV, to contradict the interview given by the deputy head of the Civil Protection (which by law has the duty of public communication) because his statements were wrongly reassuring.

An additional disquieting revelation was that the public prosecutor called as a witness the former head of Civil Protection (not involved in the trial), who, in a taped telephone call, mentioned three of the defendants as involved in a "media operation" to be held in L'Aquila. The prosecutor spent much time to quibble about "media operation in what sense" and did not ask the witness why he was naming only three out of seven defendants. The prosecutor did not ask if he really agreed with the three defendants named in the phone call to "reassure" the population, and in the case of an affirmative answer if he agreed with the other four too. Reading the minutes of the CGR meeting and ranking the comments by the commission members on a scale from really upset to "reassuring," it turns that (by coincidence?) those who made the more concerned statements are the four not mentioned in the phone call caught on tape. The judge seems not to considered individual responsibilities as an alternative to the all guilty/all innocent scheme.

IS EARTHQUAKE ENGINEERING "POINTLESS"?

Finally, the fifth and most worrisome paragraph in the sentence appears on pages 296–297: "The argument that the activity of seismic risk reduction requires the improvement of seismic standards and the reduction of the vulnerability of existing structures…according to the defendants is the predominant, if not the only, means of mitigation of seismic risk. Such a defense argument seems entirely unfounded. Regarding the evaluation and mitigation of seismic risk, the assertion that the only defense against earthquakes is to strengthen the construction and improve their ability to resist earthquakes appears as obvious as pointless. That statement is useless because it provides an indication, which is not feasible in practice. In the Italian municipalities, almost all characterized by extensive historical towns dating back over the centuries, to strengthen existing buildings…would require so large financial resources to be effectively unavailable. As it concerns the city of L'Aquila, the uniqueness of its architectural heritage…makes it clear that the intention to reduce the seismic risk is not seriously feasible." This is the first time that an official document declares that seismic engineering is pointless, since, if the resources are not readily available, it is unthinkable to reduce seismic risk. This is in striking contrast with a courageous policy that followed the L'Aquila earthquake. The Italian Government issued a law and provided funds for completing in 7 years a microzonation campaign in all the municipalities with higher seismic classification and making the resulting maps a mandatory tool for urban planning. Moreover, a fund of more than 950 millions euro is devoted to retrofit public strategic infrastructure or private buildings whose collapse could represent a serious hampering of emergency operation.

CONCLUSIONS

The L'Aquila trial was more a judgment about risk communication rather than a trial of science as it was perceived at its beginning by most of the media. The sentence tries to build a first-time case of responsibility of scientists and civil protection officers in miscommunication of risk, raising doubt about proper ethical conduct of the defendants. However the sentence misses some important points about the possibility (or not) of predicting the consequences of an earthquake. No proper consideration is given to local seismic amplification and to the importance of the fact that foreshocks are not for granted and can be recognized as such only after the main shock (thus requiring the hindsight that should be avoided in a trial). The most alarming consequence of the L'Aquila sentence could be that at a time when citizens should be given advice what to do for their safety, they may be convinced instead that since earthquakes are not predictable and seismic risk reduction is unfeasible, their only option is to run for life every time they feel shaking, hoping that the big tremors always have foreshocks.

REFERENCES

Alexander, D.E., 2014. Communicating earthquake risk to the public: the trial of the "L'Aquila Seven". Nat. Hazards. http://dx.doi.org/10.1007/s11069-014-1062-2.

Bilham, R., 2015. Mmax: ethics of the maximum credible earthquake. In: Wyss, M., Peppoloni S. (Eds.), Geoethics: Ethical Challenges and Case Studies in Earth Science. Elsevier, Waltham, Massachusetts, pp. 119–140.

Boncio, P., Pizzi, A., Cavuoto, G., Mancini, M., Piacentini, T., Miccadei, E., Cavinato, G.P., Piscitelli, S., Giocoli, A., Ferretti, G.Z., De Ferrari, R., Gallipoli, M.R., Mucciarelli, M., Di Fiore, V., Franceschini, A., Pergalani, F., Naso, G., Working Group Macroarea 3, 2011. Geological and geophysical characterisation of the Paganica-San Gregorio area after the April 6, 2009 L'Aquila earthquake (Mw 6.3, Central Italy):implications for site response. Boll. Geofis. Teor. Appl. 52. http://dx.doi.org/10.4430/bgta0014.

Del Monaco, F., Tallini, M., De Rose, C., Durante, F., 2013. HVNSR survey in historical downtown L'Aquila (Central Italy): site resonance properties vs subsoil model. Eng. Geol. 158, 34–47.

Dolce, M., Di Bucci, D., 2015. Risk management: roles and responsibilities in the decision making process. In: Wyss, M., Peppoloni S. (Eds.), Geoethics: Ethical Challenges and Case Studies in Earth Science. Elsevier, Waltham, Massachusetts, pp. 211–221.

Gallipoli, M.R., Albarello, D., Mucciarelli, M., Bianca, M., 2011. Ambient noise measurements to support emergency seismic microzonation: the Abruzzo 2009 earthquake experience. Boll. Geofis. Teor. Appl. 52 (3), 539–559. http://dx.doi.org/10.4430/bgta0031.

Mucciarelli, M., Bianca, M., Ditommaso, R., Vona, M., Gallipoli, M.R., Giocoli, A., Piscitelli, S., Rizzo, E., Picozzi, M., 2011. Peculiar earthquake damage on a reinforced concrete building in San Gregorio (L'Aquila, Italy): site effects or building defects? Bull. Earthquake Eng. 9, 825–840. http://dx.doi.org/10.1007/s10518-011-9257-3.

Wyss, M., 2015. Do probabilistic seismic hazard maps address the need of the population? In: Wyss, M., Peppoloni S. (Eds.), Geoethics: Ethical Challenges and Case Studies in Earth Science. Elsevier, Waltham, Massachusetts, pp. 239–249.

Chapter 18

Risk Management: Roles and Responsibilities in the Decision-making Process

Mauro Dolce[1] and Daniela Di Bucci[2]

[1]National Department of Civil Protection, Via Ulpiano, Roma, Italy; [2]National Department of Civil Protection, Via Vitorchiano, Roma, Italy

Chapter Outline

Abstract

Decision-making under conditions of large uncertainty is an issue that poses serious questions about the decisional chain, its complexities, the consequent responsibilities, and, ultimately, the difficulties of identifying the "right" decision to make. This paper addresses these issues from a civil protection perspective, i.e., from the point of view of an organization which often operates under highly uncertain conditions in the management of the risk cycle. Three main participants in the decision-making process are identified: scientists, political decision-makers (PDMs), and technical decision-makers (TDMs). They provide different contributions to the risk management, with frequent and intricate interactions that, however, can cause distortions in the distinction of the roles to be played, and thus of the responsibilities to be taken. In addition to scientists and decision-makers, there are other participants playing important roles in the risk cycle, and thus influencing decisions within the civil protection system, such as the technical community of professionals, the mass media, the magistrates, and the citizens. PDMs and TDMs, as well as scientists, are directly involved in the decision-making process. Professionals, journalists, magistrates, and citizens can indirectly condition decisions and their implementation. Examples are reported for all of these categories. Participants in the decisional process who understand their role and responsibilities can contribute to a more efficient and effective civil protection system, reducing the occurrence of errors and incidents, if they act within the bounds of their expertise, but avoid to adhere too rigidly to their role. How to develop a correct and fruitful interaction among all the actors is a primary target for a mature civil protection system.

Geoethics. http://dx.doi.org/10.1016/B978-0-12-799935-7.00018-6

211

Keywords: Citizens; Civil protection; Decision-makers; Judges; Mass media; Professionals; Risk cycle; Scientists; Uncertainty.

INTRODUCTION

Decision-making under large uncertainty conditions is an issue that very often arises when dealing with the management of natural and technological risks, both in the short term and in the long run. When possible, some forecast of the possible hazard evolution and of the expected relevant impacts has to be made to support well-grounded decisions. In this general frame, however, some key points need to be solved about the decisional chain, its complexities, the consequent responsibilities, and, ultimately, the difficulties of identifying the "right" decision to make, an issue that is hard to address even in an *ex post* analysis, i.e., an analysis carried out after the occurrence of the considered event.

This complex issue is practically addressed every day by civil protection organizations in relation to the management of several risks. This is especially true when, as it happens in Italy, they have in charge not only activities to be undertaken after a catastrophic event, like search-and-rescue and assistance to the population, but also prevision and prevention activities aimed at mitigating risks, which lives and property are subjected to, along the entire risk cycle.

This problem appears even more intricate when civil protection is not just a single self-contained organization, but rather a system, to which several individuals and organizations contribute with their own activities and competences to attain general risk mitigation objectives. Again, this is the case of the Italian Civil Protection, to which not only regional and local administrations and operational structures (for instance Firefighters Corps, Army) contribute, but also voluntary organizations and many other public and private organizations, as well as the scientific community, including individual citizens. The coordination of this complex system, which is called National Service of Civil Protection, is entrusted to the National Department of Civil Protection, which acts on behalf of the prime minister.

Due to some recent legal controversies directly related to the prevision/decision-making/communication process (original documents and comments can be found in the following blogs: http://processoaquila.wordpress.com/ and http://terremotiegrandirischi.com/), in which one of the authors is unfortunately involved, a lively discussion has been opened worldwide on the roles and specific responsibilities of the various actors in risk management. For example, several sessions and lectures in international conferences mainly focused on the role of scientists in this process (e.g., AGU Fall Meeting, 2012; Gasparini, 2013, in the Goldschmidt Conference). Two workshops, organized in 2011 and 2013 by the Italian Department of Civil Protection in cooperation with the International Centre on Environmental Monitoring (CIMA Foundation), the Appeals Court of Milan and the National Association of Magistrates, dealt with the different points of view of magistrates, scientists, and decision-makers ("who evaluates, who decides, who judges"—http://www.protezionecivile.gov.it/resources/cms/documents/locandina_incontro_di_studio.pdf; "civil protection in the society of risk: procedures, guarantees, responsibilities"—http://www.cimafoundation.org/convegno-nazionale-2013/). Moreover, in the field of international scientific cooperation, the Global Science Forum promoted an activity, involving senior science policy officials of the Organisation for Economic Co-operation and Development member countries in "a study of the quality of scientific policy advice for governments and consequences on the role and responsibility of scientists" (http://www.oecd.org/sti/sci-tech/oecdglobalscienceforum.htm).

In this paper, an attempt is made to point out how the different institutions, organizations, and individuals interact or should interact, in order to improve the efficiency of the system, and how the different responsibilities in the risk mitigation cycle should be distributed.

SCIENTISTS AND DECISION-MAKERS

When dealing with the decision-making process, according to a widely accepted scheme, two categories of participants are usually identified: scientists and decision-makers (e.g., Marzocchi, 2013). They are often considered as two counterparts, which dynamically interact, representing two different points of view that have to be reconciled. As described by Leroi (2010), scientists frequently model events that occurred in the past in order to understand their dynamics, whereas decision-makers need well-tested models, which are able to describe events possibly occurring in the future. The scientific approach to the risks is often probabilistic, and always affected by uncertainties. However, in most cases decisions require a yes or no answer.

Promptness can represent a not negligible issue, as well. On the one hand, scientists need a relatively long time for their work, in order to acquire more data to reduce uncertainties, preferring to wait rather than to be wrong, as underlined by Woo (2011b). On the other hand, decision-makers are generally asked to give an immediate response, often balancing low occurrence probabilities versus envisaged catastrophic consequences (Dolce and Di Bucci, 2012).

The term "decision-makers" usually encompasses, too generically, all the people involved in the main decisional process. Actually, a further distinction should be made between political decision-makers (PDMs) and technical decision-makers (TDMs), to clarify specific roles and responsibilities in the decisional process. Additional distinctions could be made, concerning for instance the PDMs who operate either at national or local level, and in relation to either general risk management policies or specific scenarios. However, in general terms, the decisional chain and the related responsibilities can be envisaged as the result of the interplay among the above mentioned three actors: scientists, PDMs, and TDMs.

In an ideal decision-making process, the following steps can be identified:

1. definition of the acceptable level of risk according to established policy (i.e., in a probabilistic framework, of the acceptable probability of occurrence of quantitatively estimated consequences for lives and property);
2. quantitative evaluation of the risk (considering hazard, vulnerability, and exposure);
3. identification of specific actions capable of reducing the risk to the acceptable level;
4. evaluation of the costs and benefits of the possible risk-mitigating actions;
5. adoption of the most suitable technical solution, according to points 1, 3, and 4; and
6. implementation of risk-mitigating actions.

Considering the three different participants defined above, it is apparently easy to assign to each of them the right place in the process of reaching a decision:

- scientists should deal with steps 2 and 4;
- PDMs should deal with step 1; and
- TDMs typically manage steps 3, 5, and 6.

There is no doubt that in many cases it can be hard to totally separate the contribution of the different actors, since some feedback is often necessary. Moreover, the contribution of different actors, with their expertise, is sometimes required also within a single step:

- scientists, in particularly difficult problems, could also provide important contributions to help decision-makers in all the other steps;
- PDMs decisions are needed also in step 5, in some complex scenarios; and
- TDMs are often required to interact with the scientific community in steps 2 and, especially, 4.

The intricate interactions just outlined, however, can cause distortions in the distinction of the roles to be played, and thus of the responsibilities to be taken. This can further happen if the participants in the decisional process do not, or cannot, accomplish their tasks or if, for various reasons, they go beyond the limits of their role. For instance:

- scientists either could not provide fully quantitative evaluations or could miss to supply scientific support in cost–benefit analyses, or could give undue advice concerning civil protection actions;
- PDMs could decide not to establish the acceptable risk levels for the community they represent (e.g., Butti, 2013). As observed in some cases, they could prefer to state that the "zero" risk solution must be pursued, which is in fact a nondecision, though appreciated by common people;

- TDMs could tend (or could be forced, in emergency conditions) making and implementing decisions they are not in charge for because of lack of scientific quantitative evaluations and of acceptable risk statements (or the impossibility to get them).

As one can easily appreciate, interaction among scientists, PDMs, and TDMs is often made even more difficult by the different languages they use. Gaining a full mutual understanding under these conditions can make the decision-making process too long. Specific expertise and capability are therefore required, in order to quickly interpret scientific information and translate it in terms of operational decisions/actions. In this framework, the example represented by the Italian Department of Civil Protection is quite specific, because it takes advantage of the inclusion, within its staff, of personnel with high-level scientific competences, who work as a link aimed at allowing for a rapid and effective integration between the scientific and the decisional level.

Countless are the examples of individuals usurping or infringing on roles not assigned to them. This probably occurred every time a decisional process had to be activated to implement countermeasures for any kind of risk, both for the short term and in the long run, as everybody involved in such decisional processes has experienced. Usurping and infringing usually results neither in a failure of the civil protection system in terms of major consequences for citizens and property, nor in negative effects on scientists and decision-makers. Nevertheless, in many cases some problems arise that could be avoided by increasing the actors' awareness.

For instance, scientists often present their own solutions in terms of emergency planning or emergency management in many newspaper interviews, as well as in some technical papers. These topics, however, typically fall within the TDMs responsibilities, and the related decisions have to take into account several pieces of information that scientists usually do not have. As a consequence, the proposed solutions (e.g., Panza, 2010) are either too generic to provide a really useful contribution to a mature system, such as civil protection is, or they are technically or economically unfeasible. This occurs especially when the level of knowledge does not allow scientists to provide information as accurate enough to support strong decisions of civil protection. Under such conditions, scientists feel, nevertheless, the need to contribute by providing suggestions on more general solutions. Most commonly, the suggested solutions are based on the idea that, given a potentially catastrophic event and without considering the actual feasibility and repeatability of such solutions, the related risk should be or tend to zero. However, it is known that this very desirable condition is unfortunately impossible to be achieved. In reality decision-makers have to balance the actions to be undertaken for a specific risk with the (limited) budget available to mitigate all the different risks affecting a given area, accounting for their respective occurrence probabilities.

An example of PDMs usurping the role of scientists is a proposal of law presented in Italy in 2013 (n. 1184) concerning a hypothesis for a national anti-seismic plan. As one out of seven "directive principles and criteria" that such a plan must fulfill, the first paragraph of the first article prescribes the scientific methods that have to be used to update the seismic zonation map (Article 1, par. 1a: "updating of the seismic classification of the national territory through the application of the probabilistic method and of the neo-deterministic method"). In this case, PDMs want to impose an undue political choice on a merely scientific question, such as the methods that should be used in a seismic hazard analysis, whereas nothing is stated on more essential political issues, such as the acceptable level of seismic risk, which is the indispensable start point for any actually feasible seismic mitigation strategy.

Another example of TDMs taking over a role of scientists is the evolution of the tsunami alert management in the Mediterranean Sea, and particularly in Italy. After the 2004 Sumatra tsunami, the UNESCO promoted a policy of tsunami alert in different parts of the world. In particular, countries surrounding the Mediterranean basin were asked to identify a national institution that in the future shall be able to manage and distribute tsunami alerts. This task is generally performed by a scientific institution, as it implies monitoring and scientific interpretation, and in fact in Italy it is presently in charge of the National Institute of Geophysics and Volcanology. Previously, however, it was in charge of the National Department of Civil Protection, which temporarily accepted the responsibility of an undue role in order to avoid interruption in the alert communication chain.

An interesting case of strategic choice in long-term seismic prevention is the assumption of a 10% probability of occurrence in 50 years (or 475-years return period) for the design earthquake in seismic codes. This choice, which establishes the accepted seismic risk for ordinary new constructions (e.g., dwellings), is not the result of a linear decision process, where first PDMs decided the acceptable level of risk, then scientists conducted scientific calculations to establish design targets to guarantee resistance of buildings to this predefined level, and finally TDMs translated the results in code rules. It was, on the contrary, the result of a "reasonable" choice, made to fulfill some rational engineering criteria for building design when the first probabilistic hazard map, which just assumed a 10% probability of being exceeded in 50 years, was released in the United States (Algermissen and Perkins, 1976). The 1976 hazard map soon became the reference for seismic design in the ATC-03 seismic code (ATC, 1978). In the premises of this code, some important principles were stated, among which the "no zero risk," or the need to balance different risks when deciding the amount of investments for the mitigation of a single risk: "While a code cannot ensure the absolute safety of buildings, it may be desirable that it should not do so as the resources to construct buildings are limited. Society must decide how it will allocate the available resources among the various ways in which it desires to protect life safety. One way or another, the anticipated benefits of various life protecting programs must be weighed against the cost of implementing such programs." The same acceptable level of seismic risk is assumed practically in every country all over the world, taking into account neither the different economic conditions nor the social consequences of possible destructive events.

OTHER ACTORS INFLUENCING DECISIONS AND THEIR RESPONSIBILITIES

Besides scientists and decision-makers, there are other actors playing important roles in the risk cycle and that thus are, or should be, involved in the civil protection system; among them, the technical community of professionals, the mass media, the magistrates, and the citizens. All of them can make decisions or adopt choices and behaviors capable of affecting the decision-making process (examples related to earthquake predictions are provided by Papadopoulos, 2015). Which is the role they play both in ordinary and emergency times, and which responsibilities, both formal and ethical, do they have?

Professionals (engineers, geologists, architects, surveyors, etc.) contribute significantly, with their everyday activities, to the long-term risk reduction, e.g., dealing with geological surveys, microzonation studies, urban planning, design of new constructions, upgrading of old constructions. Moreover, in case (or in the imminence, when possible) of an event, professionals can contribute significantly to the short-term risk reduction (before the event) or to the emergency management and the residual risk reduction (during and after the event), e.g., by helping civil protection authorities in monitoring the phenomena and assessing the damage. Skill and state-of-the-art professional capability are fundamental requirements for effective assistance, as well as an adequate knowledge of the functioning of the civil protection organization. In Italy, professionals are organized as groups of volunteers officially acknowledged by the National Service of Civil Protection. In case of an emergency, they are prepared to provide skilled support to the civil protection authorities, following a coordinated approach that has been organized and tested at times when no crisis was looming. By doing this, professionals take on the responsibility of supporting civil protection, even though this is not yet properly regulated by law.

Mass media, including not only newspapers, radio, and television, but also the Web and the social networks, are more and more rapidly increasing the importance of their role in risk prevention and risk management. There is a great need for an effective collaboration between civil protection TDMs and the media. The civil protection system could take great advantage by establishing synergies with media, especially in the dissemination of knowledge about risks and their reduction. People's awareness should be increased, and for the spreading of alerts crowdsourcing is becoming useful. However, quite often problems arise, which are probably determined by the need of media to increase their audience for commercial purposes, and to support some political orientations. The most dangerous problems are information distortion, accreditation of non-scientific ideas and nonexpert opinions, political exploitation, and false alarms.

For instance, some pseudoscientific newspapers and television programs choose to emphasize the sensational expectation of huge and sometimes almost impossible hypercatastrophes (e.g., Volcanic Explosivity Index 8—eruptions of volcanoes for which this is an unprecedented level of activity, or magnitude 8 or 9 earthquakes in regions where tectonic features indicate magnitudes hardly exceeding 7), just as some popular catastrophe movies do. Almost no self-protecting or civil protection actions can be effective against such imaginary mega disasters. This practice has severe consequences. It promotes a fatalist attitude among citizens. They will no longer be motivated to adopt (and to invest in) any form of prevention, because it would be useless in case of extreme catastrophes. This is counterproductive because the most feasible prevention actions would be extremely useful in case of medium-size and more frequent events. Instead of spreading ungrounded fear, the media should focus on more probable, yet damaging and life threatening, medium-size events. Moreover, promoting the concept of prevention allows the population to develop risk awareness and, thus, a civil protection culture.

Analogously, the decision to broadcast unverified/false yet alarmist news, often characterizing the social networks on the Web, can entail severe consequences. For instance, due to the false Web prediction of a catastrophic earthquake that should have affected Rome on May 11, 2011, about 20% of the Roman workers decided not to go to work and spend that day outdoors (Nostro et al., 2012). Reducing, or better avoiding, the occurrence of such cases should be a primary responsibility of the mass media, as they are considered part of the civil protection system.

The decisions of judges can significantly affect the behavior of the civil protection system. This has become especially true after the so-called L'Aquila trial (following the 2009 L'Aquila earthquake; e.g., Mucciarelli, 2015), which focused the attention of scientists and civil protection TDMs on properly balancing choices, actions, and responsibilities all over the world.

An interesting analysis of the relationship between civil protection and the judiciary can be found in the proceedings of one of the workshops mentioned in the Introduction (DPC and CIMA, 2013). Some passages in these proceedings provide the opinion of some judges and experts of criminal law, dealing with the bias that can affect the legal interpretation and the possible consequences of a punishing approach (i.e., an approach which looks only for a guilty party after a catastrophic event) on the decision-making process.

Concerning the possible bias in the legal interpretation, Rocco Blaiotta, Judge of the Supreme Court of cassation, says: "The idea of allowed risk...: it is difficult to establish the equilibrium point, the boundary between licit and illicit...the referee who establishes the boundary between licit and illicit is precisely the judge." Moreover, he adds: "As judges,...we are instinctively inclined to forget all the other issues occurring nearby at the same time of the considered event." And Renato Bricchetti, president of the Court of Lecco, states: "I realize...that most of the people feel the need to find a responsible, I don't want to say a scapegoat, but to know who has to be blamed for what happened. And the mass media world amplifies this demand for justice."

Dealing with the possible consequences of a punishing approach, Francesco D'Alessandro, Professor of Criminal Law at Università Cattolica of Milan, addresses the "Accusatory approach to the error: a scheme of analysis for which, in case of errors or incidents, the main effort is made to find who is the possible responsible for the event that occurred, in order to punish him. Whereas those elements of the organization that may have contributed to the adoption of a behaviour characterized by negligence, imprudence, incompetence, are left in the background." He also affirms that: "As a consequence, even if you punish a specific person, the risk conditions and the possibility to commit the same error again still continue to persist." Finally, D'Alessandro depicts the devastating effects of this approach on the risk mitigation: "The accusatory approach...induces a feeling of fear in the operators of the possible punishment...and this keeps them from reporting on the near misses, thus impeding learning by the organization. This phenomenon...is characterized by a progressive, regular adoption of behaviours that are not aimed at better managing the risk, but rather at attempting to minimize the possibility to be personally involved in a future legal controversy."

Finally, a fundamental pillar in a fully mature civil protection system is represented by the citizens. Every citizen, either as individual or organized in groups of volunteers, can play a primary role in the civil protection system, having an active—and responsible—part even in the decision-making process. At ordinary times, citizens should reduce the risks threatening their lives and property, as much as they can. In addition, they should ask for adequately safe conditions at their places of work, study, and entertainment. Moreover, they should verify that civil protection authorities prepared in advance the preventive measures that have to be adopted in case of catastrophic events, especially civil protection plans, of which citizens are primary users.

Citizens at large should be more aware of the risks which they are exposed to, and should have an adequate civil protection culture, which would allow them to adopt the aforementioned precautionary measures and induce the political representatives to carry out risk-prevention policies through both their vote and active involvement in the local political activities.

In case (or in the imminence, when possible) of an event, citizens can undertake different actions, depending on the kind of risk and on the related forecasting probabilities:

- in case of an alert, citizens should follow and implement the civil protection plans (if available) and the correct behavior learned;
- in case of very low occurrence probabilities, citizens should adopt the behavior that suits them best, more or less cautious, and calibrated on their own estimate of risk acceptability, following a nudging approach, where, according to Thaler and Sunstein (2008), a nudge "is any aspect of the choice architecture that alters people's behaviour in a predictable way without forbidding

any options or significantly changing their economic incentives. To count as a mere nudge, the intervention must be easy and cheap to avoid" (see also Woo, 2011a). When applied to the behavior that people should adopt to mitigate the risk they are concerned, both at ordinary times and during an emergency, this means that citizens are nudged to responsibly and autonomously decide to adopt some countermeasures also taking into account their own evaluation of and susceptibility to the considered risks.

FINAL REMARKS

In this paper, roles and responsibilities in the decision-making process have been addressed, looking at the management of the entire risk cycle from a civil protection perspective.

Several categories of actors have been identified. PDMs and TDMs, as well as scientists, are directly involved in the decision-making process. Professionals, journalists, magistrates, and citizens can indirectly condition decisions and their implementation. An effort has been made to highlight biased behavior of some of these categories that can negatively affect the decision process and the effective implementation of the decisions taken.

On the one hand, participants who understand their role and responsibilities can contribute to a more efficient and effective civil protection system, reducing the occurrence of errors and incidents, if they act within the bounds of their expertise. On the other hand, an excessive rigidity in limiting their roles can lead to an impossibility of making decisions, because of the interactions required when dealing with difficult and complex problems. The way to develop a correct and fruitful interaction among all the actors is therefore a primary target on which a mature civil protection system has to focus its efforts, in order to attain general risk mitigation objectives.

REFERENCES

Algermissen, S.T., Perkins, D.M., 1976. A Probabilistic Estimate of Maximum Acceleration in Rock in the Contiguous United States. U.S. Geological Survey Open-File Report 76-416. p. 45.

AGU Fall Meeting, 2012. Lessons Learned from the L'Aquila Earthquake Verdicts Press Conference. http://www.youtube.com/watch?v=xNK5nmDFgy8.

ATC, 1978. Tentative Provisions for the Development of Seismic Regulations for Buildings.

Butti, L., 2013. Il delicato equilibrio tra precauzione e causalità. Ecoscienza 4, 28–31.

Dipartimento della Protezione Civile and Fondazione CIMA (DPC and CIMA) (Ed.), 2013. Protezione Civile e responsabilità nella società del rischio. Chi valuta, chi decide, chi giudica. ETS Editor, p. 152.

Dolce, M., Di Bucci, D., 2012. Probabilità e protezione civile. Ambiente Rischio Comunicazione, quadrimestrale di analisi e monitoraggio ambientale. Decidere Nell'Incertezza 4/2012, 34–39. ISSN: 2240–1520.

Gasparini, P., 2013. Natural hazards and scientific advice: interactions among scientists, decision makers and the public. Plenary Lecture, 2013 Goldschmidt Conference (Florence, Italy) Mineral. Mag. 77 (5), 1146.

Leroi, E., 2010. Landslide risk assessment & decision making. In: Lectio Magistralis, 85th National Congress of the Italian Geological Society. Pisa (Italy), 6–8 September 2010.

Marzocchi, W., 2013. Seismic hazard and public safety. EOS 94 (27), 240–241.

Mucciarelli, M., 2015. Some comments on the first degree sentence of the "L'Aquila trial". In: Wyss, M., Peppoloni, S. (Eds.), Geoethics: Ethical Challenges and Case Studies in Earth Science. Elsevier, Waltham, Massachusetts, pp. 205–210.

Nostro, C., Amato, A., Cultrera, G., Margheriti, L., Selvaggi, G., Arcoraci, L., Casarotti, E., Di Stefano, R., Cerrato, S., 11 Maggio Team, 2012. 11 Maggio 2011: il terremoto previsto e l'Open Day all'INGV. Quad. Geofis. 98. ISSN: 1590–2595.

Panza, G.F., 2010. Verso una società preparata alle calamità ambientali: il terremoto. Geoitalia 32, 24–31. http://dx.doi.org/10.1474/Geoitalia-32-01.

Papadopoulos, G.A., 2015. Risk management: roles and responsibilities in the decision-making process. In: Wyss, M., Peppoloni, S. (Eds.), Geoethics: Ethical Challenges and Case Studies in Earth Science. Elsevier, Waltham, Massachusetts, pp. 223–237.

Thaler, R.H., Sunstein, C.R., 2008. Nudge: Improving Decisions about Health, Wealth, and Happiness. Yale University Press. ISBN: 978-0-14-311526-7 OCLC 791403664.

Woo, G., 2011a. Calculating Catastrophe. Imperial College Press, London.

Woo, G., 2011b. Earthquake decision making/processo decisionale in caso di terremoto. Amb. Risc. Comun. 1, 7–10. ISSN: 2240–1520.

Chapter 19

Communicating to the General Public Earthquake Prediction Information: Lessons Learned in Greece

Gerassimos A. Papadopoulos
Institute of Geodynamics, National Observatory of Athens, Athens, Greece

Chapter Outline

Abstract

Communicating publicly earthquake prediction results is a socially sensitive issue. In Greece important experience has been accumulated given that several public earthquake predictions have been communicated publicly in the last 33 years or so mainly by the so-called Varotsos, Alexopoulos, Nomikos group, based on geoelectric signals. A few predictions based on other precursory phenomena have also been released. In this chapter, two characteristic earthquake prediction episodes in Greece are examined as regards the circumstances under which the predictions were publicly communicated, the administrative management, the social, and other consequences as well as the role of the news media.

Keywords: Code of ethics; Earthquake prediction; Greece; VAN predictions.

INTRODUCTION

Greece occupies a particular geographic position in the tectonic setting of the eastern Mediterranean region that is characterized by a very high seismicity level. Instrumental and historical earthquake data sets (e.g., Papazachos and

Geoethics. http://dx.doi.org/10.1016/B978-0-12-799935-7.00019-8
223

Papazachou, 1997; Ambraseys, 2009) indicate the occurrence of both shallow- and intermediate-depth earthquakes with magnitude, M, up to about 8.0 or more, particularly along the Hellenic Arc (Figure 1). The main geodynamic processes that generate earthquakes are the active subduction from SSW to NNE of the Mediterranean lithosphere beneath the South Aegean area along the Hellenic Arc and Trench system, as well as the westward movement of the Anatolia block that slips along the dextral strike-slip North Anatolian Fault (e.g., Reilinger et al., 2006 and references therein).

In modern times, one of the most important seismic episodes in Greece, from both scientific and social points of view, was the earthquake sequence of February–March 1981 when three strong shocks ranging in magnitudes from of M6.3 to M6.7 ruptured the eastern Corinth Gulf on February 24 and 25 and March 4, 1981, respectively, causing extensive destruction and a death toll of 20 people. Great social concern was generated among the population in the capital city of Athens since it was the first time in modern life that the city was threatened by strong earthquakes.

It was exactly after the 1981 earthquakes that claims for predicting earthquakes in the short-term sense were released systematically by the Varotsos, Alexopoulos, Nomikos (VAN) group of solid state physicists at the University of Athens. Since

FIGURE 1 Seismotectonic elements of the Hellenic Arc and Trench (HAT) system and the four large main shocks that ruptured along HAT during 2008 *(after Papadopoulos et al., 2009)*. Dates, magnitudes, and focal mechanisms (beachball diagrams) as listed in Table 1 are shown (taken from the Harvard earthquake list, http://www.globalcmt.org/CMTsearch.html). Arrows represent the direction of plate motion, and triangles indicate volcanoes. The dashed line in the volcanic arc indicates the 150-km seismic isodepth.

then, up to writing these lines, hundreds of VAN *Earthquake Predictions Statements (EPSs)* were produced, plenty of them being communicated through the mass media. On the other hand, in this nearly 33-year time interval only very few EPSs based on premonitory seismicity patterns were publicly communicated by seismologists, while several EPSs were produced and announced by amateurs.

The experience in communicating earthquake predictions to the public in Greece is quite rich, thus providing factual material which is valuable to earthquake scientists, social scientists, public officials, and decision-makers when considering relevant ethical issues.

In another chapter (Dolce and Di Bucci, 2015) the roles and responsibilities in the decision-making process have been addressed particularly as regards earthquakes, looking at the management of the entire risk cycle from a civil protection perspective. In this chapter ethical issues are examined in relation to the handling, evaluation, and public communication of earthquake prediction results particularly in the short-term sense, approximately 2–3 months in advance. Factual material has been taken from two characteristic prediction episodes that developed in Greece in 1999 and 2008. The interest is focused not so much on the scientific basis of this or that prediction method, on the success or failure of particular EPSs, but mainly on the circumstances under which predictions were publicly communicated, as well as the social and other consequences of the public release of EPSs. This examination has been made by taking into account existing Codes of Ethics for earthquake predictions (see references in Box 1) and the Greek administration mechanism for the EPSs evaluation (Box 2). Localities referred to in the paper are shown in Figure 2.

Box 1 Codes of Ethics Concerning Earthquake Prediction

To the knowledge of the author the next items fall under the general expression "Codes of Ethics Concerning Earthquake Prediction":

1. *Responsibilities in Earthquake Prediction*, by C. R. Allen (1976). This article does not constitute a code of ethics in the strict sense. It contains personal views which, however, are of wide interest to the scientific community and to the public officials. The views expressed in that excellent article are predecessors of the guidelines organized by the Seismological Society of America (SSA) some years later.
2. *Guidelines for Earthquake Predictors*, Seismological Society of America (Anonymous, 1983). The SSA put forward for discussion a set of ethical guidelines for earthquake prediction.
3. Code for Practice for Earthquake Prediction, International Union of Geodesy and Geophysics (IUGG) (Anonymous,1984). Following an agreement reached between UNESCO and the International Association of Seismology and the Physics of Earth's Interior (IASPEI), the IUGG concluded with this Code of Practice.
4. *European Code of Ethics Concerning Earthquake Prediction*, Open Partial Agreement on the Prevention of, Protection Against, and Organization of Relief in Major Natural and Technological Disasters, Council of Europe (Anonymous, 1991).

Box 2 The Greek System for the Evaluation of Earthquake Predictions

1. The Earthquake Planning and Protection Organization (EPPO), traditionally supervised by the Ministry of Public Works (MPW), currently by the Ministry of Infrastructures, is a national service responsible to undertake actions for the mitigation of the seismic risk in Greece.

2. Due to the strong debate regarding the VAN predictions, in April 1985 the Greek government established an Earthquake Prediction Council consisting of two seismologists, one geologist, two VAN members, and the EPPO administrator.

3. For reasons of documenting better what exactly was the content of VAN messages, in the spring time of 1986 it was decided that VAN messages should be submitted to the Interministerial Committee of the Greek government and through it to the MPW and to EPPO.

4. In 1992 the MPW decided to establish the "Permanent Special Scientific Committee for the Assessment of Seismic Hazard and the Evaluation of Seismic Risk." The Chairman and the members of this Committee are appointed for a three-year term. This is the official body where EPSs should be submitted for evaluation. In the text of this chapter it is referred to as "Committee" for reasons of brevity.

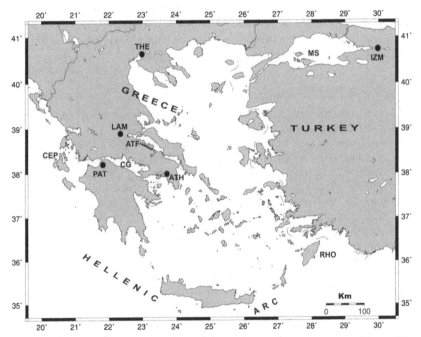

FIGURE 2 Geographic entities mentioned in the paper. Key: Atalanti Fault (ATF), Athens (ATH), Cephalonia Island (CEP), Corinth Gulf (CG), Izmit (IZM), Lamia (LAM), Marmara Sea (MS), Patras (PAT), Rhodes Island (RHO), Thessaloniki (THE).

THE VAN PREDICTIONS

The so-called VAN method is based on electric field changes, named seismic electric signals (SESs), which are recorded by electrode devices which are installed at the Earth's surface (e.g., Varotsos and Alexopoulos, 1984). The VAN group claimed that from the study of SESs they are able to predict the time of occurrence, the epicenter, and the magnitude of earthquakes. The EPSs produced by the VAN group usually were documented with the submission of telegrams directed either to VAN members or to government officials. In the last 10 years or so the documentation is made by uploading reports to the publicly accessible arXiv database of the Cornell University, USA. This second practice violates the confidentiality that the codes of ethics require, thus facilitating the public circulation of EPSs, e.g., Anon (1984, p. 28, 1991, p. 11) (see Box 1).

The VAN method caused strong controversies, initially in Greece and later internationally (e.g., see the special issue "Debate on VAN" in the Geophys. Res. Lett., May 27, 1996). Some of the claims of correlations of earthquakes with SESs published by the VAN group were based on erroneous data. For example, Wyss (1996) showed that 25% of the earthquakes that supposedly correlated with SESs were not in the authoritative bulletin of the Institute of Geodynamics, National Observatory (NOAGI) and 40% of the ones that were listed in the bulletin did not appear in the list of VAN correlations published by the VAN group.

Strong debates were also triggered because of the public release of predictions by the VAN group. The mass media, as well as administrative persons played a crucial role in the way the EPSs of VAN group were managed. In some particular instances, the public communication of EPSs by the VAN group caused great social concern. For example, in January 1991 the VAN supporter late French volcanologist Haroun Tazieff, without previous consultation with the responsible Greek authorities (see Box 2), published in the French newspaper "Le Mond" predictive information interpreted as indicating a forthcoming earthquake of M5.5–6.0 in the area of Thessaloniki city with population on the order of 1 million. This report was reproduced by Greek news media causing great concern among the population (Drakopoulos and Stavrakakis, 1996). The VAN group accepted later that it was a false alarm, but did not explained why the "prediction" was announced publicly before its submission to the government (Varotsos et al., 1996). It is quite obvious that H. Tazieff violated the code of ethics on how to handle earthquake prediction information particularly when the expected earthquake concerns a foreign country, e.g., IUGG (1984, p. 28; see Box 1).

TWO PREDICTION EPISODES

Two earthquake prediction episodes were selected for examination with the aim to cover as many ethical aspects as possible, which is from reaching to prediction results by the scientists up to communicating them via mass media and their management from the side of public officials. Scientific aspects are related to the method(s) used for the EPSs production and evaluation. Of particular interest is the selection of some scientists to communicate "predictions" which obviously are vastly incomplete, that is not specifying space, time, and magnitude intervals as a minimum information needed for the evaluation. From an administrative point of view, the interest is focused on the practice followed by decision makers for the EPSs management including evaluation. The social interest covers the role of the mass media as well as reactions of the people in the public communication of predictions. Finally, in some instances a forensic dimension was introduced when the public prosecutor intervened to investigate if EPSs expressed publicly were grounded or not.

Prediction Episode 1: The September 7, 1999 Lethal Earthquake of Athens

On September 7, 1999 the metropolitan area of Athens was hit by an earthquake ($M_w5.9$), located at a distance of only 20 km to the NW from the city center (Pavlides et al., 2002). The shock caused the collapse of several buildings and extensive damage in many communities, killing 143 people (Papadopoulos et al., 2000). The Athens earthquake possibly was triggered remotely (Papadopoulos, 2002) by the August 17, 1999 large earthquake ($M_w7.5$) of Izmit, eastern Marmara Sea, which caused a death toll of about 18,000. In view of the disastrous consequences of the Izmit earthquake, the lethal earthquake of Athens spread great anxiety among the population of the entire country. One may understand that the social situation in Greece was quite sensitive after the tragic results of the Athens earthquake.

On September 13, 1999, Prof S. Uyeda reported in the Japanese newspaper "Asahi Shinbun," having a daily tirage of about 12 million, that an anomalous SES was recorded on September 1 and 2, 1999 at the VAN station of Lamia (LAM) situated about 150 km to the north of Athens (ATH). The estimated magnitude of an expected earthquake was M5.5, but it was impossible to pinpoint the expected epicenter. Uyeda concluded that since no epicenter was available and since the VAN members had decided to announce only expected earthquakes of M6 and over, no earthquake prediction statement (EPS) was released. The report of "Asahi Shinbun" reached Athens causing unfounded rumors saying that the VAN group had predicted the earthquake. Soon after, new rumors were spread saying that the VAN group had predicted another, possibly stronger earthquake that was expected to happen in a different area. The concern of the public in the entire country increased dramatically.

On September 15, the Minister of Public Works, supervisor of Earthquake Planning and Protection Organization (EPPO) (see Box 2), revealed that after the main shock he had received a message from the VAN group and that he forwarded it to the evaluation Committee. However, because of terrific rumors circulating all over the country the Minister released the VAN message to the news media on September 15, criticizing the VAN group for the practice of not submitting the prediction directly to the Committee.

The management of the case from the side of the Minister caused strong political debate, the opposition political parties criticizing him for the way he selected to manage the seismic crisis. In a press interview of September 19, 1999 the Minister defended his decision by saying that he wanted to show everybody that there was no prediction at all since the VAN message contained uncertain and vague information. On the other hand, he wanted to confront the wave of panic spreading rapidly in the country. On September 16, the Committee examined the case and announced that the data submitted by the VAN group were inadequate for a scientific evaluation, as well as for any practical implementation. The VAN message says that after the SES of September 1 and 2, new SES of slightly larger amplitude were recorded at LAM on September 13 and 14, indicating a new epicentral area for an earthquake expected in a time window of up to 2–3 weeks. No epicentral area and no magnitude of the expected earthquake were specified.

The thriller with the VAN predictions continued the next days. In radio and TV programs of the September 28 it was revealed that on September 26 a new letter was sent by the VAN group to the Minister who on the next day verified that he received the letter. The Deputy Minister of Public Works stated in the press that the content of the new VAN letter was "silly talk." The Committee concluded the same, that the data submitted by the VAN group were inadequate neither for a scientific evaluation nor for any practical implementation. The public prosecutor asked for an urgent investigation of the radio broadcaster with the aim to determine if the information broadcast was based on facts and justified or not.

However, the agony in the country reached its zenith because of the content of the new VAN "prediction," which, although did not specify the earthquake magnitude, supported that *"Atalanti fault (which lies between LAM and ATH) is, of course, a candidate area; its southern end lies only 50–60 km far from the earlier one close to Athens."* In Greece everybody is aware of the Atalanti fault that generated two lethal earthquakes of M6.9 and M6.7 on April 20 and 27, 1894. After the new VAN message, the people did not know what to believe. Most of them considered that a new strong earthquake was possible to occur in the Athens area. In view of this, the mass media, particularly the TV channels, organized an unusual report looking for seismologists who did not stay overnight in their homes as an evidence that they may really expect a strong earthquake. The author was one of the "victims" of the safari for "dislodged seismologists."

Taking seriously into account the time window of 2–3 weeks setup in the VAN "prediction" many people expected that Sunday, October 3, 1999, was the deadline for the expected earthquake to occur. The press reported where politicians had planned to stay during that critical weekend: in Athens or elsewhere? The news was that nearly all politicians did not change their initial schedule in view of the supposed "prediction."

The dramatic evening of Sunday, October 3, became worst when an earthquake measuring $M_s 4.1$ was felt in the area of Athens at 20.03, local time.

Everybody was looking for a reliable evaluation: was that the VAN earthquake or not? Seismologists at the Institute of Geodynamics, National Observatory of Athens (NOAGI) contributed substantially to the gradual relaxation of the people emphasizing that the shock was only one of the many aftershocks that followed the main shock of September 7, 1999. No strong earthquakes occurred in Athens or elsewhere in the country in the months that followed. No VAN predictions were released publicly for a long time after this episode.

Prediction Episode 2: The 2008 "Storm" of Earthquakes and Predictions

During the first seven months of 2008, Greece experienced a "storm" of earthquakes, a "storm" of predictions and consequently a "storm" of rumors that spread throughout the country. Four main shocks all exceeding M6, ruptured along the Hellenic Arc (Figure 1). In addition, the second in line but largest main shock (M_w6.8), of February 14, 2008, was followed by two aftershocks also exceeding M6. The earthquake focal parameters are listed in Table 1 and illustrated in Figure 1. Since two smaller earthquakes occurring near Patras city on February 4, 2008 are of relevance to this discussion their focal parameters are also included in Table 1.

During the same time period, a "storm" of earthquake predictions appeared in scientific papers and reports (see Table 2). Some of the claims were circulated through the news media, thus causing a "storm" of rumors about more and possibly catastrophic earthquakes to come. As a consequence, the country experienced an unusual long-standing seismic crisis which turned into a social crisis.

TABLE 1 Focal Parameters of the 2008 Hellenic Arc and Trench Earthquakes Examined in Prediction Episode 2

Date (day/month/year)	Time (h:min:ss)	Lat (φ_N^0)	Long (λ_E^0)	h (km)	M_w	M_L
06/01/2008	05:14:24	37.11	22.78	92.4	6.2	6.1
04/02/2008	20:25:10	38.08	21.92	27	–	4.9
04/02/2008	22:15:37	38.09	21.94	25	–	5.0
14/02/2008	10:09:29	36.50	21.78	20.0	6.8	6.2
14/02/2008	12:09:02	36.22	21.75	15.0	6.5	6.1
20/02/2008	18:27:13	36.18	21.72	17.6	6.2	6.0
08/06/2008	12:25:37	37.98	21.51	24.7	6.4	6.5
15/07/2008	03:26:40	35.85	27.92	42.2	6.4	6.2

Taken from the Harvard and the National Observatory of Athens earthquake lists; http://www.globalcmt.org/CMTsearch.html and http://www.gein.noa.gr.

TABLE 2 The Chronicle of the Long-Standing Seismic and Social Crisis in Greece During 2008

No.	Date	Event	Comment 1	Comment 2
1	January 6, 2008	Strong EQ (M$_w$6.2)	No casualties, slight damage	
2	January 8, 2008	The first author of a published paper (Lagios et al., 2007) forwarded it via e-mail to the CC (this chapter's author).	The same day the CC acknowledged that the paper was received.	The EPS in this paper (p. 142) says that for the uplift identified in a radius of ≥50 km around SW Cephalonia Island (Ionian Sea) from September 2003 "the resultant magnitude for a future EQ is estimated at M≥7.0. For such M the precursory time to its occurrence is c. T=4.2–4.6 years starting from 2003."
3	January 11–13, 2008	Claims in the press that the EQ of January 8, 2008 was predicted by VAN.	No VAN or other EPSs were received by the Committee.	
4	January 13, 2008	Dr J. Rundle (UC Davis, USA) stated in the press that EQ's of M~7 are expected in western Greece, in 5–7 years ahead but this EQ does not fall within his long-term predictions.		
5	January 14, 2008	The Committee convened and evaluated the paper of event No. 2.	The paper's first author and an external expert were invited and participated. The EPPO President and Director attended ex officio.	The Committee concluded unanimously that due to problems with the measurements calibration, it is not possible to conclude with practically applicable countermeasures beyond the regular ones. The further intensification of the instrumental monitoring was recommended.

(Continued)

TABLE 2 The Chronicle of the Long-Standing Seismic and Social Crisis in Greece During 2008 — cont'd

No.	Date	Event	Comment 1	Comment 2
6	February 4, 2008	Moderate EQ's near Patras.	On February 5, 2008 VAN supporters claimed in press that these EQ's had been predicted.	These claims contradict with others expressed also by VAN supporters (see debate in Uyeda and Kamogawa, 2008, 2009 and Papadopoulos, 2010).
7	February 6, 2008	One VAN supporter from University of Patras submitted to the Committee a letter saying that "records in one of our stations indicate that the Western Hellenic Arc is excited."	On February 11, 2008 the CC replied that more data are needed as a minimum basis to go ahead with evaluation.	The CC did not convene the Committee.
8	February 10, 2008	News media, based on an arXiv report, supported that VAN predicts a strong EQ somewhere in a very large area of Greece.	EPSs were not submitted to the Committee.	
9	February 14, 2008	Large EQ (M_w6.8). Strong aftershock.	No casualties, slight damage.	
10	February 14–24, 2008	In a series of TV programs VAN supporters claimed that the February 14, 2008 EQ's had predicted and that there was evidence for more EQ's to come. Great concern was spread among the Greek population.	The CC said in TV programs that no EPSs were submitted and that the people should stay calm.	After public claims for more EQ's to come, the public prosecutor started investigation to clarify if the claims were founded or not. The CC was invited to testimony three (!) years later, i.e., on February 9, 2011. Until writing this chapter no more news were heard regarding this investigation (!!).

11	February 2008	Sometime within February or early March the paper in event 2 was uploaded in the web site of the Department of Geology, University of Athens.	Thus, the EPS related to event 2 became publicly available.	
12	March 14, 2008	The EPS in event 2 became a big issue in a TV program in the presence of four parliament members belonging to four different political parties.	In the TV program the Chairmen of EPPO and of the CC defended the unanimous evaluation mentioned in event 5. The four parliament members expressed their satisfaction for the management of the prediction by both the EPPO and the Committee.	In the time interval that followed the great public concern exaggerated due to rumors saying that the large EQ predicted in the paper of event 2 should occur in Cephalonia Island by the end of July 2008.
13	June 8, 2008	Large EQ (M_w6.4) struck Patras city and the Achaia Prefecture in general. Extensive damage was caused but no casualties were reported.	On the same day news media reported that the EQ had been predicted by an arXiv report of VAN. On June 22, same claims appeared in the press.	The next days, Prof A. Tselentis, supporter of VAN from the University of Patras, claimed in the mass media that the aftershock sequence was anomalous since no strong aftershock of M5.5–6.0 had occurred and that such an aftershock should be expected. Great concern was caused among the population.

(Continued)

TABLE 2 The Chronicle of the Long-Standing Seismic and Social Crisis in Greece During 2008—cont'd

No.	Date	Event	Comment 1	Comment 2
14	June 20, 2008 June 30, 2008	The Prefect of Achaia sent to the CC an urgent letter regarding the evaluation of the short-term seismic hazard in Patras. In parallel, apparently because of the prediction described in events 2 and 12, rumors circulated that an EQ larger than the one of June 8, 2008 should be expected in western Greece and the Ionian Sea.	The letter says that the local people got in panic particularly after the repeated statements in the news media by Prof A. Tselentis who appeared confident that an aftershock measuring M5.5–6.0 should be expected.	The largest aftershock recorded in that EQ sequence was of M4.3 (Papadopoulos et al., 2010). On June 30, 2008 the CC gave a speech and a press conference in the Scientific Park of the Patras University explaining that (1) the prediction for a larger EQ is not well-documented, and (2) the magnitude of the largest aftershock may fall in a wide range and, therefore, the aftershock sequence should not be considered as anomalous.
15	July 15, 2008	Large EQ ($M_w6.4$) offshore Rhodes, Eastern Hellenic Arc (EHA).	No casualties, no damage. Neither EPSs nor rumors appeared regarding the EHA.	On the contrary, rumors increased regarding an expected large EQ in the Ionian islands.
16	July 29, 2008	In the last two weeks of July 2008 the panic increased dramatically in Cephalonia Island due to the EPS described by the events 2 and 12. Tourists arrivals were massively canceled.	In an effort to distress people, local authorities organized a social party in Cephalonia on July 29, 2008 with the participation of ministers, parliament members, and other persons publicly recognized.	The predicted EQ never happened until writing this chapter. The largest EQ that occurred in Cephalonia until July 5, 2014 was that of 26 January 26, 2014 measuring ($M_w6.0$). The seismic and social crisis in Greece diminished after July 2008.

Key: EQ=earthquake, CC=Chairman of the Committee mentioned in paragraph 4 of Box 2, arXiv report=report uploaded at the Cornell University database without getting review, EPS=earthquake prediction statement, EPPO=Earthquake Planning and Protection Organization. The Acronym EPPO is explained in Box 2. Focal Parameters of the EQ's reported here can be found in Table 1.

In Table 2 the chronicle of the 2008 crisis in Greece is summarized in the form of sequential events. Here "event" means (1) earthquake, (2) EPS, (3) news media reports, (4) rumor or other social impact, (5) action taken by authoritative organizations or persons. The term EPS is used regardless of the quality, the completeness, the success or failure of an earthquake prediction.

The chronicle of the 2008 crisis raised several ethical issues regarding the communication of earthquake prediction results. First is the behavior of the predictors. Some selected to submit their predictions for evaluation by the Committee without circulating them publicly. However, after a not so positive evaluation, the authors of the prediction uploaded their paper to the web, thus making their prediction publicly available. Others, such as the VAN group never submitted their predictions to the Committee, preferring instead public circulation. I do not necessarily mean that the VAN group passed directly their predictions to the news media. However, by uploading the predictions to the arXiv database they made them publicly available and, therefore, easily accessible to the news media.

The role of the news media was significant during the 2008 crisis. Most of the Greek media hosted quite easily prediction stories referring either to expected strong earthquakes or to supposedly successful predictions. Media reports or broadcasting and TV programs regarding earthquake predictions as a rule were released without cross-checking. Several of the reports and programs caused great concern or even panic in the population. It is clear how the chain of panic works. After a strong earthquake, a predictor claims having predicted the earthquake that just happened and the supposed success story is reproduced uncritically by the news media. The public, being unable to evaluate the validity of the claim, tend to believe the success story. The public is convinced that the prediction of earthquakes is possible, if not a routine. Then, the people are ready to believe as true any other "prediction" that will be publicly available beforehand. This belief generates panic since the people tend not to trust the public officials who try to explain that this or that EPS is not documented.

The attitude of the politicians was neutral during the first stages of the crisis. However, as the crisis was going on toward the end of July 2008, when the supposed large earthquake was expected to occur in Cephalonia Island, they changed their behavior. Politicians participated in TV programs supporting the opinion of the Committee that the particular prediction was not documented. The Prefect of Patras city asked in writing for an evaluation by the Committee. Several politicians decided to visit Cephalonia personally with the aim to convince people that nothing will happen.

CONCLUDING REMARKS

The examination of two characteristic earthquake prediction episodes communicated publicly in Greece in 1999 and 2008 reveals a variety of administrative, social, and ethical issues. As early as 1985 the Greek administration established a Committee for the evaluation of the short-term seismic hazards including earthquake prediction results. The established Committee examined several cases, provided that the earthquake prediction statements contained the minimum information needed to evaluate.

In the socially sensitive case of 1999, however, the responsible Minister decided to communicate publicly a VAN prediction submitted to him. The objective of this violation of the codes of ethics mentioned in Box 1 was to terminate terrible rumors circulating in the country as regards the prediction. On the social level, the VAN and other predictions often caused great concern, panic, and cancellation of thousands of tourist visits e.g., documented in press reports of (1) January 1991, regarding the VAN prediction in Thessaloniki mentioned earlier, and (2) July 2008 regarding the Lagios et al. (2007) prediction analyzed in Table 2.

The role of news media in creating such negative consequences was crucial. The news media easily support "success" stories about predictions and publish supposedly valid predictions beforehand, thus causing unrest, panic, and other societal consequences. In addition, in this way they breed the belief among the general public that earthquake prediction is an operational method for the protection against earthquakes. The international status in this direction was reviewed by Jordan et al. (2011). In view of the practices of mass media, it is of particular interest to investigate a code of ethics regarding the media behavior in communicating not only earthquake predictions but also news about natural hazards in general.

The choice of the VAN group and other predictors to disseminate publicly, or to facilitate the public communication of their predictions, perhaps is motivated by different reasons in different cases. However, the VAN group is a unique case worldwide in the sense that its predictions are systematically communicated publicly during the last 33 years.

The factual material existing in Greece regarding the public communication of earthquake predictions is extremely rich and certainly deserves a more thorough study with the examination of more prediction cases.

REFERENCES

Allen, C.R., 1976. Responsibilities in earthquake prediction: presidential address to the seismological society of America. Bull. Seism. Soc. Am. 66, 2069–2074.

Ambraseys, N.N., 2009. Earthquakes in the Mediterranean and Middle East, a Multidisciplinary Study of Seismicity up to 1900. Cambridge Univ. Press, Cambridge. p. 947.

Anon, 1983. Guidelines for earthquake predictors. Bull. Seism. Soc. Am. 73, 1955–1956.

Anon, 1984. Code of practice for earthquake prediction. IUGG Chron. 165, 26–29.

Anon, 1991. European code of ethics concerning earthquake prediction. In: Council of Europe, Internat. Conf. on Earthquake Prediction. State-of-the-Art, Strasbourg, France. 15–18 October.

Debate on VAN, 1996. Special Section (Geller, R. (Ed.)). Geophys. Res. Lett. 23, May 27, 1996, 1291–1452.

Dolce, M., Di Bucci, D., 2015. Risk management: roles and responsibilities in the decision-making process. In: Wyss, M., Peppoloni S. (Eds.), Geoethics: Ethical Challenges and Case Studies in Earth Science. Elsevier, Waltham, Massachusetts, pp. 211–221.

Drakopoulos, J., Stavrakakis, G.N., 1996. A false alarm based on electrical activity recorded at a VAN station in northern Greece in December 1990. In: Debate on VAN, Special Section (Geller, R. (Ed.)) Geophys. Res. Lett. 23, 1355–1358.

Jordan, Th. H., Chen, Y.-T., Gasparini, P., Madariaga, R., Main, I., Marzocchi, W., Papadopoulos, G.A., Sobolev, G., Yamaoka, K., Zschau, J., 2011. Operational earthquake forecasting: state of knowledge and guidelines for utilization. Ann. Geophys. 54 (4), 316–391. http://dx.doi. org/10.4401/ag-5350.

Lagios, E., Sakkas, V., Papadimitriou, P., Parcharidis, I., Damiata, B.N., Chousianitis, K., Vassilopoulou, S., 2007. Crustal deformation in the Central Ionian Islands (Greece): results from DGPS and DInSAR analyses (1995–2006). Tectonophysics 444, 119–145.

Papadopoulos, G.A., 2002. The Athens, Greece, earthquake (Ms5.9) of 7 September 1999: an event triggered by the Izmit, Turkey, 17 August 1999 earthquake? Bull. Seismol. Soc. Am. 92, 312–321.

Papadopoulos, G.A., Drakatos, G., Papanastassiou, D., Kalogeras, I., Stavrakakis, G., 2000. Preliminary results about the catastrophic earthquake of 7 September 1999 in Athens, Greece. Seismol. Res. Lett. 71, 318–329.

Papadopoulos, G.A., Karastathis, V., Charalambakis, M., Fokaefs, A., 2009. A storm of strong earthquakes in Greece during 2008. EOS Trans. AGU 90 (46). http://dx.doi.org/10.1029/200 9EO460001.

Papadopoulos, G.A., 4 May 2010. Comment on "The prediction of two large earthquakes in Greece". EOS Trans. AGU 91 (18), 162.

Papazachos, B.C., Papazachou, C.B., 1997. The Earthquakes of Greece. Ziti Publ., Thessaloniki. p. 304.

Pavlides, S., Papadopoulos, G.A., Ganas, A., 2002. The fault that caused the Athens September 1999 M5.9 earthquake: field observations. Nat. Hazards 27, 61–84.

Reilinger, R., McClusky, S., Vernant, P., Lawrence, S., Ergintav, S., Cakmak, R., Ozener, H., Kadirov, F., Guliev, I., Stepanyan, R., Nadariya, M., Hahubia, G., Mahmoud, S., Sakr, K., ArRajehi, A., Paradissis, D., Al-Aydrus, A., Prilepin, M., Guseva, T., Evren, E., Dmitrotsa, A., Filikov, S.V., Gomez, F., Al-Ghazzi, R., Karam, G., 2006. GPS constraints on continental deformation in the Africa-Arabia-Eurasia continental collision zone and implications for the dynamics of plate interactions. J. Geophys. Res. 111, B05411. http://dx.doi.org/10.1029/2005JB004051.

Uyeda, S., Kamogawa, M., 23, September 2008. The prediction of two large earthquakes in Greece. EOS, Trans. Am. Geophys. Union 89 (39), 363.

Uyeda, S., Kamogawa, M., 23, September 2009. Reply to "the prediction of large greek earthquakes: a critical evaluation. EOS, Trans. Am. Geophys. Union 89 (39), 363.

Varotsos, P., Alexopoulos, K., 1984. Physical properties of the variations of the electric field of the earth preceding earthquakes. Tectonophysics 110, 73–89.

Varotsos, P., Eftaxias, K., Lazaridou, M., 1996. Reply to "a false alarm based on electrical activity recorded at a VAN station in northern Greece in December 1990 by Drakopoulos J., Stavrakakis G. Geophys. Res. Lett. 23 (11), 1359–1362.

Wyss, M., 1996. Inaccuracies in seismicity and magnitude data used by Varotsos and coworkers. In: Debate on VAN, Special Section (Geller, R. (Ed.)) Geophys. Res. Lett. 23, 1299–1302.

Chapter 20

Do Probabilistic Seismic Hazard Maps Address the Need of the Population?

Max Wyss
International Centre for Earth Simulation, Geneva, Switzerland

Chapter Outline

Abstract
Estimating the seismic hazard probabilistically (PSHA) is a standard practice. Engineering handbooks outline how this is to be done and some countries require that the method be applied to estimating the peak ground acceleration (PGA) for which the built environment must be designed. The units in which the seismic hazard is presented on maps are meters per square second (m/s^2) (or equivalently the fraction of g, the Earth's gravitational attraction) not to be exceeded with a certain probability and within a selected period, both chosen by the author of the map. I argue that there are ethical as well as technical issues with this approach. For engineers who build a high dam for a reservoir it is essential to be given a spectrum of accelerations in m/s^2 in order to design the dam to resist the accelerations deemed possible by seismologists. However, for a governor of a province with seismic activity, a map with colors representing m/s^2 is incomprehensible and therefore useless. I propose that the seismic hazard should be translated into seismic risk, expressed in numbers of casualties to be expected, such that decision-makers and the public can visualize the extent of a

Geoethics. http://dx.doi.org/10.1016/B978-0-12-799935-7.00020-4
239

disaster they may be exposed to and that they can take steps to mitigate. In addition to the problem of communication and thus service to the public, the technical problems include the facts that the PSHA maps cannot be verified directly and when they are verified indirectly, they are shown to be wrong in important regions. In the Western United States, the PSHA maps are showing accelerations twice the upper values that could possibly have happened in about the last 10,000 years, as attested to by precariously positioned rocks. Elsewhere, in the source volumes of large earthquakes (M ≥6.9), which have killed 1,000 and more during the last 20 years, the expected fatalities implied by the PSHA world map underestimate those observed by a factor of about 200. One could thus argue that the application of the PSHA method to critical structures is ethically justified because it has been formulated precisely and is accepted by the majority of practitioners, but that its extension to seismic hazard maps, which should serve the public, is immoral because the hazard is not communicated comprehensibly and the maps' implications are demonstrably wrong for important regions.

Keywords: Disaster mitigation; Seismic hazard; Seismic risk.

INTRODUCTION

In scientific thinking and for communicating, it is important to use precise language. When the popular use of a word is somewhat fuzzy, scientists sharpen its meaning to a precise definition and always use it as agreed.

Seismic hazard and *seismic risk* might not be considered distinct parameters by the public, but they mean different things to professionals. *Seismic hazard* is defined as the *probability that a given strong ground shaking* in a particular location not be exceeded by X percent during a period of Y years, whereas *seismic risk* is *the loss* that results when the strong shaking occurs. The *hazard* is usually expressed as an estimate of the *peak ground acceleration (PGA) in units of meter per square second* (m/s^2) that is expected to be exceeded with a relatively small probability during a certain exposure time. The *risk* is measured in *fatalities, injured, or monetary losses*, if the PGA occurs. Clearly, the earthquake hazard depends only on nature (as long as humans do not interfere), whereas the risk is a strong function of the exposure and its fragility. That is, in the desert, high accelerations of ground motion generate no consequences (zero risk), whereas even moderate shaking beneath a metropolis with poorly constructed buildings can produce great losses (high risk).

Seismic hazard maps are founded on knowledge of the level of seismicity of the past and the transmission properties of seismic waves (strong or weak attenuation as a function of distance from the earthquake source). *Seismic risk maps* are based on hazard maps, and the numbers of local population, as well as the properties of the built environment additionally flow into constructing them.

From this brief review it is clear that seismic hazard maps in m/s^2 are useful for engineers, who may need to design an important structure, and for scientists, who may want to construct a seismic risk map. For the public and decision-makers, seismic hazard maps are meaningless because people need to be told the losses that are possible, in order to fathom the danger they may be exposed to. The advantages and disadvantages of estimating the risk probabilistically (e.g., probability of losses in $/year) or deterministically (e.g., for a specific earthquake at a specific location in fatalities/event) forms part of the discussion in this chapter. For detailed technical discussions of methods to deal with seismic hazard and risk see Wyss (2014).

THE NEED OF THE POPULATION IN HIGHLY SEISMIC REGIONS TO KNOW THEIR RISK

Information about the *earthquake risk* is what people need, not probabilities of accelerations to occur (the hazard), which is a quantity fueling discussion in the ivory tower. Although there are many problems with disaster research (Kelman, 2015), I argue that it is only useful, and thus worth doing, if it is communicated to the users in a way they can understand. To this end, Marone et al. (2015) work toward quantifying the simplicity of texts concerning complicated scientific issues to be communicated to the public.

The need of the population with respect to the problem posed by strong earthquakes can be formulated in a nutshell: (1) People in strongly active regions should be told the worst case scenario that could be in store for them, and (2) they should be told what they can do to protect themselves.

In the world of designing high dams to withstand possible strong ground motion, the notion of Maximum Credible Earthquake (MCE) is used as a required basis for estimating the maximum peak ground accelerations, which a dam should be able to resist, PGA(MCE). The addition of calculating the possible accelerations *deterministically*, that is, as they may result in the worst case, forms an essential addition to the evaluation of the hazard for critical facilities because it serves as a reality check on the PSHA calculations, which are to some degree arbitrary because they depend on the exposure time, which has to be selected without a firm logical basis.

- The first step in this evaluation is that geologists and seismologists identify seismically active faults in the vicinity of the dam site.
- The second step is for these experts to estimate the maximum magnitude, M_{max}, of which the various faults are capable. Although this is not always easy, the principle is simple. The parameters of historical and paleoseismologically identified earthquakes that are considered to estimate M_{max} include: measured magnitudes, mapped rupture lengths and widths, as well as maximum displacements in the past.
- The third step is to calculate the expected acceleration, given the M_{max} and its distance from the project site, using an attenuation function that gives the decrease of wave amplitudes as a function of distance.

It seems obvious that people's safety deserves at least as much consideration as dam safety. Therefore, I advocate that for population centers in seismically strongly active regions estimating PGA(MCE) be made mandatory, as it is for dams. In the case of dams, engineers know how to use this information in their design of the structure. However, communities, including their leaders, do not know how to use PGA(MCE). Therefore, one must calculate for them the likely consequences, the risk. The reason I advocate using fatalities as the unit in which to quantify the risk is that the numbers of fatalities, although sometimes uncertain, is the best-known parameter describing past disasters, and it is by learning from the past that we can hope to usefully estimate what is in store in the future.

THE ASSUMPTION UNDERLYING THE PROBABILISTIC METHOD TO ESTIMATE SEISMIC HAZARD AND RISK

Perhaps a brief explanation of the standard PSHA method may be useful for the reader who is not a seismologist. After selecting a point of interest (coordinates on the globe), one divides the surrounding seismically active volumes in units that make sense. Let the unit we consider be an active fault. The seismic potential of this fault is then characterized by three parameters; a, b, and M_{max}. M_{max} is the largest magnitude the fault is thought to be capable of, and the constants a and b are obtained by counting the earthquakes as a function of magnitude. The earthquake catalogs usually span the last few decades, typically from 1973 on, when the U.S. Geological Survey started to report events systematically with high reliability worldwide. The equation (Gutenberg and Ricter, 1944)

$$\text{Log } N = a - bM \tag{1}$$

in which N is the cumulative number, defines a straight line in a logarithmic plot with the slope of b and the intercept at $M = 0$ of a (Figure 1). Therefore, a measures the productivity of the volume studied, and b controls the intercept of the extrapolation (red line in Figure 1) with the expected M_{max}. A steeply plunging line intercepts the M_{max} ordinate at a low probability, a shallow dipping line at a higher probability.

The estimate of the annual probability of M_{max} by this method requires an assumption that has never been verified: In any period of observation, the appropriate fraction of the total number of earthquakes must be accumulated that corresponds to fulfilling the law of Eqn (1) all the way from Mc to M_{max}, such that, starting from a previous M_{max}, by the day before the next M_{max} occurs one would observe a straight line from M_{max} to Mc. Thus the assumption stipulates that once the aftershock sequence has come to an end, the earthquakes in the interevent period (between two consecutive earthquakes with M_{max} on the same fault) fill the plot in Figure 1 exactly the same way the aftershocks did. It can be shown that for important fault segments the production of earthquakes during the last 40–80 years is far too low for this assumption to be correct. The conclusion is that the PSHA method is based on an unverified assumption.

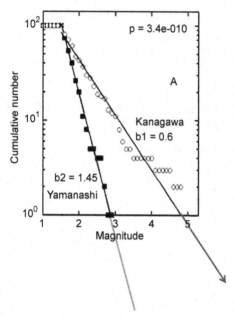

FIGURE 1   The cumulative number of earthquakes in a volume is plotted on a logarithmic scale as a function of magnitude, which is derived from the logarithm of the seismic wave amplitude. This relationship is usually well approximated by a straight line (Eqn (1), black in the figure). It is valid between the largest earthquake recorded and the smallest earthquakes reported completely, Mc. Below Mc more earthquakes occur than can be recorded with the available network, hence the relationship is not complete and no longer valid. The parameter b is the slope of the black line, a is its intercept with M=0. In the figure the seismicity of two volumes with the same productivity (the constant a is the same) are compared. In the volume with the lower b-value the extrapolation (red line) leads to a relatively high expectation of an earthquake with M7, whereas in the volume with the low b-value (green extrapolation) a large earthquake has a very low probability to occur, extrapolated by the red line to the M_{max} that has not been recorded instrumentally, thus estimating the probability of M_{max} to occur once per year.

THE ARBITRARINESS GOVERNING THE PROBABILISTIC METHOD TO ESTIMATE SEISMIC HAZARD AND RISK

The PGA values (m/s^2) are estimated as applying under certain conditions, namely the probability of exceedance in a specified period. For the values shown on the world hazard map, the probability that they may be exceeded during a period of 50 years is given as 10%. If one does not like the results under the aforementioned conditions, one can recalculate by stipulating that the probability of exceedance is only 2%, which leads to higher PGA values. An increase of the estimated PGA can also be achieved by extending the 50-year period to a longer duration. With other words, virtually any PGA value can be produced by adjusting the period and the exceedance probability. Admittedly, some values of these parameters are more reasonable than others, but the fact remains that their selection is arbitrary. This level of arbitrariness is usually not accepted in scientific analyses.

In addition to the problem highlighted here, Panza et al. (2011) edited two volumes of *Pure and Applied Geophysics* reviewing and criticizing the PSHA method and the discussion is further developed by Stirling (2014) and Panza et al. (2014).

TESTING THE PROBABILISTIC METHOD TO ESTIMATE SEISMIC HAZARD AND RISK

Testing these methods directly is not possible because the annual probability of a hazard or a risk is not measurable. However, there are two indirect methods that allow a check on the probabilistic hazard and risk calculations. Inset 1 briefly explains that precariously balanced rocks in the Western United States and elsewhere provide an upper limit for the strong ground motion to which they could have been exposed during the last approximately 10,000 years (e.g., Brune, 1999; Anderson et al., 2011). These rocks would have been toppled if

Precariously balanced rocks serve as gauges of the maximum prehistoric PGA

Above: A precariously balanced rock near active faults in Southern California being tested for the force it takes to topple it (photos curtesy J. N. Brune).

Below: The line of surviving precarious rocks half way between the San Jacinto and Elsinore faults, Southern California, limiting the possible prehistoric PGA to less than about 0.25 g (Brune et al., 2006).

The maximum PGA possible at locations of precariously balanced rocks (circles) as a function of distance from the San Andreas fault, compared to PGA values shown on the probabilistic seismic hazard map of California and calculated by similar probabilistic methods(dashed lines) (Brune, 1999).

Riverside-Aguanga Line of Precarious Rocks

PGA had exceeded approximately 0.25 g, but the seismic hazard map shows about 0.5 g as to be exceeded by a 2% probability in 50 years. By showing the value of 0.5 g on their maps, the authors suggest that community leaders should prepare for this kind of acceleration, yet this value has never been obtained in about 10,000 years. In this case the community is asked to spend unnecessary resources to protect against something that has virtually no chance to occur.

The opposite is the case for those locations on the planet where large earthquake have killed more than 1,000 people during the last 20 years. Inset 2 demonstrates that earthquakes that would generate the PGA (to be exceeded by 10% probability in 50 years), plotted on the world's seismic hazard map in the locations of large damaging earthquakes that did occur, underestimate the fatalities reported by a factor of 200 for earthquakes with $M \geq 6.9$. The world seismic hazard map

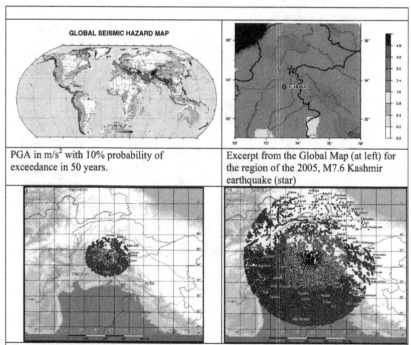

GLOBAL SEISMIC HAZARD MAP

| PGA in m/s^2 with 10% probability of exceedance in 50 years. | Excerpt from the Global Map (at left) for the region of the 2005, M7.6 Kashmir earthquake (star) |

Comparison of the settlements affected by an earthquake of M6.1 (left bottom row), as expected based on the Global Seismic Hazard map top row) at the epicenter of the 2005, M7.6 Kashmir earthquake, with the settlements actually affected (at bottom right). The size of the dots is proportional to the population, the color proportional to mean damage on a scale of 1 to 5. The method of calculation and other examples are given by Wyss et al., 2012. The mean damage represented by blue is grade 0.5, i.e. minor, thus the maps show the estimated extent of the population affected by the hypothetical (left) and the actual case (right). In this case the ratio of observed to expected fatalities is 19. On average for earthquakes with $M \geq 6.9$ that killed more than 1,000 during the last 20 years the ratio of observed to expected fatalities is 200.

INSET 2

thus conveys a false sense of security to people living near faults capable of $M \geq 6.9$ earthquakes. The power of maps and their potential to mislead consumers, who are not experts, is discussed by Heesen et al. (2015).

These two indirect tests show that the standard probabilistic seismic hazard and risk mapping method leads to incorrect results. This conclusion is also supported by a study of Zuccolo et al. (2011) who showed that in places where accelerations due to recent earthquake have been measured instrumentally, the world seismic hazard map has underestimated the accelerations by a factor of 2–3. In addition, Geller (2011) has shown maps (Figure 2) demonstrating that

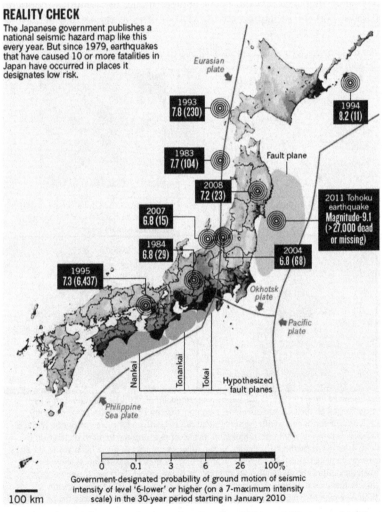

FIGURE 2 Mismatch between epicenters of earthquake which cause 10 or more fatalities and areas designated officially as high risk. *From Stein et al. (2012).*

recent large earthquakes typically have fallen into regions characterized by low seismic hazard on the probabilistic maps.

THE DETERMINISTIC APPROACH TO ASSIST THE POPULATION IN UNDERSTANDING THEIR EARTHQUAKE RISK

Deterministic is the expression used when one specific earthquake with assumed location, magnitude, and depth is taken as the worst credible case. In this approach, the estimate of how probable such an earthquake and its associated losses are is not included. We must admit that we have no reliable way to estimate the annual probability of a damaging earthquake, although the insurance industry needs to construct some kind of logic to estimate probability to conduct their business (e.g., Mitchel, 2014). When it comes to motivating the population of highly seismic regions and their leaders to take mitigating measures, I recommend stressing that the probability integrated over the lifetime of three generations of descendants is more than say 50% (or whatever the best estimate for the region in question is).

In addition, one should advise people living in poorly constructed dwellings to build themselves, or better yet have an engineer design for them, an earthquake closet in their apartment or house. Such an earthquake protection unit (EPU) would have dimensions of about 2 m by 2 m and would resist collapse, even with heavy weights falling on it. It would increase the probability of a family to survive typically several thousand-fold (Wyss, 2012).

The proposed EPU has two aspects in common with tornado shelters. (1) It is inexpensive and could be available for purchase in lumber yards, and (2) the owner understands that there is a strong likelihood that he or his descendants will be hit by a disaster.

This second point needs to be examined a little closer. In the case of tornados, the return period is measured in a few years, thus people remember. However, for earthquakes the return period is typically more than 100 years even in highly active regions, thus people forget and the annual probability of a disaster is estimated as about 1% or less. Let's consider the case of a head of household in an area close to a fault capable of M7+ earthquakes, where for several hundred years no devastating earthquake has happened. The next devastating earthquake is unavoidable, but its occurrence time is unknown. I believe that humans strongly care what happens to their descendants and that they typically know and love two to three generations after themselves. Thus, in a case where time has progressed a couple of hundred years into a 300-year recurrence interval, the probability of a hit during one of the next three generations, or the current one, is larger than 30–50%. In such a case, I definitely would build myself an earthquake closet at the cost of a few thousand dollars increasing the probability of my loved ones to survive by a few thousand times.

CONCLUSIONS

The standard probabilistic earthquake hazard assessment yields incorrect results in many areas and especially for large earthquakes causing fatalities, although it is based on a method accepted by the engineering community, thus an ethically justified method. I propose that the moral obligation to serve mankind requires that one should instead estimate the losses to be expected by the maximum credible earthquake near population centers, as is mandatory for high dams. The units of demonstrating losses should be fatalities because fatalities as a measure of a disaster is something everyone can understand and because it is a number known for about 2,500 historic earthquakes (e.g., Utsu, 2002). Constructing maps of seismic hazard in units of m/s^2 to be exceeded by some probability in a certain period is incomprehensible for decision-makers and the population. If these maps are based on the standard probabilistic method the implied consequences, the estimated seismic risk, is far from reality in many important regions.

In regions where the probability of an MCE to occur during the next 100 years is higher than a threshold (perhaps 30%) one should act as if the disaster might take place the next day and take mitigating measures. To assist the people to make a decision whether or not to invest the funds necessary for constructing an EPU one should estimate the probability that a large earthquake may happen during the next hundred years, as it has been done for the Himalayas (Khattri, 1999; Bilham et al., 2001). If the losses in the same region are also estimated (Wyss, 2005, 2006), then heads of households, as well as civil leaders, can make decisions on an informed basis whether or not to mitigate.

I propose that the standard approach to help the population to understand their seismic risk does not come up to an acceptable moral level for the following reasons. The probabilistic method to estimate seismic hazard is not based on testable assumptions. To express the seismic hazard in units of m/s^2 shows a lack of caring for the population because the units are incomprehensible. The seismic hazard and risk in urban centers does not receive as careful attention as that for high dams. And finally, the population in many areas is not informed of their earthquake risk and what they may be able to do to reduce it.

REFERENCES

Anderson, J.G., Brune, J.N., Biasi, G., Anooshehpoor, A., Purvance, M., 2011. Workshop report: applications of precarious rocks and related fragile geological features to U.S. national hazard maps. Seism. Res. Lett. 82 (3), 431–441.

Bilham, R., Gaur, V.K., Molnar, P., 2001. Himalayan seismic hazard. Science 293, 1442–1444.

Brune, J.N., 1999. Precarious rocks along the Mojave section of the San Andreas Fault, California: constraints on ground motion from great earthquakes. Seism. Res. Lett. 70, 29–33.

Brune, J.N., Anooshehpoor, A., Purvance, M.D., Brune, R.J., 2006. Band of precariously balanced rocks between the Elsinore and San Jacinto, California, fault zones: constraints on ground motion for large earthquakes. Geology 34, 137–140. http://dx.doi.org/10.1130/G22127.1.

Geller, R.J., 2011. Shake-up time for Japanese seismology. Nature 472, 407–409.

Gutenberg, R., Richter, C.F., 1944. Frequency of earthquakes in California. Bull. Seismol. Soc. Am. 34, 185–188.

Heesen, J., Lorenz, D., Voss, M., Wenzel, B., 2015. Reflections on ethics in mapping as an instrument of disaster research. In: Wyss, M., Peppoloni, S. (Eds.), Geoethics: Ethical Challenges and Case Studies in Earth Science. Elsevier, Waltham, Massachusetts, pp. 251–262.

Kelman, I., 2015. Ethics of disaster research. In: Wyss, M., Peppoloni, S. (Eds.), Geoethics: Ethical Challenges and Case Studies in Earth Science. Elsevier, Waltham, Massachusetts, pp. 37–47.

Khattri, K.N., 1999. Probabilities of occurrence of great earthquakes in the Himalaya. Proc. Indian Acad. Sci. (Earth Planet. Sci.) 108, 87–92.

Marone, E., Castro Carneiro, J., Cintra, J., Ribeiro, A., Cardoso, D., Stellfeld, C., 2015. Extreme sea level events, coastal risks, and climate changes: informing the players. In: Wyss, M., Peppoloni, S. (Eds.), Geoethics: Ethical Challenges and Case Studies in Earth Science. Elsevier, Waltham, Massachusetts, pp. 273–302.

Michel, G., 2014. Decision making under uncertainty: insuring and reinsuring earthquake risk, in *Earthquake hazard, risk and disasters*. In: Wyss, M. (Ed.). Elsevier, Waltham, Massachusetts, pp. 543–568.

Panza, G.F., Irikura, K., Kouteva-Guentcheva, M., Peresan, A., Wang, Z., Saragoni, R., 2011. Advanced seismic hazard assessment. Pure Appl. Geophys., 1–752.

Panza, G., Kossobokov, V.G., Peresan, A., Nekrasova, A., 2014. Why the standard probabilistic methods of estimating seismic hazard and risk are too often wrong, in *Earthquake hazard, risk and disasters*. In: Wyss, M. (Ed.). Elsevier, Waltham, Massachusetts, pp. 310–357.

Stein, S., Geller, R.J., Liu, M., 2012. Why earthquake hazard maps often fail and what to do about it. Tectonophysics 562, 1–25.

Stirling, M.W., 2014. The continued utility of probabilistic seismic hazard assessment, in *Earthquake hazard, risk and disasters*. In: Wyss, M. (Ed.). Elsevier, Waltham, Massachusetts, pp. 359–376.

Utsu, T., 2002. 691–717 A list of deadly earthquakes in the World: 1500–2000. In: Lee, W.K., Kanamori, H., Jennings, P.C., Kisslinger, C. (Eds.), International Handbook of Earthquake Engineering and Seismology. Academic Press, Amsterdam.

Wyss, M., 2005. Human losses expected in Himalayan earthquakes. Nat. Hazards 34, 305–314.

Wyss, M., 2006. The Kashmir M7.6 shock of 8 October 2005 calibrates estimates of losses in future Himalayan earthquakes. In: Van de Walle, B., Turoff, M. (Eds.), Proceedings of the Conference of the International Community on Information Systems for Crisis Response and Management, p. 5. Newark.

Wyss, M., 2012. The earthquake closet: rendering early-warning useful. Nat. Hazards 62 (3), 927–935.

Wyss, M., Nekrasova, A., Kossobokov, V.G., 2012. Errors in expected human losses due to incorrect seismic hazards estimates. Nat. Hazards 62 (3), 927–935. http://dx.doi.org/10.1007/s11069-012-0125-5.

Wyss, M. (Ed.), 2014. Earthquake Hazard, Risk, and Disasters. Elsevier, Waltham, Massachusetts, p. 582.

Zuccolo, E., Vaccari, F., Peresan, A., Panza, G.F., 2011. Neo-deterministic and probabilistic seismic hazard assessments: a comparison over the Italian territory. Pure Appl. Geophys. 168 (1–2), 69–83.

Chapter 21

Reflections on Ethics in Mapping as an Instrument and Result of Disaster Research

Jessica Heesen,[1] Daniel F. Lorenz,[2] Martin Voss[2] and Bettina Wenzel[2]

[1]International Centre for Ethics in the Sciences and Humanities (IZEW), Universität Tübingen, Germany; [2]Disaster Research Unit (DRU), Freie Universität Berlin, Germany

Chapter Outline

Abstract

Mapping has become an important tool in disaster research. How physical and social conditions are described by geographical methods decides the way risks, vulnerability, and resilience are evaluated and influences the efforts that are made for disaster control—as well as their fair distribution. However, the predominant methods of mapping have certain constraints concerning cultural and social factors and particularities. Addressing a comprehensive ethical consideration, these limitations and blind spots have to be taken into account. Certain aspects cannot be mapped to the full extent, such as specific forms of knowledge, risk perception, and preparedness measures. If the unmapped preconditions of mapping remain unconsidered, this can lead to problematic effects with respect to an appropriate understanding of maps on the one hand and a fair and context-related distribution of official measures to control risk and to care for resilience processes on the other hand. Map makers and map users should be aware of the problems connected to maps in order to avoid pitfalls and misinterpretations.

Keywords: Ethical considerations; Hazards; Map data; Mapping; Methodological critique; Resilience; Risk; Social process; Vulnerability.

Geoethics. http://dx.doi.org/10.1016/B978-0-12-799935-7.00021-6

PART I: THE CONCEPTS OF RISK, VULNERABILITY, AND RESILIENCE

Concepts of Risk

Usually risk is defined as the probability of an accident occurring multiplied by the expected loss in this case. How probable a loss seems and how it is ranked may depend on several factors (Douglas, 1992). There are risks that require a personal decision to take them or not and generally the individual assumes to be able to cope with these risks, for example, to manage a fixed rope route on a hiking tour. Other risks are not under the individual's control and confront him or her as a danger. For example, an unprepared single person seems to be almost helpless in the face of earthquakes, tsunamis, volcanic eruptions, storms, and also technological disasters.

Societies normally face these kinds of risks through statistical calculations of probabilities and try to react with preventive measures. Maps and increasingly geographic information systems (GIS) play an important role in the presentation of the probability of occurrence and also in the assessment of protection requirements. In this context, it has to be considered on the one hand that maps in general rely on statistical "objective" risk assessment, but on the other hand the perception of risk on an individual or a local level can strongly differ.

Concepts of Vulnerability

The concept of vulnerability emerged within different fields in the context of natural hazards and risk to help in understanding the condition or the predisposition of a system to suffer harm due to a hazard (Adger, 2006). Approaches to measure vulnerability mainly focus on selected influence factors (natural hazards, food insecurity), some of which may be mapped cartographically, while others may not. The definition of what should be understood by "vulnerability" is contested within science and economy, as well as in everyday life (Yamin et al., 2005). A rather phenomenologist view, which sees vulnerability as a measure of the probability to suffer harm in case of an event, is confronted with numerous other concepts, including further influencing factors, and means of protection referring to different timescales and spatial references (Voss, 2008).

Concepts of Resilience

Most of what has been said before with respect to vulnerability is true for the resilience approach as well: It has its background in different disciplines and its meaning is perhaps even more contested. The Latin root *resilire* is first mentioned in ancient Rome by Lucius Annaeus Seneca (Alexander, 2013), then, most prominently, reinvented within the field of ecology by C.S. Holling since the 1970s (e.g., Holling, 1973). Within the last two decades, the term ascended to the core of different discourses, e.g., disaster risk reduction, disaster education, and risk communication. Again, seen from a phenomenological and rather distant perspective, instead of questioning which hazards or other factors enhance vulnerability, the resilience approach emphasizes the factors that allow people, or groups of people, to cope with extreme events. It asks why some of the concerned people are heavily affected by natural or social hazards, while others can continue to live their "normal" life without loss and trauma, being exposed to the same stress. Obviously, culturally shaped worldviews, norms, and values play a crucial role, developed through a constant grappling with their biophysical and social environment. Seen through the resilience lens, the understanding of loss, risks, and hazards becomes correspondingly relational.

PART II: MAPPING AS SELECTIVE REPRESENTATIONS

Forms of Selection

Maps are, by definition, selective representations (Black, 1997; Raffestin, 2003) since they need to be limited to aspects for the purpose of a heuristic value (Monmonier, 1996). However, this intentional selection becomes hidden, since maps build on a specific sign system that is presented as naturally given (Pickles, 2004, p. 61). But the predominant mapping procedures are based on certain codes, standards, and knowledge (Wood and Fels, 2008), both in their production and in their reading. Since these foundations are embedded in a Western scientific culture (Harley, 2001d), non-Western or other variant countermapping cartographies such as indigenous maps (Taylor, 2008) are often not acknowledged as such (Pickles, 2004).

Furthermore, incompatible forms of knowledge and meanings or social practices are often not mapped or cannot be mapped within the Western sign system (Wainwright and Bryan, 2009). Referring to the work of Dauenhauer (1980), this is termed cartographic silence by Harley (2001c, p. 84). However, the ontologies on which GIS are based often remain invisible, and are often not questioned by the users.

Every form of mapping builds on the existence and availability of (the latest) data in the first place. The limitations of socioeconomic indicators in vulnerability and resilience mapping have been extensively discussed (King, 2001; Fekete, 2012). Choosing indicators of vulnerability or resilience from a standardized set of census data may be problematic since they usually do not encompass relevant variables such as risk perception, social networks, and preparedness measures and thus foster cartographic silence.

When analyzing and processing spatial data, several statistical and methodical problems may arise that are amplified by their visual presentation in maps. One of the most discussed issues is the definition of spatial units: The arbitrariness of the analytical unit's definition is mostly pragmatically motivated, but may sometimes be biased by political motives. Connected statistical problems are the "modifiable areal unit problem" (Openshaw, 1984; Madelin et al., 2009), as well as data aggregation and standardization, the problem of ecological fallacy, and different forms of classification.

Cartographic subtleties of visual representation such as projection, generalization, and color schemes may also lead to misinterpretations of maps. The knowledge about the sign system cannot be assumed to be globally consistent.

Reduction from Qualitative Processes to Quantitative Data

The mapping of risk, vulnerability, and resilience is carried out to prevent or mitigate disasters in terms of social impacts (casualties, loss, and trauma). The global mapping of disaster risk and vulnerability is dominated by single natural hazards like tsunamis or earthquakes, with geographic reach of waves, and the frequency of their occurrence (e.g., World Risk Report, Alliance Development Works, 2013), but do not scrutinize the heterogeneous historical, political, and cultural conditions that transform these natural processes to threats for societies and communities (Wyss, 2015). On the one hand, the magnitude or frequency of extreme natural processes alone do not bear any information about their social relevance in relation to other problems a society has to deal with. Hence, a red-colored zone displaying the risk of a tsunami does not indicate vulnerability, if people prepare themselves adequately through a culture of disaster awareness (building standards, warning systems). If such an environment-adequate culture exists, the vulnerability is reduced or may be eliminated in some cases (Figure 1).

On the other hand, the background assumption that the effects triggered by natural processes have to be solely negative as long as they potentially cause biophysical harm or casualties is a social construction underlying most forms of mapping. However, as has been shown (Macamo, 2003; Voss, 2008), the local interpretations might construe what is seen as a disaster from the outside as something that is culturally valued as positive, e.g., a fatal flood can be seen as a necessity for agriculture. What is seen as negative and dangerous depends on cultural norms and patterns (Douglas and Wildavsky, 1982). Although it seems to be beyond dispute, forms of loss are framed by different cultures, leading to different ways of dealing with it (from repression strategies to full acceptance or fatalism). To be precise, of course there are many approaches that try their best to include these social and cultural aspects into mapping. Nevertheless, cartography in general has limitations in visualizing these relativistic, complex cultural conditions, and the more global the perspective, the higher the likelihood that relevant cultural or social determinants varying the possible effects of an extreme natural event tend to be neglected (e.g., World Risk Report; Alliance Development Works, 2013). This may lead to an erroneous interpretation of the situation because it blanks out the available productive conditions and capacities of the population, even though these sources of resilience can be most important.

Without going so far as to argue for a radical cultural relativism, it should be noted that the meanings of risk, vulnerability, and resilience do not primarily evolve from recourse to single key figures, but only when linked back to social and cultural systems of meaning. This means that certain assessment criteria have to be given considerations that commonly defy quantification. Every

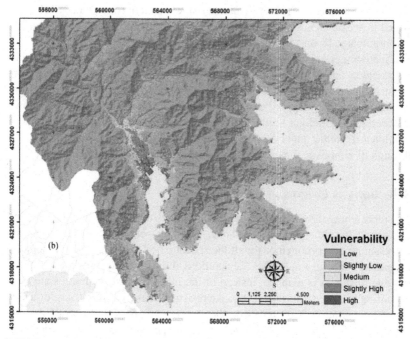

FIGURE 1   A so-called vulnerability map that does not indicate vulnerability since the map is solely based on physical parameters (elevation, slope, coastline distance, and vegetation density). Social factors, such as population, properties of the built environment, and levels of preparedness and adaptation strategies, are absent. *Source: Sambah and Miuara (2014).*

quantifying assessment is thereby limited in scope, and pure descriptive figures are proving to be insufficient when it comes to cultural significance.

Therefore, differing strategies to protect one's life are "normal" and against the background of varying cultures they are justified differently and are only rational in part, yet remain so implicitly. These rather "practical" strategies (Bourdieu, 2000, pp. 138–139) are developed in a coevolutionary process of humans and environment. Such coevolutionary and nonlinear strategies of adaptation cannot be captured by objectifying methods, for example, the local or traditional knowledge about hazards that is formed over decades defies representation in maps (Voss, 2008).

Under interpretational sovereignty of quantifying sciences (e.g., engineering, natural sciences, economics) and insurance industry, risk has for a long time and to a large extent exclusively been understood as a quantifiable and therefore objective function of probability and damage of an event (Rowe, 1977; Bonss, 1995). However, realizing that such calculations are not the foundation of risk perception, acceptance, and acting in everyday life, an alternative concept of risk being a relational construction that is highly dependent on cultural and social beliefs, values, and judgments came up in the social sciences and humanities in the 1960s (Fischhoff et al., 1984; Douglas and Wildavsky, 1982). Today, the dichotomy of so-called risk-realism and risk-constructivism is widely acknowledged in the academic literature, even though the political decision making is still dominated by an "objectivistic" understanding of risk.

Analogously, emanating from a similar positivistic position, vulnerability research opens up slowly to this complexity of different perceptions and constructivism, especially through the integration of the resilience perspective. However, the mapping of vulnerability and resilience tends to slow down this creeping paradigm shift (McEntire, 2004) from an objective-static to a relational-dynamic understanding also of the resilience of people and groups through its simplistic materialistic representation.

Selections as Origins of Interpretation

The representation of danger by mapping implies deriving factors determining disaster or loss from georeferenced data. From the perspective of its critics, this representation is synonymous with the reduction of loss to its spatial forms of appearance. For example, the accelerations due to earthquakes or the flooding due to storms is mapped, instead of the invisible fact that houses are built according to code requiring earthquake resistant construction because of a corrupt administration. With regard to the mapping of risk and vulnerability, this denotes not only that the use of GIS and their special language is problematic, but also potentially that the tacit justification links for vulnerability and insecurity can gain authority and power by using a seemingly neutral map (Glasze et al., 2005; Harendt and Sprunk, 2011). In the construction of maps, certain patterns of data interpretation are required.

The form of determination is an expression of a certain claim of validity. Maps are similar to other semantic systems in that they are a medium for subjective processes of reaching understanding for dominant ascriptions of significance (Harley, 2001b; Wood, 2010). If a context of justification with a certain semiotic system is established by maps, they become "an authority that may be hard to dislodge" (Harley, 2001a, p. 168), since alternative facts, links, and interpretations may not be displayed within the semiotic system.

With the given agency of the mapping, other concurrent interpretations are exacerbated. Therefore, maps that are seemingly only descriptive are reducing reality's complexity and become prescriptive. They establish a singular meaning and interpretations that become part of reality and the background for the interpretation of the world (Wood, 1992; Wood and Fels, 2008).

However, in the terms of media theory, actor-network theory (Latour, 1999), and affiliated approaches of critical theory or cultural studies (Pickering, 2001), a so-called objectification through mapping can be described as a "normal" process. To objectify social processes in a map means to produce a cultural code, among other, competing cultural codes. As a result of this approach, not only objectification itself has to be considered, but also the discourses or narratives that lead to it.

In this perspective, maps have to be regarded as objectifications, which media in general are. Similar to other media like television, newspapers, or books, maps show a certain perspective of the different possible views on reality in relation to a specific set of instruments. In a media-critical approach, one has to give some thought to how a sign system is used and in which way other perspectives are excluded by this special form of representation.

Moreover, maps gain authority for a broader public because they are commensurate with the demands of Internet communication and traditional broadcasting media: both ask for visualizations and a simple "language," and they both promote a trend that one can also find in the growing number of pictorial representations in online communication (e.g., icons). In media theory and practice, the reduction of reality by using numbers and statistics is an important and common method to gain attention (Luhmann, 1990). Maps are ideal for a noticeable and suggestive presentation: they combine a (putative) statistical relevance with the advantage of an attractive eye-catcher. In this way, maps are not only relevant for academia and politics, but also for the public, and it can be expected that the importance of maps will grow in the next media-dominated decades.

PART III: ETHICAL QUESTIONS OF RISK MANAGEMENT AND INTERPRETATION OF MAPS

Since mapping risk, vulnerability, and resilience is an interpretation of reality, people act upon what is presented in the maps (Thomas and Thomas, 1928). This holds especially true for mapping both in predisaster planning and crisis management. Thus, when certain areas are judged to be rendered unsafe in case of a disaster, this might lead to a self-fulfilling prophecy, because vulnerability is increased by its attribution alone. On the one hand, a world map that ascribes high vulnerability to areas or even continents by coloring them red may have practical consequences that run contrary to the intentions (Bankoff, 2003). Investors, for example, will use such maps as guidance, although they do not know much about the underlying concepts, they understand the red color as a warning sign. On the other hand, it is important to recognize that mapping risks and vulnerabilities might also have the opposite effects: once an area has been marked as "vulnerable" or "risky" this might also lead to an unjust distribution of resources when they are mainly directed toward such areas at the expense of others.

Discourses of "underdevelopment" and dependency of southern countries on industrialized countries as members of the Organisation for Economic Co-operation and Development are also partially carried forward by mapping and visualizing vulnerability, establishing stereotypical images that can lead to multiple and destructive effects in reality as discussed within postcolonial theories (Bhabha, 2004). Furthermore, for some cultures, the (scientific) mapping of risks may be perceived as a provocation of celestial powers, with the direct consequence that their feeling of security will be reduced (Voss, 2008). This holds true for matters large and small: valuation differences will be on the agenda all over the world. Every nonconsideration of different points of view—be it intended or not—leads to distortions, the valuation of which can only be made on the political level. All stakeholders should be included, such that the discussion is not restricted to experts, disregarding the needs and opinion of the people affected (Wyss, 2015).

While there are good reasons to critically reflect upon the use of maps, we should also be aware that, as with statistics (Saetnan et al., 2010), we can hardly argue in favor of giving maps up altogether, since disaster preparedness planners, emergency managers, and scientists need data on how many people in different sorts of categories may need assistance or special intervention in a crisis (King, 2001). Map makers and map users should be aware of the problems connected with maps in general in order to avoid pitfalls and misinterpretations of those maps. We also need to carefully reconsider the use of the maps we make, and be aware of the specific limitations of mapping. For a more comprehensive understanding of GIS and to avoid problematic side effects such as a misallocation of resources, a critical analysis of data sources and their relevance for the specific reason of mapping or local problem is an elementary condition. Furthermore, it is necessary to explain the meaning of symbols, signs, and colors in maps, recognizing cultural differences and to consider the possibility that people in different cultures may interpret a map differently than intended.

Basically, the abstract concepts of risk, vulnerability, and resilience need to be seen and researched with respect to their root causes and different perceptions of the relevant people, not as isolated factors that can be located and displayed easily. To fully understand the generation of risk, vulnerability, and resilience, the underlying social causes need to be regarded as complex processes, involving different spatial and temporal scales (Black, 1997). What we need is not a positivistic construction and interpretation of maps, but their foundation on a contextual and differentiated exploration of concrete situations at particular places. An inductive concept of vulnerability and resilience focuses on the given circumstances of vulnerability or resilience and the feeling of (in)security. However, this focus is not only related to geographic measurable units, but must also be widened to the social conditions that produce places and spatial entities, for example, a nation, an administrative territory, a so-called no go area, or a holy place. In other words, representations in maps should be the starting point for ongoing consideration of the underlying cause-and-effect chain, but not the final and concluding point. If we are well aware of those fundamental problems, then risk, vulnerability, and resilience maps should be seen in relation to historic developments and sociodemographic and sociopolitical parameters to make transparent complex historical (i.e., long-term) cause and effect relationships.

In a thorough interpretation of spatial mapping, a new and different approach is possible: a regionalized idea focused on explaining danger from a concrete—and not abstract—perspective (Figure 2).

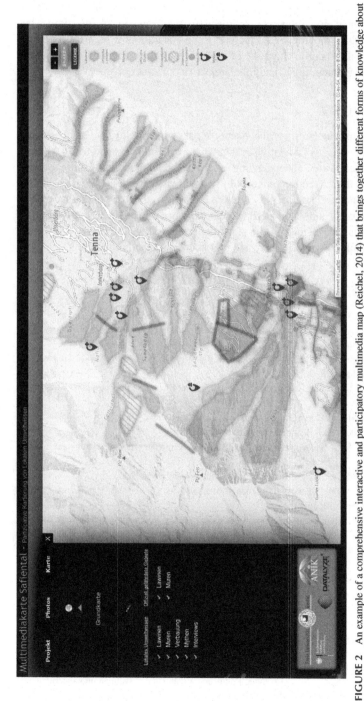

FIGURE 2   An example of a comprehensive interactive and participatory multimedia map (Reichel, 2014) that brings together different forms of knowledge about natural hazards in the Savien Valley, Alps. Scientific knowledge (SK) and traditional environmental knowledge (TEK) about avalanches (SK = violet, TEK = red) and debris flows (SK = blue gray, TEK = blue) are mapped, as well as snow slabs, housing, and avalanche barriers (gray). These mappings are combined with additional media such as interviews and videos for further explanation. The map combines a scientific expert evaluation with knowledge of the local people. It shows how valuations can differ as a result of different sources of knowledge and it therefore challenges the "objectivistic" interpretation of maps.

The regionalization of mapping can be conceived of as an indication of a context-related and inductive approach to the concepts in question. This approach leads to an augmented picture of the generation of security, a picture that includes the specific character of a place as an interplay between real factors, subjective perceptions, and intersubjective constructions.

CONCLUSIONS

This chapter introduces different concepts of risk, vulnerability, and resilience and shows the importance of a relativistic approach for their understanding. In contrast to such a context-aware understanding of societal and cultural implications, many maps are based on a reduced concept of risk and vulnerability. The article shows how maps work as selective representations and which ethical problems can evolve from mapping. The authors underline the importance of a critical interpretation of maps as media and factors for the construction of reality. The contribution ends with an example for an interactive and regionalized method of mapping.

REFERENCES

Adger, N., 2006. Vulnerability. Global Environ. Change 16 (3), 268–281.

Alexander, D.E., 2013. Resilience and disaster risk reduction: an etymological journey. Nat. Hazards Earth Syst. Sci. Discuss. 1, 1257–1284.

Alliance Development Works, 2013. World Risk Report 2012. http://www.worldriskreport.com/uploads/media/WRR_2012_en_online.pdf.

Bankoff, G., 2003. Cultures of Disaster. Society and Natural Hazards in the Philippines. Routledge, London.

Black, J., 1997. Maps and Politics. The University of Chicago Press, Chicago.

Bhabha, H.K., 2004. The Location of Culture. New York.

Bonss, W., 1995. Vom Risiko. Unsicherheit und Ungewissheit in der Moderne. Hamburger Edition, Hamburg.

Bourdieu, P., 2000. Pascalian Meditations. Stanford University Press, Stanford.

Dauenhauer, B., 1980. Silence: The Phenomenon and Its Ontological Significance. Indiana University Press, Bloomington.

Douglas, M., 1992. Risk and Blame: Essays in Cultural Theory. Routledge, London; New York.

Douglas, M., Wildavsky, A., 1982. Risk and Culture: An Essay on the Selection of Technical and Environmental Dangers. University of California Press, Berkeley.

Fekete, A., 2012. Spatial disaster vulnerability and risk assessments: challenges in their quality and acceptance. Nat. Hazards 61 (3), 1161–1178.

Fischhoff, B., Lichtenstein, S., Slovic, P., Derby, S.L., Keeney, R., 1984. Acceptable Risk. Cambridge University Press.

Glasze, G., Pütz, R., Rolfes, M., 2005. Die Verräumlichung von (Un-)Sicherheit, Kriminalität und Sicherheitspolitiken – Herausforderungen einer Kritischen Kriminalgeographie. In: Glasze, G., Pütz, R., Rolfes, M. (Eds.), Diskurs – Stadt – Kriminalität. Städtische (Un-)Sicherheiten aus der Perspektive von Stadtforschung und Kritischer Kriminalgeographie. Bielefeld: Transcript, pp. 13–58.

Harendt, A., Sprunk, D., 2011. Erzählter Raum und Erzählraum: (Kultur)Raumkonstruktionen zwischen Diskurs und Performanz. Social Geogr. 6, 15–27.

Harley, J., 2001a. Deconstructing the map. In: Harley, J. (Ed.), The New Nature of Maps. John Hopkins University Press, Baltimore, pp. 150–168.

Harley, J., 2001b. Maps, knowledge and power. In: Harley, J. (Ed.), The New Nature of Maps. John Hopkins University Press, Baltimore, pp. 51–81.

Harley, J., 2001c. Silences and secrecy. The hidden agenda of cartography in early modern Europe. In: Harley, J. (Ed.), The New Nature of Maps. John Hopkins University Press, Baltimore, pp. 83–107.

Harley, J., 2001d. Text and contexts in the interpretation of early maps. In: Harley, J. (Ed.), The New Nature of Maps. John Hopkins University Press, Baltimore, pp. 33–49.

Holling, C.S., 1973. Resilience and stability of ecological systems. Annu. Rev. Ecol. Syst. 4, 1–23.

King, D., 2001. Uses and limitations of socioeconomic indicators of community vulnerability to natural hazards: data and disasters in Northern Australia. Nat. Hazards 24 (2), 147–156.

Latour, B., 1999. Politiques de la nature. Comment faire entrer les sciences en démocratie. La Découverte, Paris (Engl. Politics of Nature).

Luhmann, N., 1990. Gesellschaftliche Komplexität und öffentliche Meinung (Societal Complexity and Public Opinion). In: Soziologische Aufklärung 5. Konstruktivistische Perspektiven. Westdeutscher Verlag, Opladen, pp. 170–183.

Macamo, E., 2003. Nach der Katastrophe ist die Katastrophe. Die 2000er Überschwemmung in der dörflichen Wahrnehmung in Mosambik. In: Clausen, L., Geenen, L.E., Macamo, E. (Eds.), Entsetzliche soziale Prozesse. Theorie und Empirie der Katastrophen. Lit, Münster, pp. 167–184.

Madelin, M., Grasland, C., Mathian, H., Sanders, L., 2009. Das 'MAUP': modifiable areal unit – problem oder Fortschritt? (The "MAUP": Modifiable Area unit – Problem or Progress?) Inform. Raumentwick. 10/11, 645–660.

McEntire, D., 2004. Development, disasters and vulnerability: a discussion of divergent theories and the need for their integration. Disaster Prev. Manage. Int. J. 13 (3), 193–198.

Monmonier, M., 1996. How to Lie with Maps. The University of Chicago Press, Chicago.

Openshaw, S., 1984. The modifiable areal unit problem. In: Concepts and Techniques in Modern Geography, vol. 38. Geo Books, Norwich.

Pickering, A., 2001. Practice and posthumanism: social theory and a history of agency. In: Schatzki, T., Knorr Cetina, K., von Savigny, E. (Eds.), The Practice Turn in Contemporary Theory. Routledge, London/New York, pp. 172–183.

Pickles, J., 2004. A History of Spaces: Cartographic Reason, Mapping and the Geo-coded World. Routledge, London.

Raffestin, C., 2003. Statistik, Raum, Macht (Statistics, Space, Power). In: Klauser, F. (Ed.), Claude Raffestin – zu einer Geographie der Territorialität [Raffestin, Claude: Statistique, espace, pouvoir. In: Swiss Statistical Society Bulletin (47/2003), S. 7–10]. Steiner, Stuttgart, pp. 117–123.

Reichel, C., 2014. Multimediakarte Safiental. Online: http://medien.cedis.fu-berlin.de/cedis_medien/projekte/safiental/#home.

Rowe, W., 1977. The Anatomy of Risk. Wiley, New York.

Saetnan, A., Lomell, H., Hammer, S. (Eds.), 2010. The Mutual Construction of Statistics and Society. Routledge, London.

Sambah, A.B., Miura, F., 2014. Integration of spatial analysis for Tsunami Inundation and impact assessment. J. Geogr. Inform. Syst. 6, 11–22.

Taylor, J.J., 2008. Naming the land: San countermapping in Namibia's West Caprivi. Geoforum 39, 1766–1775.

Thomas, W., Thomas, D., 1928. The Child in America: Behavior Problems and Programs. Knopf, New York.

Voss, M., 2008. The vulnerable can't speak: an integrative vulnerability approach to disaster and climate change research. Behemoth. J. Civiliz. 1 (3), 39–71.

Wainwright, J., Bryan, J., 2009. Cartography, territory, property: postcolonial reflections on indigenous counter-mapping in Nicaragua and Belize. Cultural Geogr. 16, 153–178.

Wood, D., 1992. The Power of Maps. The Guilford Press, New York.

Wood, D., 2010. Rethinking the Power of Maps. The Guilford Press, New York.

Wood, D., Fels, J., 2008. The Natures of Maps. The University of Chicago Press, Chicago.

Wyss, M., 2015. Do probabilistic seismic hazard maps address the needs of the population? In: Wyss, M., Peppoloni, S. (Eds.), Geoethics: Ethical Challenges and Case Studies in Earth Science. Elsevier, Waltham, Massachusetts, pp. 239–249.

Yamin, F., Rahman, A., Huq, S., 2005. Vulnerability, Adaptation and Climate Disasters: A Conceptual Overview. Climate Change and Disasters Group, Institute of Development Studies (IDS), University of Sussex.

Chapter 22

Geoethics, Risk-Communication, and Scientific Issues in Earthquake Science

Robert J. Geller

Department of Earth and Planetary Science, Graduate School of Science, University of Tokyo, Tokyo, Japan

Chapter Outline

Abstract

I consider risk-communication and geoethics issues in earthquake science. The former basically arise due to the demand from the public and stakeholders for deterministic information about what is inherently a stochastic process. The latter basically arise due to the tendency to exaggerate claims of immediate applicability to secure or increase research funding and the incentives to provide answers regarding earthquake and tsunami hazards that will please clients and governments. A common feature of the cases considered in this chapter is the need to clearly communicate the uncertainties of the present scientific state of the art.

Keywords: Risk-communication; Earthquakes; Tsunamis; Earthquake prediction; Nuclear safety; Funding; L'Aquila.

In my view anyone who is not very confused all the time about ethical issues is out of touch with the frightful (and joyous) richness of the world. But at least being confused means that we are considering our ethical stance and that of the institutions we associate with. That's a good start.

Tom Peters, "The Pursuit of Wow!," Vintage, 1994, pp. 87–88.

Geoethics. http://dx.doi.org/10.1016/B978-0-12-799935-7.00022-8

INTRODUCTION

Geoethics is a relatively new term denoting the ethical issues that arise in geoscience. Presumably this term was inspired by bioethics, which denotes the ethical issues arising in biomedical research.

Geoscientists have well-developed and generally accepted ethical standards that govern the conduct of scientific research and its publication, which are matters concerning conduct within the scientific community (e.g., Mayer, 2015). They are also subject to guidelines for prevention of improper practices (fabrication, falsification, and plagiarism) in scientific research; most nations seem to model their definitions on those published by the U.S. National Institutes of Health (http://grants.nih.gov/Grants/research_integrity/definitions.htm). Conduct of externally funded scientific research is governed by the laws and government regulations of each nation, which tend to be similar in general, or by contracts with nongovernmental sponsors. Applied geoscientists (e.g., exploration geophysicists) are usually subject to rules for dealing with clients set out by their professional societies. Although violations occur from time to time regarding the above matters, there is little or no dispute (excepting possible gray area issues) regarding what constitutes proper conduct.

In contrast, much less attention has been paid to the question of how geoscientists should deal with the public or government on matters that involve the intersection of science and public policy. Recently, however, researchers in many fields of geoscience have found themselves embroiled in controversies regarding such questions. Controversies involving climate change and fracking are two well-known instances. In this chapter, I consider cases arising in earthquake science. These cases involve scientific, risk-communication, and geoethics issues, which are entangled in a complex way.

The cases considered in this chapter arise fundamentally due to the strong desire of the public and other stakeholders for deterministic answers to scientific questions, such as the occurrence of future earthquakes, which are inherently stochastic. This is the reason that risk-communication issues arise. Scientific issues arise because models or theories used to make forecasts of natural hazards are often subject to large uncertainties, may not have been sufficiently verified, and, indeed, may in some cases be incorrect. Geoethics issues arise due largely to the incentives that act to encourage geoscientists to make overly certain deterministic statements that may not be sufficiently supported by the present state of the art, primarily to obtain funding and to respond to the demands of their clients.

L'AQUILA-LIKE HYPOTHETICAL CASE

Much recent discussion has been prompted by the controversy surrounding the 2009 L'Aquila, Italy, earthquake (e.g., Jordan et al., 2011; Cartlidge, 2012) and the subsequent criminal prosecution of members of Italy's National Commission for the Forecast and Prevention of Major Risks. The story at first glance sounds simple: the Commission members were accused of making false assurances of safety before a magnitude (M) 6.3 quake that hit L'Aquila on April 6, 2009, which killed over 300 people. However, as one looks into the details more closely the story becomes more complicated, and the details become harder to confirm, especially for someone like me who lives almost halfway around the globe and is not fluent in Italian. So rather than discuss L'Aquila I will follow a device used in law schools, and discuss a hypothetical case (hypo, as they are called), which is loosely based on L'Aquila.

My hypo is set in the mythical town of Quakeville, a small town in California in the desert on the west side of the Central Valley near Interstate Highway 5, about halfway between Los Angeles and San Francisco. A swarm of small earthquakes, none greater than M3.5, occurs in a previously quiescent region, and a disk jockey (DJ) from the local AM radio station calls Professor Stanley Foghorn, a seismologist from the (mythical) University of California at Fresno, to ask if the residents should be worried about the possibility that a much larger quake might occur. What should Professor Foghorn tell the radio station?

Professor Foghorn is fully conversant with recent theoretical work on the way earthquake occurrence probabilities vary with time, so he quite accurately tells the DJ that there always is some chance of a quake at any time—it is never "completely safe." He further, quite accurately, tells the DJ that the swarm has increased the instantaneous probability of a large quake by a factor of, say, over 10, but that the absolute probability is still quite small. Unfortunately, the DJ, a college dropout with a liberal arts background, is incapable of following this explanation and starts yelling. "Professor, cut out the fancy double talk and just answer a straight question. Are we safe or not? Just answer 'Yes or no'." What will Professor Foghorn say? Of course what he should say is that earthquake occurrence is inherently a stochastic process, and that he is very sorry, but the present state of science does not permit a straight answer of the safe/unsafe type. But what is likely to actually happen is that in order to get the DJ off the phone the professor will say, "OK if you want a straight 'safe or unsafe' answer, I'll say 'safe'." After all, even after the swarm, the chance of a large quake in, say, the next week is still much closer to 0% than 100%, so in this sense the professor's answer is correct. The problem, of course, is that there is still some nonzero chance of a large quake. And if that improbable but not impossible event occurs, the same DJ who demanded the "safe or unsafe" answer in the first place will be calling for the professor's head on a platter.

The above hypo, in my opinion, captures the "essence of L'Aquila." The hypo does not involve any ethical issues, just issues of risk-communication. It is very hard for scientists to explain accurately to the public the facts about potentially devastating events that have a very small, but nonzero, probability of occurrence, and it is even harder to explain the uncertainties in those probabilities, even though this is highly desirable (Stein and Geller, 2012). It may also be almost impossible for scientists who are inexperienced in dealing with the media to avoid giving "safe or unsafe" answers when pressed hard by reporters under near-panic conditions.

One lesson to be learned is that, because it is very difficult to try to educate the media on what science can and cannot say when a crisis is already upon us, earthquake scientists should be making ongoing efforts to educate the public and the media under "peacetime conditions." The problem though is that turnover among reporters is rapid, and they are busy covering a variety of breaking stories, so the only time they will pay attention to earthquakes is in times of crisis. Since this discussion is taking us away from geoethics I will not pursue it further, but I do think it is necessary to recognize that risk-communication issues are also important and to separate them from geoethics issues.

FUNDING AND ETHICS

Ethical issues in every field of research arise from considerations involving funding. Much larger sums are typically available for applied research promising direct benefits to the public than for basic research. This leads scientists in every field to tend to oversell the prospects for applicability of their research (Ghiselin, 1989; pp. 74–75). For example, a U.S. government program combining research in fundamental biological science with medical applications was sold in 1971 as the "War on Cancer" (Anonymous, 1971; Sporn, 1996; Hanahan, 2014).

The dilemma is that exaggerations of prospects for immediate applicability are ubiquitous, so scientists who refuse to stretch the truth are at a disadvantage in securing funding. This situation was summed up by the pithy remark of a senior scientist to two graduate students who questioned the propriety of such exaggerations in a small-scale grant application: "You guys don't understand. First we have to get the money."

The problem is somewhat akin to the famous "prisoner's dilemma" (e.g., Nowak and Sigmund, 1993). If an individual researcher or the leaders of some given field of science unilaterally refrain from overselling the applicability of their work they will lose funding to other individuals or other fields. So even though almost everyone recognizes the ethical issues, almost no one will stop fibbing. And even if, hypothetically, the global leaders of all fields of science were to hold a summit conference and agree to stop overselling the immediate applicability of their research, the result might be an across the board drop in funding for science. So we are probably stuck with this ethical problem for the foreseeable future.

EARTHQUAKE PREDICTION IN JAPAN

Japan's national earthquake prediction project has been in existence since 1965 (see Geller (1997) for details; I have also written about this topic extensively in Japanese, but omit references here). The establishment of an earthquake prediction *research* project was first proposed by a group of leading Japanese earthquake scientists in 1962. Their proposal had a detailed discussion of the observational networks they were proposing to establish, but only a vague discussion of how data from these networks would be used to predict earthquakes—essentially they were just proposing to collect a lot of data and hope that something would turn up, which it has not to date. The initial proposal promised that 10 years after the program started they would give an answer to the question of whether or not earthquakes were predictable. That answer was due in 1975, but still has not been delivered.

The initial Japan earthquake prediction research proposal was an exaggeration on the level of the War on Cancer, but the nature of the problems escalated twice in the first 15 years of the project. First, in 1969, the word "research" was dropped from the project name, so it became the national earthquake prediction program, thereby securing increased funding at the cost of further leading the public to hold unrealistically positive expectations of what was possible.

Second, in the mid-1970s, it was mooted (based on arguments that earthquakes repeat more or less periodically, see Section 5 of Geller (1997) for details) that an M8 earthquake was imminent in the Tokai district between Nagoya and Shizuoka. (No such event has yet occurred, whereas the M9 Tohoku earthquake occurred in March 2011.) On the basis that if a network of observatories was established the government could implement a program for detecting precursors and issuing alarms, an operational earthquake prediction system administered by the Japan Meteorological Agency (JMA) was set up, and remains in force today. Under this system the Prime Minister is empowered to declare a state of emergency and order businesses, schools, roads, public transport, and many other activities in the Tokai area to cease if the JMA notifies him that "precursors" have been detected. The JMA is not tasked with advising the Prime Minister to issue alarms for earthquakes outside the Tokai district and has no legal authority or responsibility for doing so.

Even though the system for making an imminent prediction of a large earthquake in the Tokai district remains in operation today, there seems to be a tacit consensus that it will never actually be used. Even so, it seems unwise to allow this system to continue in effect, especially given the fact that, despite great progress in seismology since 1978, no reliable and accurate precursors have ever been confirmed to exist. The continued existence of the Tokai earthquake prediction system misleads the public and media into excessive belief that prediction is possible, since the government cannot openly admit that the system is being continued primarily to justify ongoing funding for public works projects (e.g., seawalls and roads) related to the "Tokai earthquake" (the budget for which far exceeds the budget for earthquake prediction) and for the JMA, and to avoid losing face.

The prediction system for the "Tokai earthquake" involves an expert committee of six seismologists who are tasked with being on call to the director of the JMA. These six scientists will be summoned to the JMA's headquarters if and when candidate precursors are observed and they will advise the JMA director whether or not to tell the prime minister to declare a state of emergency. Given the lack of scientific evidence for the existence of reliable and accurate precursors one might question whether scientists should accept appointment to such a committee, but, in practice, it is probably difficult for researchers with other connections to the government to refuse such an appointment.

Aside from the Tokai earthquake prediction operation, funding for Japan's earthquake prediction program has recently been cut back, while another program operated by "the Headquarters for Earthquake Research Promotion" (http://www.jishin.go.jp/main/index-e.html), which was established after the 1995 Kobe earthquake, has much larger funding. One of the problems with the 1965–present earthquake prediction program was that the allocation of funds among a limited number of institutions was essentially fixed, with most of the funding being used for large-scale observational projects in seismology and other fields of geophysics not directly related to earthquake prediction. I have written in detail about this problem in Japanese, but omit details here. A search of the database of stories in the Asahi Shimbun newspaper from about 1980–present found 197 articles containing both the phrases "zishin yochi" ("earthquake prediction" and "yakudatu" ("to use for the purpose of"), which demonstrates the frequency of such claims.

THE FUKUSHIMA NUCLEAR ACCIDENT

One of the major consequences of the 2011 Tohoku earthquake was the Fukushima Dai-ichi nuclear power station (NPS) accident. As is widely known, the combination of the earthquake and tsunami caused a complete loss of electric power (station blackout), which led to meltdowns and hydrogen explosions, rendering a zone with an approximate radius of 20 km around the plant uninhabitable. I have written or coauthored several articles about the accident and about nuclear safety issues going forward (including Nöggerath et al. (2011), Geller (2013a,b), Geller (2014)). Based on the discussion in these articles, let us consider the geoethics aspects of the Fukushima accident and of nuclear power in general. A classic book on seismic aspects of the nuclear safety issue (Meehan, 1986) also provides valuable insights.

The basic problem is that nuclear power plants, or, for that matter, all other engineered structures, must be designed to withstand some particular level of ground acceleration, some particular height of tsunami, and so on. On the other hand, there is no upper limit to these values imposed by Earth science. As we raise the bar higher, the probability of exceedance decreases, but it will never reach exactly zero. In other words, the probability of an earthquake- or tsunami-induced nuclear accident can be reduced by spending more money on more stringent safeguards, but it can never be reduced to zero. The decision on whether or not society is prepared to accept these risks, and, if so, at what level, is inherently political, and should, in my opinion, be made as transparently as possible.

However, what actually happened before Fukushima was both nontransparent and scientifically indefensible. A group of experts (which did not include any specialists in paleotsunami deposits) was convened by the Japan Society of Civil Engineers (JSCE). As shown clearly and explicitly in the report cited below, the members of this group had close ties to the government and/or the electric power industry. Their report (http://committees.jsce.or.jp/ceofnp/system/files/ JSCE_Tsunami_060519.pdf) was based on basically the same assumptions used in making hazard maps (i.e., that characteristic earthquakes repeat more or less periodically), and did not refer to the possibility of a magnitude 9 earthquake. Recent research has strongly questioned the characteristic earthquake/ earthquake cycle hypothesis (e.g., Kagan et al., 2012), and hazard maps based on this model perform poorly in forecasting future earthquakes (Stein et al., 2012; Wyss, 2015; see also comment by Frankel (2013); reply by Stein et al. (2013)). Furthermore, several studies (e.g., Stein and Okal, 2007; McCaffrey, 2008) pointed out that M9 quakes could occur at any subduction zone.

As discussed by Interim Report (2011, Part VI) and Nöggerath et al. (2011), the original design basis tsunami of about 3 m for the Fukushima Dai-ichi NPS was based on tsunamis known to have struck that specific site from about 1600 to 1965, so the original standard was based on the tsunami from the 1960 Chile earthquake, despite the fact that much larger tsunamis were known to have struck the coast of Tohoku farther to the north (in 1896 and 1933), and that 22 tsunamis higher than 10 m have been historically documented to have struck various parts of Japan in the past (see database maintained by the Siberian Division of the Russian Academy of Sciences: http://tsun.sscc.ru/htdbpac/). In the event, based mainly on an application of the methods presented in the JSCE report, a tsunami height of about 6 m was estimated by the plant operator, Tokyo Electric Power Company (TEPCO), as the safety basis for the Fukushima Dai-ichi NPS after 2002. As discussed by Interim Report (2011, Part VI) and Nöggerath et al. (2011), several chances to further upgrade the tsunami design basis at various points in time from 2002 to 2011 were not acted on by the plant operator or the regulator. A tsunami much higher than 6 m struck the Fukushima Dai-chi NPS on March 11, 2011, and the subsequent level-7 nuclear accident occurred. The aftermath is notable. The government (both the previous Democratic Party of Japan administration and the present Liberal Democratic Party administration) took the position that the actual size of the tsunami was unforeseeable ("so-tei-gai" in Japanese), and no one has been held responsible for the accident.

We thus can see clearly the de facto role of geoscience consultants in the nuclear industry. The JSCE expert committee made their best judgment of the methods that should be used to estimate the height of the expected tsunami. No one can know whether they may have been subconsciously influenced by their ties to the industry and government, and also no one can know whether they were picked (and other potential members rejected) for their views. However, once they promulgated a method for obtaining the value for the tsunami design basis, this somehow was treated by the government and the nuclear power industry as an ironclad safety standard—it was somehow assumed that merely by ensuring the Fukushima Dai-ichi NPS met the 6 m or so tsunami standard obtained by TEPCO using the JSCE methodology with input parameters chosen by TEPCO (which it probably did), that tsunami safety was absolutely assured. Furthermore, the design basis estimated using the JSCE methodology became, in practice, an ironclad alibi, as any larger tsunami was treated as an act of God.

The events leading up to the Fukushima accident illustrate a more general theme, namely the division of responsibility between engineers and their geo-science consultants. The latter play a vital role in studies of natural hazards safety issues involved in nuclear power plants and other key infrastructure, but it should be clearly acknowledged by all parties that responsibility for safety issues lies with the plant operator, not the consultants.

Looking back at the Fukushima Dai-ichi accident, we can see the familiar themes of risk-communication issues and geoethics issues. A tsunami design basis is no better than the reliability of the earthquakes assumed as the tsunami sources, so it should have been obvious to everyone involved in the design and safety review process that any promulgated value for the design tsunami was just a guess, and that there was a finite (and nonnegligible) probability of exceedance. The power company and government should have made contingency plans for exceedance, but did not. Turning to the JSCE panel itself, its members should have, but did not, made clear in their report the degree to which their estimates of the tsunami design basis obtained using their methodology were subject to uncertainty. Finally, the various members of the JSCE committee should not necessarily have been disqualified by their close ties to the government and nuclear power industry, but it would have been highly desirable to have also included other members who were not so closely connected to the nuclear/government establishment.

As discussed above, the Fukushima accident also raises fundamental scientific questions. The plant operator based its tsunami countermeasures on past tsunamis at the site, rather than much larger tsunamis at tectonically similar sites (subduction zones) inside and outside Japan. Since the Earth's age is 4.6 billion years but we only have at most on the order of 1000 years of historical data, and since even paleotsunami data go back perhaps at most 10,000 years, it might be wise to trade space for time by averaging over tectonically similar environments throughout the world. This type of thinking might also be applicable to hazard estimates for the Central and Eastern U.S. For example, rather than placing a "bullseye" around New Madrid, it might be more appropriate to average this over the entire Central and Eastern U.S. (Stein, 2010).

CONCLUSIONS

As shown in this chapter geoethics issues arise in many contexts in earthquake science. In many cases the issues arising are not pure geoethics issues, but also involve thorny scientific and risk-communication aspects. Three cases are considered in detail in this chapter. The first was a "L'Aquila-like" scenario in which the issue was how seismologists should inform the public about the likelihood of a possible imminent large earthquake when a swarm of small earthquakes is occurring. The second was the issues involved in securing funding for earthquake prediction research and in ongoing government programs in Japan for operational earthquake prediction in the Tokai region. Finally, the issues involved in the tsunami countermeasures at the Fukushima Dai-ichi NPS were discussed.

A common theme in these cases is that the public, government, and industrial clients have a strong desire to receive specific and deterministic information on future events (earthquakes and tsunamis) that the current scientific state of the art cannot provide. Under these circumstances it behooves scientists to clearly communicate the uncertainty of their statements regarding future natural hazards. It also behooves clients (plant operators and governments) to obtain advice from a broadly representative sample of the opinions of the geoscience community rather than only a limited subset. Finally, while advice from geoscience consultants is important, all stakeholders must recognize that it is only advice and is uncertain, and that the final responsibility for safety issues rests with the plant operator. Similarly, the public and government must recognize that the fact that a plant meets some specific design level does not imply "absolute safety," and that in the end the decision on which risks to accept and which to reject is inherently political.

REFERENCES

Anonymous, 1971. Nixon and Kennedy peddle rival cancer cures. Nature 229, 590–592.

Cartlidge, E., 2012. Aftershocks in the courtroom. Science 338, 184–188.

Frankel, A., 2013. Comment on "Why earthquake hazard maps often fail and what to do about it". Tectonophysics 592, 200–206.

Geller, R.J., 1997. Earthquake prediction: a critical review. Geophys. J. Int. 131, 425–450.

Geller, R.J., 2013a. A Seismologist's View of Nuclear Safety Issues in Japan — Part I. Forum on Energy http://forumonenergy.com/2013/08/15/robert-j-geller-a-seismologists-view-of-nuclear-safety-issues-in-japan-part-i/.

Geller, R.J., 2013b. A Seismologist's View of Nuclear Safety Issues in Japan — Part II. Forum on Energy http://forumonenergy.com/2013/09/04/robert-j-geller-a-seismologists-view-of-nuclear-safety-issues-in-japan-part-ii/.

Geller, R.J., 2014. Back to the Future: Restarting Japan's Nuclear Power Plants. Nikkei Asian Review http://asia.nikkei.com/Viewpoints/Perspectives/Back-to-the-future-Restarting-Japans-nuclear-power-plants.

Ghiselin, M., 1989. Intellectual Compromise—The Bottom Line. Paragon, New York.

Hanahan, D., 2014. Rethinking the war on cancer. Lancet 383, 558–563.

Interim Report, 2011. Investigation Committee on the Accident at the Fukushima Nuclear Power Stations of the Tokyo Electric Power Company, December 26, 2011. http://www.cas.go.jp/jp/seisaku/icanps/eng/interim-report.html.

Jordan, T.H., Chen, Y.T., Gasparini, P., Madariaga, R., Main, I., Marzocchi, W., Papadopoulos, G., Sobolev, G., Yamaoka, K., Zschau, J., 2011. Operational earthquake forecasting: state of knowledge and guidelines for utilization. Ann. Geophys. 54, 315–391.

Kagan, Y.Y., Jackson, D.D., Geller, R.J., 2012. Characteristic earthquake model, 1884–2011, R.I.P. Seismol. Res. Lett. 83, 951–953.

Mayer, T., 2015. Research integrity: the bedrock of the geosciences. In: Wyss, M., Peppoloni, S. (Eds.), Geoethics: Ethical Challenges and Case Studies in Earth Science. Elsevier, Waltham, Massachusetts, pp. 71–81

McCaffrey, R., 2008. Global frequency of magnitude 9 earthquakes. Geology 36, 263–266.

Meehan, R., 1986. The Atom and the Fault—Experts, Earthquakes, and Nuclear Power. MIT Press.

Nöggerath, J., Geller, R.J., Gusiakov, V.K., 2011. Fukushima: the myth of safety, the reality of geoscience. Bull. At. Sci. 67 (5), 37–46.

Nowak, M., Sigmund, K., 1993. A strategy of win stay, lose shift that outperforms tit-for-tat in the prisoner's dilemma game. Nature 364, 56–58.

Sporn, M.B., 1996. The war on cancer. Lancet 347, 1377–1381.

Stein, S., 2010. Disaster Deferred: A New View of Earthquake Hazards in the New Madrid Seismic Zone. Columbia Univ. Press, New York.

Stein, S., Geller, R.J., 2012. Communicating uncertainties in natural hazard forecasts. EOS Trans. Am. Geophys. Union 93, 361–372.

Stein, S., Okal, E.A., 2007. Ultralong period seismic study of the December 2004 Indian Ocean earthquake and implications for regional tectonics and the subduction process. Bull. Seismol. Soc. Am. 97, S279–S295.

Stein, S., Geller, R.J., Liu, M., 2012. Why earthquake hazard maps often fail and what to do about it. Tectonophysics 562–563, 1–25.

Stein, S., Geller, R.J., Liu, M., 2013. Reply to comment by Arthur Frankel on "Why earthquake hazard maps often fail and what to do about it". Tectonophysics 592, 207–209.

Wyss, M. 2015, Do probabilistic seismic hazard maps address the need of the population? In: Wyss, M., Peppoloni, S. (Eds.), Geoethics: Ethical Challenges and Case Studies in Earth Science. Elsevier, Waltham, Massachusetts, pp. 239–249.

Chapter 23

Extreme Sea Level Events, Coastal Risks, and Climate Changes: Informing the Players

Eduardo Marone,[1,2,3] Juliane C. Carneiro,[1] Marcio M. Cintra,[1]
Andréa Ribeiro,[1] Denis Cardoso[1] and Carol Stellfeld[2]
[1]Graduate Program in Coastal and Oceanic Systems, Federal University of Paraná, Brazil;
[2]Graduate Program in Geology, Federal University of Paraná, Brazil; [3]International Ocean
Institute – OC, Federal University of Paraná, Brazil

Chapter Outline

Abstract

Floods are probably the most devastating natural hazards that the local society faces nowadays, being a problem that concerns both the local population and the management authorities. With the expected sea level rise during the next years and tropical storms that would become stronger and more frequent, the scenarios of local impacts of sea level rise and storm surges in coastal areas need to be addressed. In this chapter, we present some results of our studies related to coastal floods in Southern Brazil and the subsequent setups according to the main scenarios for 2099 of the Intergovernmental Panel on Climate Change. The objective of this work is also to show different language approaches to communicate these results to all the players involved in these problems: the decision makers and the general public, not only academics. The results are shown using the traditional scientific format accompanied by "translations," located in boxes, in a more simple language. The translation we mentioned is not just the need of putting in plain words our scientific results when communicating outside the academic world, nor just a need of not being "cryptic." This reaches beyond from the obvious, because we, scientists, had not been formally trained to explain things other than in a scientific language. We hope this chapter will also

Geoethics. http://dx.doi.org/10.1016/B978-0-12-799935-7.00023-X

serve as an alert for training young scientists in the use of appropriate language and the need of being socially sympathetic in doing so. One of our ethical responsibilities as geoscientists is to ensure that our discoveries are properly disseminated, particularly when they could affect the livelihood of other human beings and/or the environment. By doing so, we construct bridges between academics and stakeholders, because the potential solutions for the problems that could affect the region in the near future are in the hands of this social component.

Keywords: Sea level rise; Climate changes; Informing the public; Geoethics.

INTRODUCTION

Occurring commonly in Brazilian coastal and continental areas, floods are probably the most devastating natural hazards that the local society faces nowadays, being a problem that concerns both the local population and the management authorities. With the expected sea level rise during the next years and tropical storms that would become stronger and more frequent, causing loss of lives, environmental damage, and socioeconomic stress, the scenarios of local impacts of sea level rise and storm surges in coastal areas need to be addressed.

In this chapter, we present some results of our studies related to coastal floods in southern Brazil and the subsequent setups according to some of the Intergovernmental Panel on Climate Change (IPCC)'s scenarios developed for the year 2099. The results are shown using the traditional scientific format and followed by "translations" located in boxes. Although the "scientific" results are being presented in the traditional academic form, the main objective of this work is to show them with different language approaches to communicate with all the players involved in these problems: the decision makers and the general public, not only academics.

To illustrate the need for "translation" for other players, we would like to mention a real case of a troublesome misunderstanding caused by us, scientists, in the coastal society of the Paraná state, Southern Brazil. Some years ago, one of our colleagues at our university in the capital city, a much-respected scientist, informed the population through a press declaration that, on a given day, "we will experience the highest astronomical tide of the century." That statement (scientifically true and accurate) caused some panic in the coastal communities, which consulted us, at the coastal research campus of the same university, if they were supposed to take all their belongings and move up to the hills.

Thus, the translation we mentioned is not just the need of putting in plain words our scientific results when communicating outside the academic community, nor just a need of not being "cryptic." It goes far away from the obvious, because we, scientists, had not been formally trained to explain things other than in a scientific language (usually).

To create the "translation" introduced in boxes along the "scientific" text, we used the Up-Goer Five Text Editor (http://splasho.com/upgoer5/), which allows writing text using only the 1000 most used English words. We allowed ourselves to use a maximum of five other simple words per box not present at the Up-Goer dictionary, not considering geographical names or units in the count. That was necessary because simple words such as sea, beach, sand, and storm are not among the 1000 words present in the Up-Goer dictionary, and they are simple enough anyhow. On the other hand, the non scientific public we targeted speaks Portuguese, not English, and we do not have an Up-Goer tool for that language. Anyhow, each box was also produced in Portuguese, as simply as possible, to disseminate our results to the local community.

We hope this chapter will also serve as an alert to the need of training young scientists in the use of appropriate language and the need of being socially sympathetic in doing so, not just publish relevant outcomes in the academic world, in spite of the PoP (Publish or Perish) environment. In particular, this chapter was the result of a class and fieldwork with a group of graduate students of Oceanography and Geology, in both MSc and PhD levels, who we hope will develop their careers with this extra skill.

Why?

In older times, the sea was up more than 100 m, and many times higher than today. That was, usually, because the ice over land disappeared as an answer of the world to a warmer air. Today, the most important study group of people in the world, called the IPCC, is saying that, for the year 2099, the sea possible grow figures could be of the order of half a meter or higher than at present time. The reason for this raise is because we are warming the air by burning very old things found deep in the ground.

Even if we do not know the very right figures of this grow for any single place, IPCC gives us some numbers that can be used to have an idea of the problem we may face in 2099 at any given place.

Many people (and other life things) lives, works and have fun at the sea side, and beaches of Paraná and some places of Santa Catarina, and will be hurt by this sea raise (SLR).

In this work we would like to show to these people that there is a problem at these places and that we need to be ready to face them.

The idea was thought up by the very bright people of IPCC, whose language makes it harder for normal people to 'get' it. Sometimes though (and this is the worst) it is just because no one thought to explain it in the right way.

We will try to make everyone understand it, saying how much the sea could grow by 2099 at these places, showing it is not too hard to study with good figures, and being clear about what to do. By knowing the right numbers, we could be ready with plans, which will help us to make the problems less important, allowing us to continue to enjoy the *seaside*.

Coastal erosion is a global problem: at least 70% of sandy beaches around the world are recessional (Bird, 1985). There must be a worldwide cause for such ubiquitous beach erosion. The three possible candidates are sea level rise (SLR), change in storm climate, and human interference. Erosion of sandy beaches is a physical process that involves a redistribution of sand from the beach face to offshore and it is the most common effect of coastal storms (Zhang et al., 2004).The scientific interest in the study of climate change and its associated impacts has increased significantly in recent decades (Todisco and Vergni, 2008). Much of this progress is due to the efforts of the IPCC, which is the main international group of experts for monitoring and evaluating the consequences that climate change will bring in the future (Meehl et al., 2007).With the dissemination of IPCC's Fourth Report in 2007 (IPCC, 2007), the discussion of global climate change has acquired great proportions, leaving the strictly scientific sphere, not without some troubles. Showing projected scenarios for the end of the century, the best estimate is of an overall increase between 0.18 and 0.59 m for mean sea level (MSL) at the end of the twenty-first century (Meehl et al., 2007; Chust et al., 2010). However, it is important to consider a high regional variability of these rates, since there are a number of other dynamic elements and associated geomorphological changes driving the ocean levels and acting at different spatial and temporal scales.The projections of SLR and its applications in regional impact studies are an important source of information for flood risk assessment and socioenvironmental effects on coastal habitats and populations (Marengo and Valverde, 2007). In addition, the risk assessment in areas that could be affected because of the rise in MSL is potentially useful for risk management and local management of coastal environments, where most of the population lives.The Brazilian shore has been studied for the past 50 years and the data set produced by these observations indicates a trend of SLR of 4 mm/year (de Mesquita et al., 2005; Marengo and Valverde, 2007; Santos et al., 2010), thus falling outside of the IPCC figures. In the southeastern and southern regions (where Paraná and Santa Catarina states are located), an increase in relative sea level would be enough to potentially flood significant areas of coastal plains, causing changes in the dynamics of sandy beaches and also affecting the biological zonation of marshes and mangroves, potentially promoting their total elimination (Arasaki et al., 2008). The combined effects of climate conditions and natural evolution of coastline may be visible in coastal regions using the proper tools.The issues of global warming scenarios concerning erosion and coastline position are related, not only because of the projected SLR but also because of the combined effect of the long-term rise of water levels, which are secularly increasing, and coastal storms, which are also expected to increase in frequency and power, promoting further erosional impacts.Due to the global character of the IPCC results, monitoring and estimating regional projections of SLR are extremely important to elaborate local management plans and take preventive mitigation actions against the potential impacts that coastal areas may suffer in the future.

COASTLINE CHANGES IN RESPONSE TO SEA LEVEL RISE

The scientific objective of this work is to investigate the possible changes in coastal environments of Paraná and northern coast of Santa Catarina states arising from the increase in the MSL (based on some scenarios predicted by IPCC), as well as the morphodynamics of sandy beaches. In addition, considerations regarding the increase in intensity and frequency of extreme events, as well as human local interference, have to be taken into account.

The Area Studied

The coasts of Paraná and North Santa Catarina (Figure 1) are characterized by small extensions compared to other Brazilian states, with sandy shores facing the open ocean and two well-developed estuaries. It is possible to identify three main types of shores on the coast of Paraná and northern Santa Catarina: estuarine or protected (~1300 km), open sea or oceanic (~60 km), and estuarine mouths (~100 km) (Angulo and Araújo, 1996). The oceanic coast has approximately a rectilinear form and northeast orientation and can be divided into three main areas: the northern one in Superagüí/Ararapira, with length around ~14 km; a central one between Pontal do Sul and Caiobá, with ~31 km; and one to the South, between Guaratuba and Saí bar, with ~11 km. The coast is characterized by the presence of medium to fine sand beaches, of dissipative type, and predominantly dominated by waves and drift currents, showing seasonal erosion and sedimentation processes (Angulo and Araujo, 1996).

To the south of this area lies the Bay of Guaratuba (area 60 km²), with shallow sandy areas in its mouth as part of an ebb tide delta (Bucci, 2009). Most activities there are linked to leisure and nonindustrial fishing. At the north-central portion is located the Paranaguá Estuarine Complex (PEC), with an approximate surface area of 600 km² (Lana et al., 2000) and where two main features can be observed. At the north–south oriented branch of PEC are located the bays of Laranjeiras and Guaraqueçaba, with a use similar to that of the Bay of Guaratuba; at the east–west axis are located the bays of Paranaguá and Antonina, where in addition to the leisure-time activities and artisanal fisheries, intense port and some industrial activities are undertaken.

The coasts facing the open ocean are mostly sandy beaches, with very low human occupation at the northern part of the studied area (Superagüí/Ararapira) and a pristine ecosystem with protected areas. The southern sandy beaches facing the open ocean (from Pontal do Sul to Babitonga) are mostly disorderly occupied for human leisure activities and housing, with an almost continuous line of touristic villages.

Where?

The study was done in Paraná and north Santa Catarina coasts, Brazil (Figure 1). It is possible to find three types of sandy coasts: not open (~1300 km),

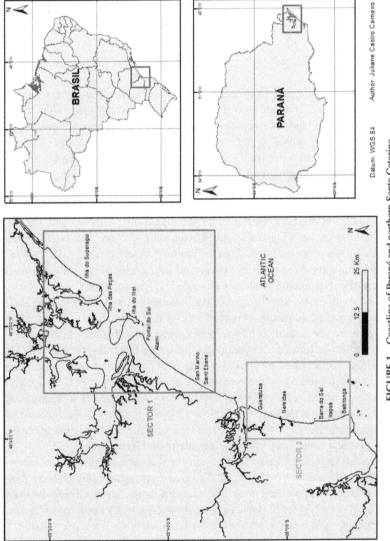

FIGURE 1   Coastline of Paraná and northern Santa Catarina.

open (~60 km), and mouths (~100 km). The open and mouth coasts are those focused on here because they are the ones to be hurt the most being open to the power of waves and storms. The coast runs northeast with three different areas: the Superagui part at north, with a reach of around ~14 km; the middle part, between Pontal do Sul and Caiobá, with ~31 km; and the last between Guaratuba and Saí bar, with ~11 km, at South.

The houses along the coast are very different from place to place, having areas with many people, and grow figures, and others still without human use. Besides, at the South coast of the State of Paraná the houses are too close to the coast or even on the beach, with people not considering enough the normal changes that use to happen on those places by normal causes.

Our work was done cutting the area in two. *Sector 1* has no great human uses with well kept grounds made of *dunes* and *marshes* at *north*, and, in the middle, some small towns by the *beach*. On the other hand, *Sector 2* at *Paraná* and on the *north* of *Santa Catarina*, we have middle towns.

The Local Dynamic

The wave systems and storm surges that hit the coast are linked mostly to the atmospheric generation centers in distant oceanic areas of Southern Atlantic Ocean, regardless of the directions of local winds. On the coast of Paraná, the predominant waves are from the south and south-southeast, shaping a north-oriented littoral drift (Veiga, 2005; Angulo et al., 2006). The coast is dominated by semidiurnal microtides, with spring tidal ranges up to 2.0 m. The spring tidal currents also have a great potential to transport fine sediments that are exported, for instance, from the PEC to the shallow platform (Noernberg, 2001).

The coasts around estuary mouths are dominated by waves and tidal currents and have large temporal variability, with advances and retreats of the coastline related to minor changes in the configuration of the tidal deltas (Angulo and Araújo, 1996). Weather events of great intensity (extreme storm surge events) are important factors in the modification of the coast and the beaches adjacent to the mouths (Lamour et al., 2003).

The land occupation along the coastlines is very irregular, with some areas being heavily inhabited and others not being used for any human activity. From the southern premises of PEC to the northern coasts of Santa Catarina, the coast is limited only by few natural barriers (such as bays, channels, and mangroves) and its urbanization presents heterogeneous characteristics, with the presence of modern resort areas and spots of permanent housing for the low-income population, both in the process of intensification (Moura and Werneck, 2000). The southern coast of the state of Paraná is characterized for a human land use too close to the coastline or even on the beach, promoting the destruction of the foredunes. It is also impacted by construction works, dredging, and urban infrastructure that do not consider adequately the natural dynamics.

METHOD

The basic assumptions for this work were

1. The main scenarios of the IPCC for the increase of MSL are the best estimates available at the time.
2. The coastline will vary mainly for two reasons:
   a. The vertical height variation of the water column (MSL).
   b. Morphodynamic processes derived from the variation of the MSL.
3. The actual position of the coastline was defined following the main debris lines, which correspond to the mean distance the seawater penetrates inland on mean extreme events.

Other causes that may contribute significantly to the coastline variations (such as changes in freshwater flow and continental sediments, in atmospheric dynamics, and in the currents and waves, arising from climate change) have not been the object of this preliminary study and, even if they may eventually make a significant contribution, they were not considered because they are difficult to estimate with the proper error intervals. As a result, estimates of this work should be considered as a baseline, a starting point, for creating scenarios related to the variation of the Paraná and North of Santa Catarina coastline, and not as a prediction.

We consider both static and dynamic progression of the coastline. The first one responds to the single event of the sea level rising and allows calculating the distance the sea will penetrate landward using a very simple trigonometric relationship based on the slope angle α of the beach:

$$R' = S \tan \alpha$$

R'=static indentation of the coastline due to SLR (meters)
S=SLR (meters)

However, the rise of the sea level leads to dynamic consequences, since the erosional processes promoted by waves could also produce further landward progression of the coastline. Thus, to determine the full width of the range of potential flooding in the sandy beaches, we considered the effects of the high MSL rise under both assumptions (a) and (b) above, by applying the Bruun's law (Bruun, 1954, 1962).

Bruun (1954, 1962) first constructed a simple two-dimensional cross-shore model based on the equilibrium beach profile concept to estimate the erosion of sandy beaches in response to rising sea level, based on the following assumptions (Zhang et al., 2004):

1. The equilibrium active beach profile is defined as an idealized statistical average profile over seasonal and storm-induced fluctuations, and includes an underwater part that makes up most of the profile.
2. If other conditions remain unchanged and sea level rises, the active beach profile will achieve equilibrium with the new sea level by shifting landward and upward.

3. Sediment eroded from the upper beach is equal to sediment deposited on the nearshore bottom.
4. The increase in elevation of the nearshore bottom resulting from deposition of sediments from the beach face and dune is equal to the rise of sea level.

The derivation presented by Bruun is based on the conjecture that the resulting indentation is not dependent on the shape of the beach profile, the point of intersection of the new and old profiles, or the position of the seaward slope angle of the offshore bar (Zhang et al., 2004).

Thus, the Bruun model will indicate the erosive advance (R) of the coastline according to the equation:

$$R = \frac{SLG}{H}$$

where

R = dynamical indentation of the coastline due to SLR (meters)
S = SLR (meters) = (MSL variant)
L = length of the active profile (meters)
H = height of the active profile (meters)
G = ratio of eroded material that remains in the active profile

The height of the active profile (H) was determined by adding the height of the last active feature (the top of the beach ridge) with the depth of closure (DOC) (taken as constant and equal to 7.25 m for the area and its wave climate). The DOC is the water depth (typically around 10 m) at which ocean surface waves no longer significantly transport bottom sediments. In addition, L is the horizontal distance between the maximum elevation of the active profile and the DOC. It is appropriate mathematically to use the Bruun model to compute the underlying long-term beach response to SLR when the rise is small compared to the active beach profile height (H) (Zhang et al., 2004).

The Bruun rule along open-ocean coasts is an approximation. It is recommended not to be applied, or to be used carefully, in regions of shoreline change induced by gradients in longshore sediment transport and variations in the active beach profile slope. Shoreline segments influenced by tidal inlets and coastal engineering projects (such as seawalls or beach replenishment) obscure the relation between SLR and beach erosion because such segments depart radically from the assumptions of the Bruun model (Zhang et al., 2004).

For further details, a review of the Bruun rule and other extended models concerning the response of beaches to SLR has been presented by SCOR (1991).

The actual coastline was established based on the mean range of the debris lines resulting from the action of the "mean" extreme events affecting the

MSL (mean storm surge level). The high water line (HWL) used by other authors (Zhang et al., 2004) as the shoreline indicator was not the approach in this study because the debris lines we used are clearer in the images we analyzed or at field, and do not depend on higher frequency phenomena (tidal cycle and spring and neap tides, for instance). Mainly, we assume the debris lines integrate better the reach of the seawater landward due to mean extreme events.

Finally, high-resolution satellite imagery (GeoEye IKONOS, 1-m resolution) were georeferenced on Arcgys using the World Imagery Extension. Only the best quality images are presented for brevity, defining the actual coastline as in assumption (c) above. The projected coastlines were plotted over the actual images.

How?

We begin this work accepting that the *IPCC's* book is our best guess to what is going to happen with the *coast* on the next years. The *coast* line will shift its position because of changes of the sea level (which position at present is the point that *sea* water reaches during high water attacks), that will result in other changes that happen at the *coast*. Because of this we used a form called *Bruun*, that considers how much *sand* is taken off and brought to the *beach* when the *sea* level changes.

We use this simple way to know how much the *sea* will cover the land around year 2099, using *sand* type, *beach* reach, edge and climb, and the *sea* position going up (SLR) at Paraná and some Santa Catarina *beaches*. In this way we were able to get the number that the *sea* can reach toward land for a given *sea* grow. We used three *sea* grow numbers following the work of *IPCC* (small SLR, mean SLR, and highest SLR). Following these numbers we can guess how high in land the *sea* can be in 2099. Final figures were done using *satellite* pictures in order to make it more easy to see the problem we could have in some years because of *sea* grow. On the final pictures one can see the actual coast line and how it should get with the lowest and bigger sea level changes expected by IPCC.

IPCC SCENARIOS

Three distinct scenarios were chosen from the IPCC findings, selected considering the relative elevation of MSL for the end of the century, with different rates of atmospheric warming and increases in MSL. The projections were obtained from the studies presented by the IPCC in its 2007 report, which confirms the trends, incorporates new knowledge to the previous document, and presents as a special feature the expansion of certainty from 60% to 90% on the SLR and on the resulting impacts (Nacaratti, 2008).

The first scenario chosen was the B1, which represents a more optimistic expectation about the greenhouse gas emissions and the relative elevation of

MSL, and estimates an increase of 0.18–0.38 m for the year 2099. The scenario B2 was also selected because it represents a more realistic projection, consistent with the actual processes nowadays, with estimated rates of relative elevation from 0.20 to 0.43 m. The last scenario was the high-emission scenario, or the worst-case scenario regarding the relative elevation of MSL (A1FI scenario), with an estimated rate of 0.26–0.59 m.

What IPCC Says?

IPCC, a group of very bright people, has been studying the forces of *climate changes* all over the world. In the middle of these changes, the warming of the air will finish, with others things, in the rise of the *sea* reach, all around, up to 4 mm/year. This bright people do not make *predictions* as people who try to see what will come from cards, cause we do not know the changes that well. Instead, they build what we call *scenarios*, which are ways to look at the times that will come with up to 90% of trust for the figures on 2099. As the warming of the air could be different standing on how humans will burn their stuffs in the next years, *IPCC* said that if we burn less, the raise would be up to 32 cm (*scenario B1*). If we continue to burn the stuffs at the same hurry as today, the sea could raise up to 42 cm (*scenario B2*). But, if we do not take care of the world, burning more and more every day, the worst case *scenario* (*A1F1*) suggest a grow of up to 58 cm. Also, they say that the waves and *storms* will become more usual and hard until 2099.

We used these *scenarios* (*B1*, *B2*, and *A1FI*) to take a look on what could happen, with a 90% of trust, on the *coasts* of *Paraná* and *North* of *Santa Catarina*, by 2099.

SCIENTIFIC RESULTS

Field measurements and data compilations of topography and slope along the coast were used to obtain the projections about potential scenarios of indentation of the coastline due to the variations of the MSL following the main IPCC scenarios and the coastal erosion dynamic.

In the very few regions that did not present historical data, field survey was carried out with precision level (Leica Geosystems, Jogger 20/24 model).

The results found for the length of the coastal indents (minimum and maximum) were calculated by applying the law of Bruun using the historical and new field data, according to the three projections of the IPCC MSL rise above, and are presented in Table 1.

We mostly analyzed shoreline sections where longshore sediment transport is in equilibrium (i.e., there is no gradient) and areas with human interferences on natural morphodynamics were not included. Except for very few beaches in the estuary mouths in the study area, most of them are in the requested long-term equilibrium condition and even those have shown a seasonal equilibrium (Angulo and Araújo, 1996), which allows us to apply the Bruun rule to them for longer time scales.

TABLE 1 Indents of the shoreline on the Coast of the State of Paraná and north of Santa Catarina.

			Indentation of Coastline According of IPCC Scenarios (B1, B2, A1FI) (m)					
			B1 (S_{min} = 0.18 m S_{max} = 0.38 m)		B2 (S_{min} = 0.2 m S_{max} = 0.43 m)	A1FI (S_{min} = 0.23 m S_{max} = 0.51 m)		
	H (m)	L (m)	B1 Min	B1 Max	B2 Min	B2 Max	A1FI Min	A1FI Max
Sector 1								
Ararapira	11.25	197.20	3.16	6.66	3.51	7.54	4.03	8.94
Peças Island	9.22	185.40	3.62	7.64	4.02	8.64	4.62	10.25
Mel Island: Farol	9.25	278.90	5.43	11.46	6.03	12.97	6.93	15.38
Mel Island: Praia Grande	10.45	229.00	3.94	8.33	4.38	9.42	5.04	11.18
Mel Island: Isthmus	9.31	533.00	10.31	21.76	11.45	24.62	13.17	29.20
Mel Island: Ponta Oeste	9.49	369.90	7.01	14.81	7.79	16.75	8.96	19.87
Pontal do Sul	9.55	437.70	8.25	17.42	9.17	19.71	10.54	23.37
Atami	8.75	1617.00	33.26	70.22	36.96	79.46	42.50	94.25
San Marino	8.25	1818.00	39.67	83.74	44.07	94.76	50.68	112.39
Saint Etiene	9.82	163.90	3.01	6.34	3.34	7.18	3.84	8.52
Sector 2								
Guaratuba	11.15	194.80	3.14	6.64	3.49	7.51	4.02	8.91
Nereidas	12.25	175.60	2.58	5.45	2.87	6.16	3.30	7.31
Coroados	10.95	251.00	4.13	8.71	4.58	9.86	5.27	11.69
Saí bar	11.55	223.60	3.48	7.36	3.87	8.32	4.45	9.87
Itapoá	9.96	223.80	4.04	8.53	4.49	9.66	5.17	11.45
Figueira do Pontal	9.38	169.20	3.25	6.86	3.61	7.76	4.15	9.20
Babitonga	10.03	76.90	1.38	2.91	1.53	3.30	1.76	3.91

The dotted red line indicates the values of the more conservative and the most extreme coastline setups for B1, B2, and A1FI scenarios of IPCC.

The following figures (Figs 2-12) show the projected and compared coastlines for 2099 according to B1 (optimistic scenario) and the maximum values of the A1FI (worst-case scenario), and the actual coastline for some selected places. Here, we compared both situations: the most conservative and the most aggressive in relation to the advance of the sea toward the land (B2 scenario falls always between the other two). Thus, projections indicate the minimal increase in the more optimistic scenario and the maximum increase in the most pessimistic, determining a range of potential variation of the coastline where, according to the IPPC error estimates, the coastline position at the year 2099 in the area of study must be within a 90% of certainty level.

We divided the area into two different sectors (Figure 1). Sector 1 comprises the Restinga de Ararapira, the Peças and Mel Islands, and the southern premise of PEC (Pontal do Sul, Atami, Saint Etiene, and San Marino). The less urbanized portion of the Paraná coast in the north presents maximum setups ranging from 3 to 9 m at Ararapira, larger variations on the isthmus of Mel Island (10–30 m), and the maximum values on the southern coasts of PEC (from a minimum of ~8 m to a maximum setup of ~112 m). Thus, the northern part of Sector 1, a sparsely populated region covered by well-preserved native vegetation of dunes and marshes, will suffer modest advances of sea level (Table 1).

Considering the islands on the mouth of PEC, the indents on Mel Island indicate that the main morphological change would be the disappearance of the isthmus connecting the east–west portions of this island (Figure 2). In the northern part of Sector 1, in particular at the Ararapira and Peças Islands, is found a well-preserved native forest and low human demography, with the exception of the Peças village, which would be significantly affected by the advance of the sea since the houses are on the water line. Therefore, in those areas, these coast setups, as the Praia Grande case (Figure 3), are likely to be not so critical due to the absence of urbanization and the presence of fixed structures near the current coastline, except in the Peças village and the Mel Island isthmus.

On the southern part of the PEC mouth, i.e., at Pontal do Sul (Figure 4), Atami (Figure 5), Saint Etiene (Figure 6), and San Marino (Figure 7), we have the greatest advances of the sea over the coastal plain, ranging from ~8 m (minimum at Pontal do Sul for IPCC/B1) to ~112 m (maximum at San Marino for IPCC/A1FI), with decreasing values at Saint Etiene, which presents a greater beach slope resulting in smaller indentations (3–8 m).

The values of the maximum setup at Atami (Figure 5) were ~70 m for scenario B1 and ~95 m for scenario A1FI. Although the values were high in relation to those of the other beaches, the presence of an extensive sandbank in that beach could ensure less trouble, since the main urban structure (streets and homes) is far enough from the MSL advances presented in these projections. The beach of San Marino (Figure 5) will retreat from 40 to 112 m for the scenarios B1 and A1FI, respectively. Note that even with the more conservative projection (scenario B1), the current beach area will be submersed and when the most extreme projection (A1FI) is taken into consideration, the urban range nearest the sea (composed of streets and several buildings) will be destroyed by the advance of the coastline, while extreme storm surges will affect further inland.

Sector 2 (Figure 1) is composed of touristic beaches on the southern coastline of Paraná (Guaratuba, Nereidas, Coroados, and Saí bar) plus a few beaches on the northern coast of Santa Catarina (Itapoá, Figueira do Pontal, and Babitonga) (Table 1).

In Sector 2, the beaches Guaratuba, Nereidas, Coroados, the Saí bar, Itapoã, and Babitonga present modest setups (Figures 8–12, respectively), with the lowest value in Nereidas (3–7 m). With these setups (maximum <11 m in Coroados or Itapoã), these beaches do not present a high flood risk in relation to the advancement of the MSL toward the land, although storm surges could impact occupied areas. These areas have a strong urban presence along the beaches, but the declivity of their profiles does not allow the sea level to reach the buildings.

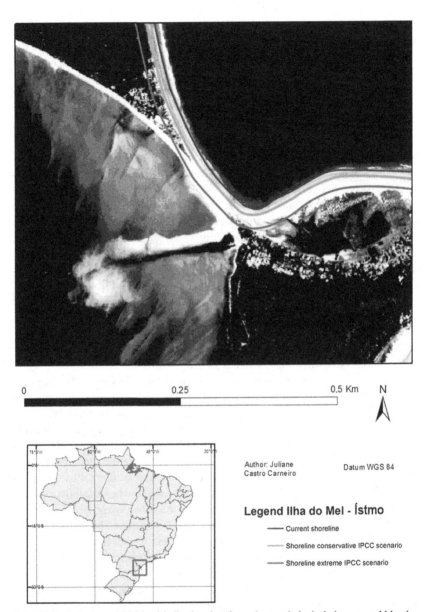

FIGURE 2 Indents on Mel Island indicating that the main morphological change would be the disappearance of the isthmus. Note that lines are presented only for the northern side of the isthmus. Sector 1. Blue line indicates the actual coastline, green line the expected coastline under IPCC scenario B1, and red line the coastline under A1FI extreme sea level scenario.

However, it must be considered that such modifications could compromise the morphological structure of the beach system due to the absence of sandy preserved areas (dunes or marshes), which allows the maintenance of the system in cases of severe beach erosion.

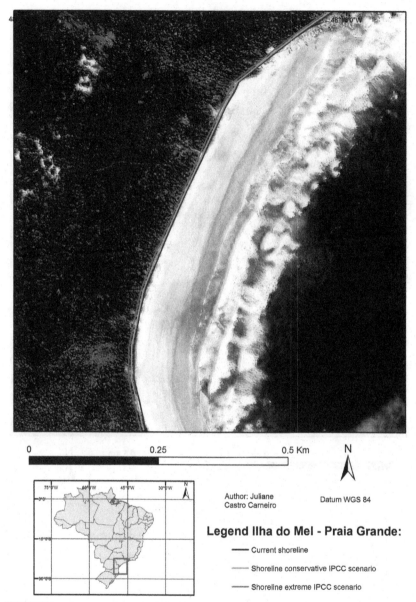

Author: Juliane
Castro Carneiro

Datum WGS 84

Legend Ilha do Mel - Praia Grande:

—— Current shoreline

—— Shoreline conservative IPCC scenario

—— Shoreline extreme IPCC scenario

FIGURE 3 Indent scenarios on Praia Grande, Mel Island. Sector 1. Blue line indicates the actual coastline, green line the expected coastline under IPCC scenario B1, and red line the coastline under A1FI extreme sea level scenario.

The beaches farther south of Sector 2 (Figueira do Pontal, Itapoá, and Babitonga) showed a higher risk of coastline indentation, not because the setup values (4 m to 11 m in Itapoá) but due to urban constructions (quite advanced toward the coastline and leaving little room for changes in the MSL) and because

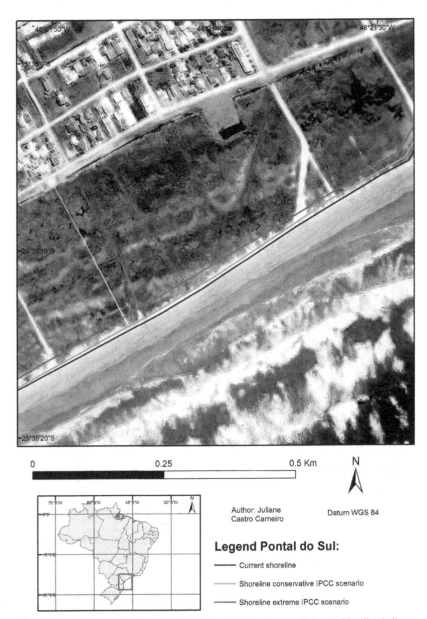

Legend Pontal do Sul:

——— Current shoreline

——— Shoreline conservative IPCC scenario

——— Shoreline extreme IPCC scenario

Author: Juliane Castro Carneiro

Datum WGS 84

FIGURE 4 Indent scenarios for Pontal do Sul, at PEC mouth area. Sector 1. Blue line indicates the actual coastline, green line the expected coastline under IPCC scenario B1, and red line the coastline under A1FI extreme sea level scenario.

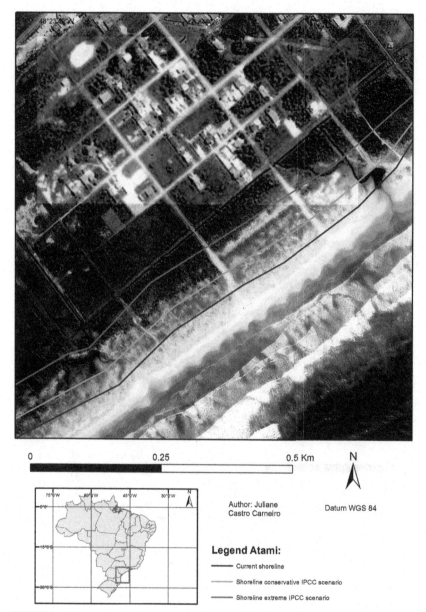

FIGURE 5 Indent scenarios for Atami beach, south of PEC mouth area. Sector 1. Blue line indicates the actual coastline, green line the expected coastline under IPCC scenario B1, and red line the coastline under A1FI extreme sea level scenario.

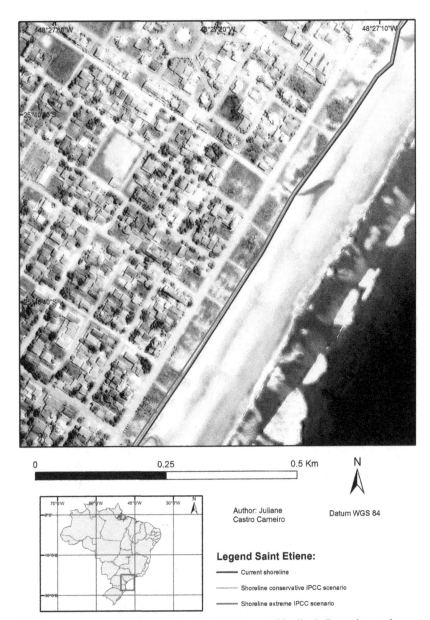

FIGURE 6   Indent scenarios for Saint Etiene beach. Sector 1. Blue line indicates the actual coastline, green line the expected coastline under IPCC scenario B1, and red line the coastline under A1FI extreme sea level scenario.

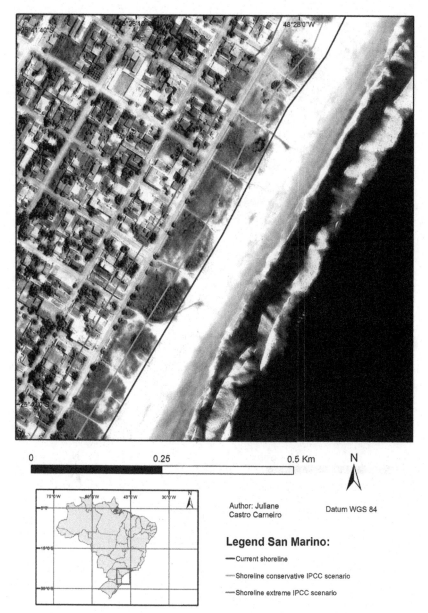

FIGURE 7   Indents scenarios for San Marino beach. Sector 1. Blue line indicates the actual coastline, green line the expected coastline under IPCC scenario B1, and red line the coastline under A1FI extreme sea level scenario.

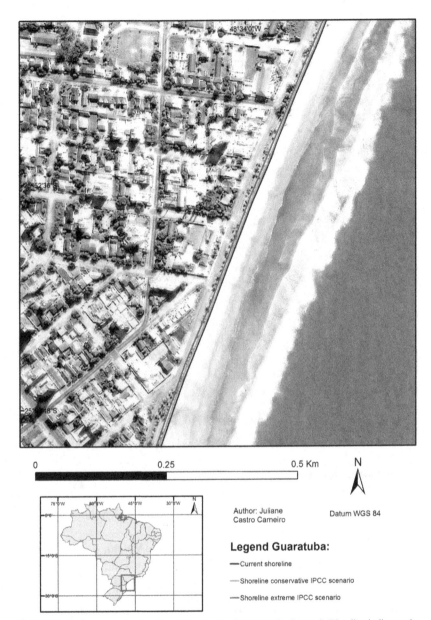

FIGURE 8 Indent scenarios for Guaratuba, south of bay mouth. Sector 2. Blue line indicates the actual coastline, green line the expected coastline under IPCC scenario B1, and red line the coastline under A1FI extreme sea level scenario.

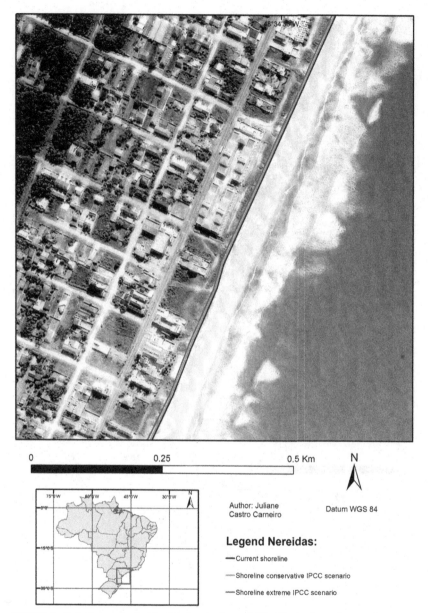

FIGURE 9 Indent scenarios for Nereidas beach. Sector 2. Blue line indicates the actual coastline, green line the expected coastline under IPCC scenario B1, and red line the coastline under A1FI extreme sea level scenario.

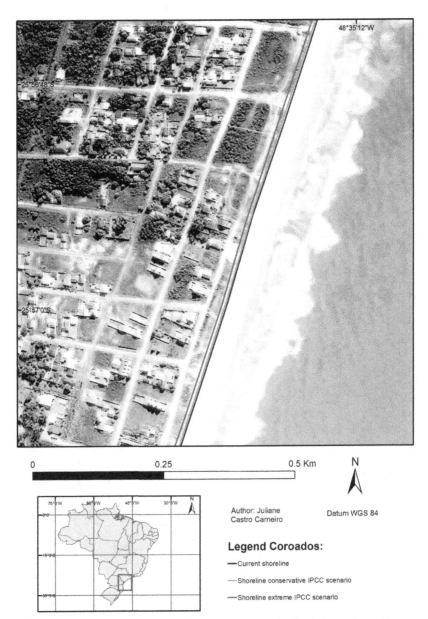

Author: Juliane
Castro Carneiro

Datum WGS 84

Legend Coroados:

—Current shoreline

—Shoreline conservative IPCC scenario

—Shoreline extreme IPCC scenario

FIGURE 10 Indent scenarios for Coroados beach. Sector 2. Blue line indicates the actual coastline, green line the expected coastline under IPCC scenario B1, and red line the coastline under A1FI extreme sea level scenario.

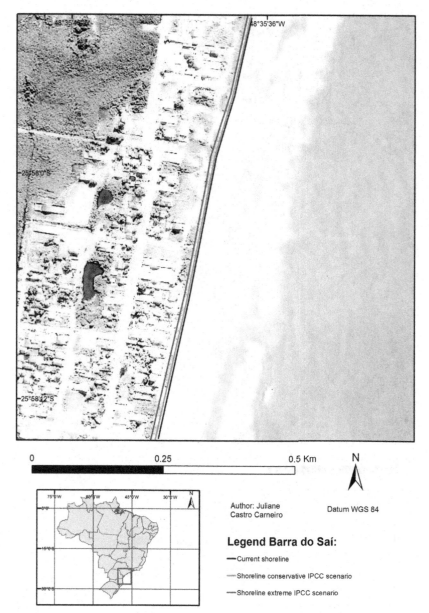

Author: Juliane
Castro Carneiro

Datum WGS 84

Legend Barra do Saí:

—Current shoreline

—Shoreline conservative IPCC scenario

—Shoreline extreme IPCC scenario

FIGURE 11   Indent scenarios for Saí bar. Sector 2. Blue line indicates the actual coastline, green line the expected coastline under IPCC scenario B1, and red line the coastline under A1FI extreme sea level scenario.

FIGURE 12   Indent scenarios for Itapoã. Sector 2. Blue line indicates the actual coastline, green line the expected coastline under IPCC scenario B1, and red line the coastline under A1FI extreme sea level scenario.

these beaches are reported as showing a state of coastal erosion (Gupta, 1999). On these beaches, some buildings and urban tessellations will be affected, in particular on the beach of Figueira do Pontal, where the current position of the urban constructions in relation to the shoreline has the lowest margin for adaptation. Severe storm surges will certainly critically affect these areas.

DISCUSSION

Even considering the worst-case IPCC scenario of SLR for the next decades, the estimated rates would affect coastal constructions only in few places of Paraná and northern Santa Catarina coast. Still, morphological changes like narrowing of the beaches and decrease in beach slope may occur due to changes in sea level (Angulo and Araújo, 1996), exposing areas considered protected in other forecasting models. The extra reach of sea intrusion due to severe storm surges needs also to be taken into account.

Another IPCC result is that the number of extreme events (cyclones, hurricanes, storm surges, and maritime agitation) will increase in the coming years and that usual weather events (such as cold fronts, storms, agitation, drift, and return currents) will intensify. Knowing that the morphology of estuaries and beaches in the region is highly influenced by meteorological phenomena (Lamour et al., 2003), there is a need for further investigation of the action of these extreme events in the case of an increase of the mean sea level on the coast of Paraná.

Not only the coastline would be affected by changes in sea level, but also ecotones such as dunes, mangroves, and marshes, which would not be resilient enough to migrate at the same speed or, most probably, they will not have enough room to do so. Aside from the ecological loss (since such environments are important habitats and nurseries for animals and plants in the coastal zone), the loss of such systems would be an extra damage for the new coastline, which will be able to migrate easily without the physical barrier that those ecosystems represent. The resilience of these systems should be better established, in particular that of ecotones located at the points where the projections of this work indicate critical setups.

The projected scenarios become more worrisome for some areas of Paraná's coast that are in actual erosional process (Souza, 1999). These coasts would become more vulnerable to future erosive process, since there would be an increase in the frequency and intensity of coastal flooding.

In addition, looking at Figures 4 and 7, corresponding to Pontal do Sul and San Marino, respectively, it is possible to note the "scars" in the permanent vegetation produced, instead, for "extreme" extreme events. These extreme storm surges could penetrate more than 100 m landward, and this buffer distance should also be considered in any risk assessment study.

The erosion promoted by extreme storm surges is the most important cause of coastal retreat at small time scales, and the size of the indentations produced could be greater than the ones projected using the IPCC scenarios and the Bruun law. The closest (and most alarming) example of this coastal erosion can be found at the southernmost Brazilian state, Rio Grande do Sul, in the Hermenegildo beach. The erosional processes at this beach were not studied until 1998, when a series of studies were released on this topic (Tomazelli and Dillenburg, 1998; Calliari et al., 1998a,b). Dillenburg et al. (2004) compiled these studies and suggested that the main cause of erosion on this part of the Brazilian shore is a long-term negative balance on the sediment availability. Even most of the Paraná and northern Santa Catarina beaches have shown a seasonal equilibrium (Angulo and Araújo, 1996); some of them are losing sediments, as those beaches south of Sector 2 (Gupta, 1999), and it is necessary to pay particular attention to these cases. The problems at the Hermenegildo beach forced the coastal manager to remove houses and buildings because of the coast line transgression, a solution that seems the most cost-effective in these cases.

CONCLUSIONS

We examined what will happen, in an average sense, to sandy beaches of Paraná and northern Santa Catarina, Brazil, using well-accepted IPCC scenarios to determine the risks of the area under SLR projections and sandy beach erosion. The main tool of this study was the Bruun rule, which shows the potential reach of the ocean on landward direction as a relation of the rates of SLR and using shoreline position and shape. The Bruun rule describes how beach profiles respond to SLR if other conditions (e.g., sediment supply) remain unchanged, and this process will occur as long as there is a rise in sea level. We used the best quality data available in both short- and long-term time scales, removing or minimizing the influence of inlets and human interference on sediment supply.

The SLR will induce further beach erosion, and the rate of coastal erosional dynamic retreat R will be of a magnitude greater than the rate R' of coastal indentation for a static SLR. It has to be clear that SLR does not cause long-term erosion directly because just the SLR has too little energy associated with it. Higher sea levels act as an enabler of erosion because higher water levels allow waves to act further up the beach profile and move sediment seaward.

The variability of the results of the sea intrusion landward between the studied areas is large (from 7.3 m at Nereidas to 112 m at San Marino, in the worst-case scenario) and seems to be related to some particular parameters such as the beach's slope and local morphodynamics.

We assumed that the debris lines integrate better the actual reach of the seawater landward due to mean extreme events, and used them to define the position of the actual coastline. Great storms, which we called "extreme" extreme events, capable of overwashing some extra 100 m in some areas at present times, would invade larger areas of the fore beach in the future. If the "extreme" extreme events will become more frequent and intense, the buffer area that the coastal planners have to leave free of human occupation should have to be increased proportionally. Even if there is substantial evidence that the effect of storms on shoreline position is episodic, rather than secular, the increase in their activity and intensity has to be considered in coastal management.

From this perspective, SLR is ultimately responsible for long-term beach erosion on the Paraná and northern Santa Catarina beaches, as well as sandy beaches everywhere.

It is also necessary to mention the potential losses with the destruction of aquifers and river mouths, which would be "contaminated" by the saline wedge, as well as the coastal ecosystems that will need room to move landward.

However, we do not need to run right now up to the hills and, probably, our grandchildren will have to move a little landward in our coast. Anyhow, coastal managers have to define, under the light of these figures, a long-term policy for land use and solve some focal problems on those locations where human construction is already misplaced. Moreover, we would like to con-clude offering an educated guess: we do not recommend the use of coastal protection constructions (either hard or soft) on this area. Most of these works only promote the loss of landscape value and tourism potential by the construction of hard coastal defenses aiming to contain the advance of the sea, and induces major cultural and economic losses for the populations that reside or are established seasonally to leisure. There is enough room, yet, to properly place the human settlements landward, and in the time scales we are presenting our scenarios, it will be certainly a more cost-effective way of action.

FINAL REMARKS

In this chapter we tried to show that even when writing a scientific work we could make some effort to explain the methods and results in as simple words as possible, allowing some educated reader to follow what we are saying without losing scientific rigor. In addition, we may use tools to write with the most basic words without being scientifically inaccurate but at the level in what even a child could get the message.

We avoided as much as possible using complicated words to explain scientific concepts and results along the scientific text, which was accompanied by boxes we wrote using the 1000 most common English words. Also, in Anex I, we show a poster we use to exhibit in local fairs and touristic events in our coast, in the most plain Portuguese, to divulge our results for the locals, who will be affected in the future as our projections show. Also, we construct bridges between the academic community and the stakeholders, because the potential solutions for the problems that could affect the region in the near future are in the hands of this social component.

One of our ethical responsibilities is to ensure that our discoveries are prop-erly communicated, particularly when they could affect the livelihood of other human beings and/or the environment.

O mar vai subir! Temos que correr?

Eduardo Marone; Denis Cardoso, Juliane Castro Carneiro; Marcio Cintra; Andréa Ribeiro; Carol Stellfeld - GFM-PGSISCO/PosGeo/UFPR

Em tempos antigos, o mar chegou a subir mais de 100 metros acima do nível atual e, muitas vezes, até dezenas de metros acima do que está hoje. Isso ocorreu, no passado, porque o gelo que está sobre a terra desapareceu como resposta natural da terra a um ar mais quente. Hoje, o grupo de pessoas mais importante de mundo que estuda o clima, chamado de IPCC, está dizendo que, para o ano de 2099, o mar poderá crescer em números que poderiam ser da ordem de meio metro ou superior acima do nível atual. A razão desse aumento é porque estamos esquentando o ar pela queima de coisas que pegamos da terra para aquecer nossas casas e fábricas, mover carros, aviões e barcos, e até "limpar" campos para a agropecuária.

Mesmo se nós não sabemos os números exatos deste crescimento para um dado lugar, o IPCC dá-nos alguns valores que podem ser utilizados para se ter uma ideia do problema que podemos ter que enfrentar em 2099 em qualquer lugar.

Muitas pessoas (e animais e plantas) vivem, trabalham e se divertem ao lado do mar nas praias do Paraná e do norte de Santa Catarina, e vão ter se preparar para este aumento do nível do mar.

Com este trabalho gostaríamos de mostrar a este povo que há um problema nesses locais e que precisamos estar preparados para enfrentá-los.

A ideia foi pensada pelas pessoas muito brilhantes do IPCC, mas que falam uma linguagem meio difícil de entender pelas pessoas normais. Às vezes, porém (e este é o pior caso), é simplesmente porque ninguém pensou em explicá-lo com as palavras certas.

Vamos tentar fazer com que todos entendamos, dizendo o quanto o mar pode crescer até 2.099 nesses locais, mostrando que não é muito difícil estudar o problema e ser claros sobre o que fazer. Ao conhecer os números certos, nós poderemos estar prontos com os planos de ação que nos ajudem a tornar os problemas menos importantes, o que nos permitirá continuar a desfrutar do mar.

Como calculamos:

Usamos uma maneira simples de saber o quanto o mar vai invadir a terra em torno do ano 2099, usando o tipo de areia, o alcance, a borda e a inclinação da praia, junto com o quanto o mar pode subir no Paraná e em algumas praias de Santa Catarina. Desta forma conseguimos obter quantos metros o mar pode atingir em direção à terra para um determinado local. Usamos três possíveis valores do crescimento nível do mar de acordo com o trabalho do IPCC (pequeno, médio e alto). Seguindo esses números podemos imaginar o quanto terra adentro o mar pode chegar em 2099 . As figuras finais foram feitas usando fotos de satélite, a fim de tornar mais fácil ver o problema que poderíamos ter, em alguns anos, por causa do aumento do nível do mar.

ONDE

VEJAMOS ALGUMAS:

SETOR 1:

SETOR 2:

QUANTO

E DAÍ?

Nossos resultados para as projeções de elevação do nível do mar foram divididos em dois setores: O Setor 1 compreende uma área pouco povoada, coberta por vegetação nativa e bem preservada de dunas e pântanos marinhos ao norte e a região com balneários menos densos no centro. Por outro lado, o Setor 2 é composto por praias turísticas no litoral sul do Paraná, mais três praias no litoral Norte de Santa Catarina. O Setor 1 vai sofrer modestos avanços do mar sobre a planície costeira até 2099, no entanto, nessas áreas as configurações da costa tendem a não ser tão críticas, devido à ausência de urbanização e à ausência de estruturas fixas perto da costa atual, exceto na Vila das Peças e no istmo da Ilha do Mel. Nessas áreas, mesmo com a projeção mais otimista, a área de praia atual será submersa e, se usada a projeção mais extrema, a faixa urbana mais próxima do mar será destruída pelo avanço da linha de costa e o istmo se romperá. Em San Marino, onde o avanço máximo pode chegar a mais de 100 m, a atual área de restinga ficará submersa e tormentas extremas poderão atingir a área urbanizada.

No setor 2, as praias mais ao norte não apresentam um alto risco de inundação em relação ao avanço do mar para a terra. Estas áreas têm uma forte presença urbana ao longo das praias, mas a declividade de suas costas não permite a ascensão do mar até alcançar os prédios. No entanto, nas praias mais ao sul do setor 2 algumas edificações e ruas serão afetadas.

Nós não precisamos correr, mas nossos netos vão ter que se mexer no futuro, e seria bom se nossos governos se mexem agora.

CONTER O MAR NÃO É BOM NEGÓCIO MELHOR SERÁ NOS MEXER UM POUCO!

ANEX I   Poster with the results of the scientific work used to divulge the results among the common citizens, who speak Portuguese.

What We Mean?

We do not have to be worry about the *sea level* grow in 2099 at *Paraná* and *north* of *Santa Catarina beaches*. But our children will have to move a little up the land. *Sector* 1 will feel small *sea raise* on land direction up to 2099, in almost all places.

Villa da Peças and *Mel* Island, however, will feel the raise at a hurt level, while at *San Marino* the houses and streets could be hit.

In *Sector 2*, the *beaches* further *north* do not present a high chance of being hurt because the sea grow will not be great. These areas have strong housing places, but the *slope* of the *beaches* does not allow the grow of the *sea* to reach the houses. However, the *beaches* further *south* of Sector 2 have houses and streets too near to the water and would be hit during *storms*, which will become more usual and hard. As soon as we move our houses and streets up to the land, the better. To do so, we have to get those who have the power to be more kind with our coast.

ACKNOWLEDGMENTS

This research was partially funded by the International Ocean Institute (OC-Brazil) and by Lloyd's Register Foundation, which invests in science, engineering and technology for public benefit, worldwide.

REFERENCES

Angulo, R.J., Araújo, A.D., 1996. Classificação da Costa Paranaense com base na sua Dinâmica, como Subsídio à Ocupação da Orla Litorânea. Bol. Paranaense Geoci. 44, 7–17.

Angulo, R.J., Lessa, G.C., Souza, M.C., 2006. A critical review of Mid- to Late Holocene sea level fluctuations on the eastern Brazilian coastline. Quat. Sci. Rev. 25, 486–506.

Arasaki, E., Alfredini, P., Do Amaral, R.F., Lamparelli, C.C., 2008. Os efeitos no ambiente marinho da elevação do nível do mar em regiões da Baixada Santista, Brasil. Revis. Brasil. Recurs. Hídricos 13 (2), 165–175.

Bird, E.C.F., 1985. Coastline Changes. Wiley & Sons, New York. 219 pp.

Bruun, P., 1954. Coast Erosion and the Development of Beach Profiles. U.S. Army Beach Erosion Board Technical Memorandum 44. U.S. Army Engineering Waterways Experiment Station. 79 pp.

Bruun, P., 1962. Sea-level rise as a cause of shore erosion. J. Waterways Harbors Div. 88, 117–130.

Bucci, R., 2009. Modelagem numérica da dinâmica da baía de Guaratuba. Dissertação de Mestrado em Sistemas Costeiros e Oceânicos – UFPR. 83 pp.

Calliari, L.J., Speranski, N., Boukareva, I., 1998a. Stable focus of wave rays as a reason of local erosion at the southern Brazilian coast. J. Coast Res. SI 26, 19–23.

Calliari, L.J., Tozzi, H., Klein, A.H.F., 1998b. Beach morphology and coastline erosion associated with storm surges in southern Brazil – Rio Grande to Chuí, RS. Acad. Bras. Cienc. 70, 231–247.

Chust, G., Caballero, A., Marcos, M., Liria, P., Hernández, C., Borja, A., 2010. Regional scenarios of sea level rise and impacts on Basque (Bay of Biscay) coastal habitats, throughout the 21st century. Estuarine Coastal Shelf Sci. 87, 113–124.

Dillenburg, S.R., Esteves, L.S., Tomazelli, L.J., 2004. Coastal erosion in Southern Brazil. Acad. Bras. Cienc. 76 (3).

Gupta, Barun K. Sen (Ed.), 1999. Modern Foraminifera. Kluwer Academic Publishers, Dordrecht, The Netherlands; 200 pp. ISBN: 0-412-82430-2.

IPCC – Intergovernmental Panel on Climate Change, 2007. Climate Change 2007: The Physical Science Basis Summary for Policymakers. Geneva, 21 pp.

Lamour, M.R., Noernberg, M.A., Quadros, C.J.L., Odreski, L.L.R., Soares, C.R., 2003. Erosão na desembocadura sul da baía de Paranaguá e sua relação com o assoreamento do canal da Galheta. In: Proceedings IX Congresso da Associação Brasileira de Estudos do Quaternário.

Lana, P.C., Marone, E., Lopes, R.M., Machado, E.C., 2000. Ecological Studies, Coastal Marine Ecosystems of Latin America. The Subtropical Estuarine Complex of Paranaguá Bay, Brazil, vol. 144, Editora Springer-Verlag, Berlin Heidelberg. pp. 131–145.

Marengo, J.A., Valverde, M.C., 2007. Caracterização do clima no Século XX e Cenário de Mudanças de clima para o Brasil no Século XXI usando os modelos do IPCC-AR4. Revis. Multici. 8, 5–28.

Meehl, G.A., Stocker, T.F., Collins, P., Friedlingstein, A., Gaye, J., Gregory, A., Kitoh, R., Knutti, A., 2007. Global climate projections. In: Solomon, S., Manning, D.Q.M., Chen, Z. (Eds.), Climate Change 2007: The Physical Science Basis. Cambridge University Press, Cambridge, United Kingdom, New York. Contribution of Working Group I to the Fourth Assessment Report of the Intergovernmental Panel on Climate Change.

de Mesquita, A.R., França, C.A.S., Ducarme, B., Venedikov, A., de Abreu, M.A., Vieira Diaz, R., Blitzkow, D., de Freitas, S.R.C., Trabanco, J.A.L., 2005. Analysis of the Mean Sea Level from a 50 Years Tide Gauge Record and GPS Observations at Cananéia (São Paulo – Brazil). International Association for the Physics of the Ocean, Volume Abstracts, Cairns, Australia.

Moura, R., Werneck, D.Z., 2000. Ocupação contínua litorânea do Paraná: uma leitura do espaço. Revis. Paranaense Desenvolvimento 99, 61–82.

Nacaratti, M.A., 2008. Os cenários de mudanças climáticas como novo condicionante para gestão urbana: as perspectivas para a população da Cidade do Rio de Janeiro. In: Anais do XVI Encontro Nacional de Estudos Populacionais, Caxambu, MG.

Noernberg, M.A., 2001. Processos morfodinâmicos no Complexo Estuarino de Paranaguá – Paraná – Brasil: um estudo a partir de dados in situ e LANDSAT-TM. Tese de Doutorado, Pós Graduação em Geologia Ambiental. UFPR. 179 pp.

Santos, D.N., Silva, V.P.R., Sousa, F.A.S., Silva, R.E., 2010. Estudo de alguns cenários climáticos para o Nordeste do Brasil. Revis. Brasil. Engenharia Agrícol. Ambiental 14, 492–500.

SCOR, 1991. The response of beaches to sea-level changes: a review of predictive models. J. Coastal Res. 7, 895–921.

Souza, M.C., 1999. Mapeamento da Planície Costeira e Morfologia e Dinâmica das praias do município de Itapoá, estado de Santa Catarina: Subsídios à ocupação. Dissertação de Mestrado, Pós Graduação em Geologia Ambiental. UFPR. 169 pp.

Todisco, F., Vergni, L., 2008. Climatic changes in Central Italy and their potential effects on corn water consumption. Agric. For. Meteorol. 148, 1–11.

Tomazelli, L.J., Dillenburg, S.R., 1998. O uso do registro geológico e geomorfológico na avaliação da erosão de longo prazo na costa do Rio Grande do Sul. Geosul 14, 47–53.

Veiga, F.A., 2005. Processos morfodinâmicos e sedimentológicos na plataforma continental rasa paranaense. Tese de Doutorado, Pós-graduação em Geologia Ambiental. UFPR. 193 pp.

Zhang, K., Douglas, B.C., Leatherman, S.P., 2004. Global Warming and Coastal Erosion. Climatic Change, vol. 64, Kluwer Academic Publishers. pp. 41–58.

Section V

NATURAL AND ANTHROPOGENIC HAZARDS

Chapter 24

Thoughts on Ethics in Volcanic Hazard Research

Jurgen Neuberg
School of Earth & Environment, The University of Leeds, UK

Chapter Outline

Abstract

The comfort zone in which a researcher in the natural sciences usually operates does not necessarily include societal aspects, direct communication with members of the public, or involvement with the media. Unless a research team wants to grab the media's attention, outreach activities and interaction with the public are often considered a "necessary evil" to fulfill obligations, which are nowadays written routinely into research contracts. A different approach is essential when dealing with research surrounding natural hazards, risk, and its mitigation. Not only is the ultimate aim of such research the mitigation of individual and societal risk, saving lives, and reducing impact on infrastructure and assets; the participation of the public is, in most cases, an essential contribution to its success. Effective communication is the *condicio sine qua non*. In this article, I shall reflect on several aspects of communication, the clash between the academic and the real world, as well as the conflict of interest between individual scientists and research groups. I base these thoughts about ethical behavior on more than 20 years of experience in volcanological research at different universities worldwide, in several volcano observatories, and from being a member of a scientific advisory committee for volcanic risk assessment for the British Government in Montserrat, West Indies. The examples used here are taken from the field of volcano monitoring and forecasting, and are loosely subdivided into three categories: the problem of forecasting and predictability, the minefield of communication with the public, and tension and stress in the academic world.

Keywords: Expert elicitation; Risk mitigation; Scientific advice; Uncertainty and policy; Volcanic hazard.

Geoethics. http://dx.doi.org/10.1016/B978-0-12-799935-7.00024-1
305

APPROPRIATE METHOD FOR FORECASTING AND VOLCANIC RISK ASSESSMENT

The ultimate aim of volcanological research is to be able to forecast volcanic behavior in time and space. This includes an estimate of location, strength, and start of an eruption, but also the duration, and often the most difficult of all, the end of volcanic activity. Such forecasts have been attempted on many different timescales ranging from a few hours to allow rescue teams to operate, to many years for strategic planning purposes in volcanic areas. Due to major advances in computer power and software development, forecasts are nowadays based on numerical simulations with which many volcanic phenomena can be modeled. The calculation of run-out distances of pyroclastic flows (highly mobile, glowing clouds of hot gas and rocks), lava and mudflows, and the dispersion of volcanic ash in the atmosphere are examples that help to delineate dangerous areas from safe zones.

Interpreting the results of such numerical simulations is the tricky and often misleading part: while a sophisticated graphics output, including formal error estimates, gives an impression of predictability, the results actually show a much stronger dependence on the particular numerical model used for the simulation. During the eruption of Iceland's Eyjafjallajökull volcano in 2010, monitoring agencies employed several well-tested climate models to forecast the ash dispersion in the eruption cloud. The resulting high uncertainty of the predictions caused major disruptions in air traffic, grounding flights around the world. This was mainly due to the lack of volcanological input parameters to the models, such as the size distribution of ash particles, the clustering behavior of ash once airborne, and a detailed eruption chronology. The climate models had been generated for use in a completely different context, and never applied to volcanic eruptions.

No model will ever precisely represent natural phenomena; the more complex the natural system, the higher the chance to encounter nonlinear, essentially unpredictable processes. Furthermore, complex computer codes—particularly in the hands of inexperienced users—can be overparameterized or operate outside the range they were originally designed for. This can sometimes result in a computational behavior even more erratic than the system they are meant to represent. Unconstrained numerical models out of research kitchens are often not helpful at all, and furthermore, not everything that is scientifically interesting is necessarily useful for hazard modeling. The danger lies in the unconditional belief of some scientists in their own models. A set of conceptual models to guide data interpretation and decision making is essential, but in a volcanic crisis, robustness and simplicity are preferable to mathematical elegance and sophistication. Furthermore, while I believe in the merit of an open academic discussion of even the most controversial topics, a volcanic crisis is not the right time or place to have these discussions, as tensions are already running high.

Regarding models and calculations: once a resident of Montserrat, West Indies, asked me why scientists with all their computer power could not calculate the precise run-out distance of pyroclastic flows of the island's Soufriere Hills volcano. I asked him in turn, if he would trust the car manufacturer's specifications on fuel consumption (in miles/gallon) and would fill his tank with a minimum amount of fuel to reach a certain destination. He replied, that he would not do so, of course, for the distance one can cover with a gallon of fuel depended on driving style, road surface, weight, and even tyre pressure. He finally understood my point when I told him that many different conditions affected the run-out distance of a pyroclastic flow as well, and in addition, we did not even know how much "fuel" we had.

Where a volcanic process or an entire chain of processes cannot be modeled adequately, one has to rely on the collective experience and expertise of scientists. *Expert elicitation* is an approach to quantify the opinions or sometimes merely "gut feelings" of a group of experts. An open discussion where scientific opinions are aired is followed by an anonymous voting exercise among the scientists on probable scenarios. The result is a single Figure 1 given as a probability, including uncertainties, that a certain scenario will occur. Once this probability is checked for its credibility and robustness, it can be passed on to decision makers. Through the anonymity in the voting procedure, this approach delivers a relatively unbiased result as dominant personalities among the experts are prevented from controlling the end result. On the other hand, a certain weighting of experts is sometimes necessary to take different levels of expertise and experience into account.

No matter if scientific advice is given as colorful hazard maps (with error bars) or as elicitation results (with uncertainties), it is absolutely essential that

FIGURE 1 Massive ash clouds due to pyroclastic flows on Soufriere Hill volcano, Montserrat, West Indies. Photo courtesy Montserrat Volcano Observatory.

the decision makers, politicians, or civil defense personnel on the receiving end gain an understanding of the concept of probability and uncertainty. Any attempt to reduce a result further to a "yes" or "no" statement, often demanded by decision makers, could lead to a deceptive illusion of a clear line between safety and danger, and must be avoided under all circumstances.

In most cases, volcanologists have gained expertise in one or more areas of volcanology, which is already a very wide and multidisciplinary field. Usually this does not include general risk management like evacuation planning which lies in the competence of civil defense personnel. Experts should stick to their expertise when giving advice, even though it might appear that their advice has been ignored. An untimely or hasty evacuation might result in more casualties than no evacuation at all. I experienced an evacuation exercise in Colima, Mexico, and was very surprised how complex of an operation it really is: provision of shelters and food, medical supplies, transport planning, police and army support must all be thought of, so the operation is smooth in a real scenario. Most volcanologists will restrict themselves to giving advice on hazard issues, but with reliable population and other data available, advice may be extended to a quantitative risk analysis. An increasing number of volcanologists add this expertise to their portfolio.

BIASES IN COMMUNICATING VOLCANIC HAZARD AND RISK

As volcanologists, we are sometimes asked to leave our *ivory tower* and slip into the shoes of the people who are directly affected by the advice we give—but should we? Surely, we should be aware of further implications and perhaps even hardship, that results as a consequence of our scientific advice—but should we? Is it possible to cut emotional ties and focus on the science only? Or could a closer contact with the affected public bias expert opinion? In theory—no, but in reality the lines are much more blurred.

Independence from the funding body is the first rule for minimizing bias. It does not come as a surprise that hygiene experts found out that paper towels in toilets are the best way forward to improve food safety, if the study is sponsored by the paper industry. However, when dealing with people directly during volcanic risk assessment, there are more subtle dependences at play: a governor of a small volcanic island who might have a say in the future existence of an advisory panel you are on; a hotel owner, and good friend, whose property is located at the boundary of an exclusion zone; the owner of the villa who saved up money all his working life for the retirement home in the Caribbean, which he cannot reach now because it is located just beyond the evacuation line; or the resident of Montserrat who aggressively accused me, the scientist, not the volcano, of ruining the island! How tempting and easy would it be to please these people by shifting the results of a risk assessment ever so slightly. Hence, it is sometimes not that straight forward to stick to an unbiased opinion and not to care about the consequences; but we must, because the volcano does not care about those either.

I once met a European colleague during an eruptive phase at Tungurahua volcano in Ecuador where we discussed volcano–seismic data that he had just recorded. He tried to convince me that the intensity of volcanic tremor, as well as the frequency of small explosions, was steadily decreasing. We engaged in an enthusiastic discussion, but after analyzing the seismic data set, I could not share his opinion. Not only did our opinions differ, so did the purpose of our visits to Tungurahua volcano. While I was in Ecuador to negotiate a research project and was just a casual visitor to that volcano (attracted by the eruption), my colleague was on a mission to deploy further seismic stations closer to the volcanic edifice for his own research project. The level of activity dictated whether he could achieve this goal or not—and he had only two weeks left.

At some institutions, the occupational risk level for volcanologists is determined by elicitation (see above) by the same scientists that monitor the volcano and carry out the fieldwork. While on sabbatical leave in New Zealand, I was involved in an interesting discussion among the scientists of GNS Science (*Geological & Nuclear Sciences*, the agency charged with the monitoring of New Zealand volcanoes). During this discussion, a potential conflict of interest was identified between keen researchers who wanted to make progress in the understanding of a particular volcano by acquiring more data in a risky zone. However, the health and safety measures inflicted by them through the elicitation process prevented them from undertaking fieldwork. While my colleague on Tungurahua was probably not even aware of his own conflict of interests, the scientists at GNS Science identified and therefore mitigated this problem. As long as elicitation results remain merely a guideline for an independent decision made by a superior line manager (as it is the case at GNS Science), this process to determine occupational risk is appropriate and probably as good as it can get. However, the final liability that lies with the line manager remains a critical issue and can push opinion bias in the opposite direction: better safe than sorry.

Long gone is the era of the hero volcanologist who had to risk his life to acquire data; improved risk perception and mitigation measures have led to the development of many remote sensing techniques through satellites, remotely operated gas spectrometers, drones equipped with sensors, as well as unmanned helicopters to deploy monitoring equipment. All these measures have reduced the individual risk of volcanologists. If liability issues differ between the private sector and government agencies, it can lead to bizarre situations. For example, cunning tour operators bring unassuming tourists into areas, declared too hazardous for volcanologists to work in.

In some cases, external factors can cause biased communication of scientific opinion. One such factor can be funding if the level of governmental support is directly linked to the volcanic alert level. During an eruption of a volcano in the Canary Island, emergency agencies were tempted to keep the alert on a level higher than justified to maintain the degree of financial support. This is a direct consequence to the fact that governments and politicians are usually operating in a *responding* mode toward natural disasters, rather than switching to a *preventing* mode concerning the resulting risk.

Another strong factor lies in the power of the media—press, radio, and television stations alike. It is impressive and should not be underestimated. Universities and other research institutions offer media training to their staff to spare the unassuming scientist from getting into unpleasant and sometimes embarrassing situations. Live radio or television broadcasts are the most critical, particularly when conducted by a cunning presenter. Recorded interviews and even written statements can be broken up, and parts taken out of context can convey a completely different message. Insisting on proof reading an article or checking the final broadcast is desirable but often not possible. On the other hand, for a scientist well trained and accustomed to media procedures, media coverage offers fantastic opportunities to explain research results, and to interact with the public. This is particularly important in hazard and risk research, where the education and preparedness of the wider public is part of the risk mitigation strategy (e.g., De Lucia, 2015; Dolce and Di Bucci, 2015; Mucciarelli, 2015; Papadopoulos, 2015).

VOLCANIC RESEARCH ASSETS: HOARDING OR SHARING

Volcano observatories or institutions charged with monitoring differ from other volcanological research establishments regarding their principal function. Observatories monitor a specific volcano or group of volcanoes to serve local or regional communities, while other institutions, mainly universities, conduct general volcanological research without the burden of acquiring and archiving huge amounts of data. This leads to a certain discrepancy affecting research staff in both types of organization: On the one hand, observatory staff control a large quantity of multidisciplinary volcanological data, but do not have the time for detailed data analyses. On the other hand, researchers in universities have the time and funding to analyze data in detail, but have limited data access, often under severe restriction and red tape.

Another inequity concerns the availability of new research tools and methods to observatories, when created by scientists outside. Many of these innovative creations promise major advances in volcano forecasting, but are not tested under operational conditions in observatories. The offshoot is that observatories often sit on their data, while universities are reluctant to share their intellectual property.

It is obvious that researchers in both organizations need each other to make their respective organizations work effectively, and produce the adequate research output essential for the development of their own academic career, for both are under the same pressure: "publish or perish." Several models exist to accomplish this synergy. French volcano observatories are operated through the *IPGP* (*Institut de Physique du Globe de Paris*), a research institute with university status. Other organizations running observatories seek close cooperation with universities and cosupervise students on different levels. *MVO* (*Montserrat*

Volcano Observatory), like other observatories, have put a data policy in place that regulates data sharing and ownership through a *Memorandum of Understanding* between participating organizations. These different cooperations and setups work reasonably well, but are above all highly dependent on trust between individuals. In fact, all academic institutions share the pressure to produce research output, obtain research funding, push the frontiers of knowledge, and deal with challenge and competition. The big difference in hazard research is that lives can depend on such academic rivalry, hence, by and large different rules ought to apply.

In one rather controversial case, and in contrast to what has already been mentioned above, an individual university researcher blocked data access to a network of deformation sensors for a period of ten years. This network was jointly operated by a university and an observatory, and as a consequence, the observatory was not able to pass data on to other researchers. While the individual provided merely some rudimentary analysis, other research groups were prepared to and could have analyzed the data sets in much more detail, using more advanced methods. It is understandable that the individual researcher wanted to protect his research interests given that much time and funding had been spent to install the network, however, the fact that a thorough data analysis could not be performed demonstrates that this conflict of interests prevented the observatory from fulfilling its function. Again, a volcano observatory is not merely a place to acquire research data, but in the first place is an organization to carry out monitoring, risk assessment, and mitigation.

CONCLUSIONS

Recent developments of new research tools, improved methods, and interpretation techniques have advanced the field of volcanic hazard research. Schemes such as expert elicitation have complemented this development and have had an important positive impact on the communication aspects of hazard research. Even though different countries have different rules and various setups to organize the chain from volcanic monitoring to implementing risk mitigation measures, by and large, both science and civil defense agencies serve the public in a much improved manner.

Often politicians demand that science should show a united front, for arguments in public would undermine public trust in science. However, at the end of the day science is not different to any other human endeavor that one should not suppress debate and in a hazards context in particular one should not avoid alternative conceptual models but embrace them. Furthermore, in the long run the danger of undermining public trust by suppressing debate is much greater. Alternative models and explanations do not need to confuse, different opinions can be explained and expressed as a form of uncertainty, but the essential difference is to avoid debating in the full glare of the media in an ego-serving, allegedly scholarly discourse.

Those are a few random thoughts and examples from several years of experience working as a scientist in the somewhat risky field of research into natural hazards. Scientists in any research field encounter some of the difficulties and dilemmas brought up here, as well. Nonetheless, research into hazard and risk, their mitigation or prevention contains this extra factor, which puts both importance and consequence a notch higher. For the scientist this is very rewarding, on the other hand, a much more cautious approach needs to be followed. I hope these thoughts contribute to develop awareness toward this aim.

REFERENCES

De Lucia, M., 2015. "When will Vesuvius Erupt?" why research institutes must maintain a dialogue with the public in a high risk volcanic area: the Vesuvius Museum observatory. In: Wyss, M., Peppoloni, S. (Eds.), Geoethics: Ethical Challenges and Case Studies in Earth Science. Elsevier, Waltham, Massachusetts, pp. 335–349.

Dolce, M., Di Bucci, D., 2015. Risk management: roles and responsibilities in the decision-making process. In: Wyss, M., Peppoloni, S. (Eds.), Geoethics: Ethical Challenges and Case Studies in Earth Science. Elsevier, Waltham, Massachusetts, pp. 211–221.

Mucciarelli, M., 2015. Some comments on the first degree sentence of the "L'Aquila trial." In: Wyss, M., Peppoloni, S. (Eds.), Geoethics: Ethical Challenges and Case Studies in Earth Science. Elsevier, Waltham, Massachusetts, pp. 205–210.

Papadopoulos, G., 2015. Communicating to the general public earthquake prediction information: lessons learned in Greece. In: Wyss, M., Peppoloni, S. (Eds.), Geoethics: Ethical Challenges and Case Studies in Earth Science. Elsevier, Waltham, Massachusetts, pp. 223–237

FURTHER READING

Aspinall, W., 2010. A route to more tractable expert advice. Nature 463, 294–295. http://dx.doi.org/10.1038/463294a.

Cooke, R.M., 1991. Experts in Uncertainty: Opinion and Subjective Probability in Science. Oxford Univ. Press.

Donovan, A., Oppenheimer, C., 2013. Science, Policy and Place in Volcanic Disasters: Insights from Montserrat, Environ. Sci. Policy. http://dx.doi.org/10.1016/j.envsci.2013.08.009.

Gluckman, P., 2014. The art of science advice to government. Nature 507, 163–165.

Chapter 25

Ethical Issues in the Decision-making for Earthquake Preparedness, Prediction, and Early Warning: A Discussion in the Perspective of Game Theory

Zhongliang Wu[1,2]

[1]*Institute of Geophysics, China Earthquake Administration, Beijing, People's Republic of China;*
[2]*Key Lab for Computational Geodynamics, Chinese Academy of Sciences, Beijing, People's Republic of China*

Chapter Outline

Abstract

Game theory is applied as a conceptual framework for the discussion of ethical issues in the decision-making for earthquake preparedness, prediction, and early warning. Evolutionary game theory explains the "seismo-illogical cycle." Game tree is used to discuss the policy preference in different countries, that some countries put earthquake forecast as a priority for the reduction of earthquake disaster risk, while others do not. The conceptual framework of repeated games is used to solve the paradox that in the context of ethical discussions, it seems inappropriate to think in terms of cost-effectiveness when lives are at stake.

Keywords: Earthquake early warning; Earthquake forecast; Earthquake preparedness; Game theory.

Geoethics. http://dx.doi.org/10.1016/B978-0-12-799935-7.00025-3
313

INTRODUCTION

In the explanation, prediction, and evaluation of social behavior in the context of ethics, game theory and decision theory play an important role (Harsanyi, 1992; Verbeek and Morris, 2010). Although ethical problems related to earthquake countermeasures have been defined and discussed for a long time (e.g., Wu, 1989; Sol and Turan, 2004), in the seismological community, game theory has not been a widely used theoretical tool to understand the problems associated with the reduction of earthquake disaster risk. Nevertheless, it is natural to apply game theory when considering the scientific, technical, and social factors related to earthquake disasters. In studying the decision-making issues related to earthquake engineering, there was suggestion of applying game theory to discuss the interaction among political actors (Ellwood and Ellwood, 1998). In "pure" seismological studies, the game theory approach has been applied to formulate the problem allowing a whole class of optimal methods of aftershock identification, in which each method is optimal depending on the goals and gives the best trade-off between the number of missed aftershocks and the number of incorrectly identified ones (Molchan and Dmitrieva, 1992).

This chapter tries to apply game theory as a conceptual framework to the discussion on the ethical issues in the decision-making for earthquake preparedness, prediction, and early warning. Because of the present limited capability of earthquake science to predict earthquakes, and because of the controversies in the definitions, the two concepts, "earthquake forecast" and "earthquake prediction," are not defined and differentiated in a sophisticated way, in this chapter[1].

THE "SEISMO-ILLOGICAL CYCLE"

The "seismo-illogical cycle" in seismic risk and earthquake disaster management deals with the embarrassing phenomenon that a large investment is made when a big disaster due to an earthquake occurs, but the investment decreases till the next large earthquake (Ismail-Zadeh, 2009). Ironically, if a seismic cycle is described by an increase of tectonic stress toward an earthquake (in which the earthquake causes a stress drop), the "seismo-illogical cycle" is characterized by a decrease of investment toward the next large seismic event.

---

1. Sometimes "forecast" is defined as the statement that "some earthquakes would happen probably," and "prediction" is defined as the statement that "an earthquake will happen almost definitely." On the other hand, sometimes "forecast" indicates the statement on the likelihood of future earthquakes made by scientific community, and "prediction" indicates the message conveyed to the public based on the decision-making of some functional organizations. In some countries/regions, this decision-making load is on the shoulder of the government.

Game theory can provide some conceptual understandings of this phenomenon in the perspective of evolutionary game, in which ethical choice is understood as the unintended side effect of the interactions of agents, emerging from a series of repeated interactions between different agent groups. For simplification, most models of game theory deal with two-person interaction. The role of such simplified models is somehow similar to the "toy models" in physics, such as the Ising model in statistical physics[2] to understand critical phenomena.

Consider a population with two kinds of choices or two kinds of members with different choices, namely "active" or "inactive," in the investment to the reduction of earthquake risk. The mode of the discussion is similar to the hawk-dove evolutionary game (Harrington, 2009), but considers only the simplest case, aiming at the formulation of the problem. For two members in the community, the case of both being active leads to the sharing of the load of investment (or the cost) and the sharing of the benefit for the reduction of earthquake disasters. If one is active and the other is inactive, then the active member takes the load of the investment, but still both share the benefit of such investment. On the other hand, if neither is active, then both the investment and the benefit become zero. The benefit and cost are represented by E and C, respectively, and this game can be represented by the game matrix as follow:

	Active	Inactive
Active	$(E-C), (E-C)$	$(E-2C), E$
Inactive	$E, (E-2C)$	$0, 0$

The choice is represented in the rows for one member and the columns for another. Each cell in the matrix represents the outcome of each possible pair of choices.

Consider the situation that after a disastrous earthquake when all the members in the community are active, with the probability of changing to inactive (using the language of ecology, the invasion mutant) being p (in which p is between 0 and 1). Consider an active member in the community. When she/he meets another active member (with probability $1-p$), her/his expected utility is $(E-C)$; when she/he meets an inactive member (the invasion mutant, with probability p), her/his expected utility is $(E-2C)$, that is, she/he has to take the whole load while sharing the benefit. Then the evolutionary fitness of this active member (or the active strategy), is

$$(1-p) \cdot (E-C) + p \cdot (E-2C) = E - (1+p) \cdot C$$

---

2. http://en.wikipedia.org/wiki/Ising_model.

In the same way of thinking, the evolutionary fitness of the inactive strategy is

$$(1-p) \cdot E + p \cdot 0 = (1-p) \cdot E$$

That is, for an inactive mutant, when she/he meets an active member (with probability $1-p$), her/his expected utility is E; when she/he meets another inactive member (the invasion mutant, with probability p), her/his expected utility is zero. Comparing the two strategies, the active and the inactive, the condition for the active strategy to survive in the evolution requires

$$E - (1+p) \cdot C > (1-p) \cdot E$$

that is,

$$p \cdot (E-C) > C$$

If $(E-C)$ is small, or even negative (when people think that the investment cannot be paid back since the next earthquake is still "long in the future"), then the above inequality is impossible, and thus the active strategy is not "evolutionarily stable," which leads to the "seismo-illogical cycle," that with time lapses the active community will evolve to inactive.

THE ETHICAL PARADOX

More sophisticated manipulation of the terms E, C, and $(E-C)$, the portions of the members being active, q, and those changing to inactive, p, may lead to further discussion on the above mentioned problem (for the method, see, Dixit et al., 1999; Harrington, 2009). Finding the evolutionarily stable strategies, it is also necessary to differentiate the terms E and C for the active and the inactive, respectively. This may, or may not, give hints to the solution of the "seismo-illogical cycle." But one of the more critical key issues is related to the paradoxical comparison of E and C, in which C is related to risk management and represented by an amount of money, and E is related to disaster management and represented by both property and lives saved.

In the perspective of ethics, it is always a difficult problem to consider property and lives at the same time, although in some cases such considerations could be formulated in a quantitative way (Zuo and Zhang, 1989; van Stiphout et al., 2010).[3] The paradox lies in that apparently it is inadequate to compare property and lives directly, since in principle lives should be the "singularity" in the calculation of property.[4] On the other hand, however, the

---

3. For example, an analysis of the L'Aquila earthquake pointed out that "We consider an evacuation of all people in vulnerable buildings (EMS-98 class A), costing $500/person/day on average and the willingness to pay for a life saved by the government is $1M…" (van Stiphout et al., 2010).
4. As a matter of fact, one of the conditions for the inequality $p \cdot (E-C) > C$ to exist, in which p is between 0 and 1, is to take E as infinite, that is, a singular point. Explanation of this extreme situation is straightforward simple, that when a society considers to save lives in the next earthquake at any cost, then the "seismo-illogical cycle" does not appear.

majority of society accepts that in the decision-making related to earthquake preparedness and prediction, the potential loss of property and lives has to be considered at the same time.

One solution of this paradoxical social choice is based on the theory of repeated games. Although the value of lives cannot be compared directly with property in a single game, in the repeated games, lives can be compared with lives. In the decision-making related to the dilemma whether an earthquake prediction, with known uncertainty, is to be announced to the public, for a single game, miss-to-predict has to be avoided, even if it entails a high cost of too many false alarms, since to save lives at any cost should be the principle. On the other hand, however, if repeated games are considered, then both the misses-to-predict and the false alarms have to be minimized at the same time, since too many false alarms will cause the loss of confidence of the public in the prediction, and this may potentially affect the next round of game in the game sequence, leading to more lives lost in the future earthquakes (even if it is correctly predicted). In this regard, to consider the effect of both misses-to-predict and false alarms, in the language of economic loss, is ethically rational in the perspective of repeated games. On the other hand, the simple explanation, that potential losses of misses-to-predict and false alarms (represented by losses of property and lives) need to be considered at the same time, although operationally practical, is not an exact statement.

The paradox of comparing property and life also exists for an earthquake early warning system (EEWS).[5] Related to an EEWS, one of the critical comments is that because of the "blind zone," in which differential time between the P and S wave arrivals does not permit a practical early warning, the EEWS is actually of little use in saving lives and property, questioning the values of the EEWS in the reduction of earthquake disasters (e.g., Wyss, 2012). Indeed this comment is to a large extent reasonable, since generally high intensity areas are often the "blind zone" of the EEWS. However, for destructive earthquakes, considering both the preparedness stage and the emergency rescue and relief stage (in which to save lives "at any cost" is the guiding principle), the value of the EEWS can be understood, even if the situation of inland great earthquakes with the extended rupture larger than the "blind zone" of the EEWS (Ma et al., 2012) is not considered.

POLICY SELECTION RELATED TO EARTHQUAKE PREDICTION

Now we turn to another application of game theory to the problems of earthquake disaster reduction. In the perspective of game theory, regarding earthquakes, beyond seismological agencies and the public, the Earth, as a critical player in the game, plays an important role. Sometimes in game theory, nature is treated as a "dummy player."

---

5. Doughton, S., 2011. Warning system worked, but is it worth the cost? *The Seattle Times*, http://seattletimes.com/html/localnews/2014489762_quakewarning14m.html.

Playing with the Earth who produces earthquakes, the game tree for the policy selection related to earthquake prediction can be formulated as follows:

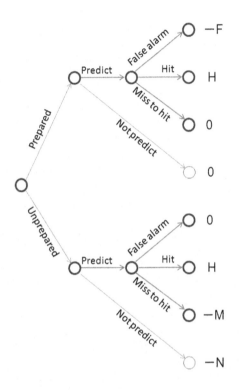

The game is between two players, the Earth (as a dummy player), and the human being. Thus in the game tree, only the payoff of the human being are marked. The above game tree considers the simplest (may be oversimplified) case, that for a prepared community, miss-to-predict and not to predict are tolerated by the public due to the engineering countermeasures, but false alarms are criticized by the public due to the unnecessary loss of property. On the other hand, in an unprepared community, miss-to-predict and not to predict are blamed by the society because of the tremendous loss of lives, but false alarms are tolerated by the society. Note that unprepared community is generally related to the developing economies in which the loss of property due to the false-alarm-based evacuation is still acceptable because of the low level of economic development. This is absolutely an oversimplified model, since in real life a community is to a large extent a middle point between these two extremes. Another assumption is that, to make the quantification simple, the utility or payoff of all the game nodes are linearly proportional to the hit rate, the false-alarm rate, or the rate of the misses-to-hit.

For a prepared community, the utility for prediction is

$$(H - F) > 0$$

Here we simply take H as the hit rate, and F as the false-alarm rate, which is assumed to be proportional to the "cost/benefit of the action." Taking this form is partly because that in the field of earthquake prediction, the relative operating characteristic (ROC) test (Swets, 1973; Molchan, 1997) has been applied to evaluate earthquake prediction schemes for a long time. The Chinese seismological community has been using the "R-value" to evaluate the performance of earthquake predictions (Xu, 1989; Shi et al., 2001).

$$R = H - F$$

The R-value corresponds to the difference between the vertical and the horizontal axes in the ROC diagram. In view of the ROC test, the condition $(H - F) > 0$ indicates that in the community with good preparedness, only in the case that the prediction outperforms random guess, the prediction is considered as a strategy with priority.

For an unprepared community, in a similar way, one may consider the rate of miss-to-hit, M, and the rate of not-taking-actions-for-prediction, N, although they are not commonly used in the study of earthquake prediction. Then the utility for prediction is

$$(H - M) > - N$$

That is

$$H > (M - N)$$

When M = N, one has

$$H > 0$$

This indicates that in the community without good preparedness, successful predictions, or "hits," play the role of stimulating the preference of prediction. An interesting example might be the earthquake prediction program in China. Despite limited successes, the Chinese seismological community has been putting earthquake prediction as one of its priorities for a long time. Chen and Wang (2010) discussed in detail the relation between earthquake prediction and the engineering measures for mitigating earthquake disasters before the 2008 Wenchuan earthquake. They envisaged that the Wenchuan earthquake will bring about some transitions of the predominant measures for reducing earthquake disasters. Among the envisaged changes, one of the most important ones might be putting more emphasis on the engineering and preparedness component. Six years have passed since the Wenchuan earthquake. It is true that the preparedness has been strengthened in many aspects. However, earthquake prediction is still kept as one of the research priorities in China, and the anticipated "paradigm transition" seems not clear. One of the plausible reasons for this situation is the above mentioned H > 0 criterion. In China, successful predictions, although with limited numbers and controversial evaluations, have been playing an important role in encouraging studies on earthquake prediction. Among these successes was the unique prediction-based evacuation of the 1975 Haicheng earthquake, a marvelous success due mainly to the occurrence of an unusually pronounced foreshock sequence (Ma et al., 1985; Wyss, 1991). Retrospective scenario analysis shows that significant lives were saved due to the evacuation (Wyss and Wu, 2013). In this regard, it can be estimated that unless a significant progress has been accomplished in the preparedness component so that China becomes "prepared," it seems unlikely that earthquake prediction studies as a research priority in Chinese seismological studies will change its position. When getting "prepared", more likely this priority will still be kept due to the criterion R > 0 as mentioned above.

CONCLUDING REMARKS AND DISCUSSION

Game theory deals with the competition among different rational agents, while ethics deals with the common values within a community of such players. Decision-making is a result of the selection based on the comparison of different strategies and the understanding of the basic values, while game theory discusses the outcome of the competition between different agents, generally it evaluates an equilibrium of the strategies of different players. When considering the three issues, decision-making, game theory, and ethics, at the same time, in the context of earthquake preparedness, prediction, and early warning, some dilemmas and/or paradoxes can be understood to some extent.

With cautions in mind about the limitation of such a conceptual discussion, ethical consideration in the decision-making for the reduction of earthquake risk, in the perspective of game theory, may provide some insights into the social processes related to the study and application of earthquake safety. On the other hand, the earthquake problem itself, with rich experiences and lessons during its long history, may also be potentially able to contribute to the study of game theory and decision theory.

ACKNOWLEDGMENTS

Thanks to Editor Max Wyss for the invitation and valuable suggestions for the revision, to Shaoxie Xu for stimulating discussion on the application of game theory, and to Alik Ismail-Zadeh for discussion on the "seismo-illogical cycle." Tengfei Ma helped in collecting the references.

REFERENCES

Chen, Q.F., Wang, K.L., 2010. The 2008 Wenchuan earthquake and earthquake prediction in China. Bull. Seismol. Soc. Am. 100, 2840–2857. http://dx.doi.org/10.1785/0120090314.

Dixit, A., Skeath, S., Reiley, D., 1999. Games of Strategy. W. W. Norton & Company.

Ellwood, J.W., Ellwood, L.G., 1998. Political engineering for earthquakes. In: Comerio, M., Gordon, P. (Eds.), Defining the Links between Planning, Policy Analysis, Economics and Earthquake Engineering. Pacific Earthquake Engineering Research Center, University of California, Berkeley, pp. 45–60. Report No. PEER-98/04.

Harrington, J.E., 2009. Games, Strategies, and Decision Making. Worth Publishers.

Harsanyi, J.C., 1992. Chapter 19: game and decision theoretic models in ethics. In: Handbook of Game Theory with Economic Applications, vol. 1, Elsevier, pp. 669–707.

Ismail-Zadeh, A., 2009. Computational geodynamics as a component of comprehensive seismic hazards analysis. In: Geophysical Hazards, International Year of Planet Earth. Springer Science + Business Media B.V., http://dx.doi.org/10.1007/978-90-481-3236-2_11.

Ma, X.J., Wu, Z.L., Peng, H.S., Ma, T.F., 2012. Challenging the limit of EEW: a scenario of EEWS application based on the lessons of the 2008 Wenchuan earthquake. In: Advances in Geosciences, vol. 31, World Scientific, pp. 11–22.

Ma, Z.J., Fu, Z.X., Zhang, Y.Z., Wang, C.M., Zhang, G.M., Liu, D.F., 1985. Earthquake Prediction—Nine Major Earthquakes in China (1966–1976). Seismological Press, Beijing (in Chinese, with English edition published in 1990 by Springer-Verlag, Berlin).

Molchan, G.M., 1997. Earthquake prediction as a decision-making problem. Pure Appl. Geophys. 149, 233–247.

Molchan, G.M., Dmitrieva, O.E., 1992. Aftershock identification: methods and new approaches. Geophys. J. Int. 109, 501–516.

Shi, Y.L., Liu, J., Zhang, G.M., 2001. An evaluation of Chinese annual earthquake predictions, 1990–1998. J. Appl. Probab. 38A, 222–231.

Sol, A., Turan, H., 2004. The ethics of earthquake prediction. Sci. Eng. Ethics 10, 655–666.

Swets, J.A., 1973. The relative operating characteristic in psychology. Science 182, 990–1000.

van Stiphout, T., Wiemer, S., Marzocchi, W., 2010. Are short-term evacuations warranted? Case of the 2009 L'Aquila earthquake. Geophys. Res. Lett. 37, L06306. http://dx.doi.org/10.1029/20 09GL042352.

Verbeek, B., Morris, C., 2010. Game theory and ethics. In: Stanford Encyclopedia of Philosophy. http://www.science.uva.nl/~seop/entries/game-ethics/.

Wu, Z.L., 1989. Seismoethics? An introduction to the ethical discussion on earthquake countermeasures. In: Proceedings of the Symposium on Preparedness, Mitigation & Management of Natural Disasters, New Delhi, India, vol. 1, pp. 23–24.

Wyss, M. (Ed.), 1991. Evaluation of Proposed Earthquake Precursors. American Geophysical Union, Washington, DC.

Wyss, M., 2012. The earthquake closet: rendering early-warning useful. Nat. Hazards 62, 927–935.

Wyss, M., Wu, Z.L., 2013. How many lives were saved by the evacuation before the M7.3 Haicheng earthquake of 1975? Seismol. Res. Lett. 85, 126–129.

Xu, S.X., 1989. The evaluation of earthquake prediction ability. In: State Seismological Bureau (Ed.), The Practical Research Papers on Earthquake Prediction Methods (Seismicity Section). Seismological Press, pp. 586–589 (in Chinese).

Zuo, Q.J., Zhang, S.Q., 1989. Social and economic responses to earthquake prediction and countermeasures in place of prediction issuance. Acta Seismol. Sinica 11, 191–197 (in Chinese with English abstract).

r

Chapter 26

Geoethics and Risk-Communication Issues in Japan's Disaster Management System Revealed by the 2011 Tohoku Earthquake and Tsunami

Megumi Sugimoto

Institute of Decision Sciences for sustainable Society, Kyushu University, Fukuoka, Japan

Chapter Outline

Abstract

This chapter discusses issues in Japan's disaster management system revealed by the 2011 Tohoku earthquake and tsunami, and by the Fukushima Dai-ichi nuclear power station accident. Many important decisions were based on scientific data, but appear not to have sufficiently considered the uncertainties of the data and the societal aspects of the problems. The issues that arose show the need for scientists to appropriately deal with geoethics and risk-communication issues.

Keywords: Ethical decisions; Nuclear power station; Risk-communication; Tsunami; Uncertainties.

Geoethics. http://dx.doi.org/10.1016/B978-0-12-799935-7.00026-5

JAPAN'S TSUNAMI COUNTERMEASURES BEFORE THE TOHOKU EARTHQUAKE

The March 11, 2011 Tohoku earthquake (with a magnitude estimated to be either M9.0 or M9.1, depending on which value is used; abbreviated as "Tohoku earthquake," below) and its tsunami killed 18,508 people, including the missing (National Police Agency report as of April 2014). The 1896 Meiji-Sanriku earthquake and the tsunami it generated killed 21,915 people, almost the same number of casualties as the 2011 Tohoku earthquake and tsunami.

During the period from 1896 to 2011 Japan developed a tsunami early warning system, earthquake early warning system, extensive networks of seismological and GPS observatories, a stringent building code for earthquake-resistant construction, disaster communication networks, disaster education programs, and has conducted frequent evacuation drills. In this chapter I consider why these efforts failed to reduce the casualty level significantly over that of 1896, I also examine some of the ethical issues involved.

Japan's preparations for earthquakes and tsunamis are based on the magnitude of the anticipated earthquake for each region. The government organization coordinating the estimation of anticipated earthquakes is the "Headquarters for Earthquake Research Promotion" (HERP), which is under the Ministry of Education, Culture, Sports, Science and Technology (MEXT). Prior to 2011, HERP stated that earthquake and tsunami countermeasures for the Pacific coast of the Tohoku region should be based on the assumption of a magnitude 8.2 earthquake, rather the M9.0 (or 9.1) megathrust earthquake, which actually occurred in 2011. Was this reasonable, in view of what was known before 2011?

After the 2004 Sumatra earthquake (with magnitude estimated to be between M9.1 and 9.3) and tsunami, several seismological studies (e.g., Kanamori et al., 2006; Umino et al., 2006) had pointed out the possibility of an earthquake larger than M8.2 in Tohoku. Geologists found tsunami deposits from the 869 Jogan earthquake in the Sendai plain along the Tohoku coast (Minoura and Nakaya, 1990), which suggested that the Jogan earthquake had been extremely large. Several geoscientists suggested, based on the tsunami deposits along the Tohoku coast and nearby rice fields, that the Jogan earthquake had been larger than M8.4. However, the above research results, some of which date back as early as 1990, were not reflected in HERP's estimates of the anticipated earthquake magnitude in Tohoku. Since disaster countermeasures in Japan are based on HERP's estimate of the anticipated magnitude, the above research results unfortunately were thus not taken into account in actual disaster countermeasures.

The situation in Japan is in great contrast to that in the Pacific northwest of the USA. Atwater (1992) found coastal tsunami deposits, which were consistent with several M9-class earthquakes in the past 2000 years, and by 2006, several state governments were basing their disaster mitigation plans on the anticipation of an earthquake magnitude of 9.0 or 9.1. However, geologists studying tsunami deposits were significantly underrepresented on the various committees advising HERP, and so the results of the paleotsunami studies were not reflected in Japanese government policy. Omitting details, note in passing that one seismologist argued, based on simulations, that the Jogan tsunami deposits were consistent with a magnitude of 8.4. However, even if one accepts these results, they do not show that the magnitude of the Jogan earthquake could not have been larger than 8.4. Nevertheless, prior to 2011 Japanese disaster countermeasures planning assumed that the magnitude of the Jogan event was 8.4. This served to exacerbate the impact of the 2011 disaster.

Hazard maps made before 2011 by Japanese civil engineering researchers calculated tsunami heights and inundation estimates on the basis of the anticipated earthquake with M8.2 specified by HERP. As shown in Figure 1, these hazard maps greatly underestimated the actual values observed for the 2011 tsunami. Local governments built tsunami evacuation buildings and specified hazardous areas based on HERP's anticipated earthquake and the tsunami simulations, and distributed tsunami hazard maps based on these values to each home. The result was that many people died in 2011 due to failure to evacuate or due to taking refuge in insufficiently high tsunami evacuation buildings. Also, Tokyo Electric Power Company (TEPCO) partly based its tsunami countermeasures on these values, which contributed to the severity (INES Level 7) of the accident at the Fukushima Dai-ichi nuclear power station (IAEA, 2011). The International Nuclear and Radiological Event Scale (INES) is rated by the International Atomic Energy Agency (IAEA). It is able to prompt communication of safety-significant information in case of nuclear accidents divided by 7 levels. Level 7 shows major accident in Fukushima event.

Japan's disaster mitigation system is depicted schematically in Figure 2 as consisting of three layers: seismology, civil engineering, and disaster mitigation planning. This clearly shows the pitfalls. Enormous efforts are made to mitigate

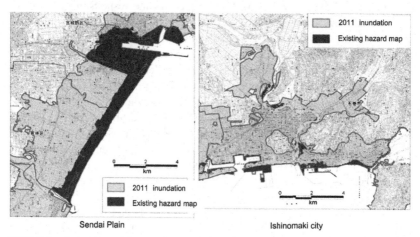

Sendai Plain   Ishinomaki city

FIGURE 1 Comparison of the actual tsunami inundation in 2011 and tsunami hazard maps based on numerical models based for an M8.2 earthquake. *Japan cabinet office (2011)*.

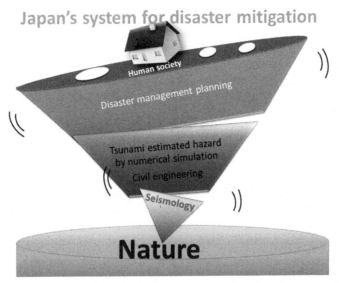

FIGURE 2 Schematic depiction of Japan's system for disaster mitigation. *(Sugimoto and Dengler, 2011).*

disasters, but all of these efforts are based on the magnitude of the anticipated earthquake supplied by HERP. If that value, which is obviously subject to large uncertainties due to the not very advanced seismological state of the art, is too small, then a disaster like that in 2011 is fated to happen.

OFFICIAL WARNING INFORMATION IN 2011

The Japanese Meteorological Agency (JMA) has the responsibility for issuing tsunami early warnings in Japan (Dengler and Sugimoto, 2011). After the 1993 Hokkaido Nansei-Oki earthquake and the tsunami, which struck Okushiri Island, the JMA developed a tsunami early warning system, which can issue alerts within 3 min after an earthquake (Japanese Meteorological Agency, 2011).

On March 11, 2011, the JMA issued a series of alerts assessing the likely tsunami hazard posed by the earthquake (Table 1). The initial bulletin, issued 3 min after the earthquake, estimated an M7.9 earthquake and forecast 3 m surges along the Iwate and Fukushima coasts, and waves as high as 6 m of bulletin 6 in the Table 1 in Miyagi. This information was disseminated to emergency officials and to the public via television, radio, and text messages to mobile phones. Succeeding bulletins expanded the warning areas and increased the expected water height; however, it took nearly 4 hour before the warnings stopped changing.

The JMA recently changed its protocol for very large earthquakes because of its underestimation of the tsunami height on March 11, 2011. In the early stages of its development of the tsunami warning system, the JMA prioritized rapid warnings after earthquakes. The JMA has gradually changed its priorities

TABLE 1 Japan Meteorological Agency Tsunami Bulletins

Earthquake Magnitude		JMA Bulletins				
		Bull 1 14:50	Bull 6 15:14	Bull 10 15:31	Bull 16 16:09	Bull 31 18:47
		7.9	7.9	7.9	8.4	8.8
Tsunami forecast region	Arr. Time	Forecast tsunami heights (meters)				
Iwate Prefecture	*1	3	6	>10	>10	>10
Miyagi Prefecture	15:00	6	>10	>10	>10	>10
Fukushima Prefecture	15:10	3	6	>10	>10	>10
Kujukuri and Sotobo area, Chiba Prefecture	15:20	2	3	>10	>10	>10
Izu Islands	15:20	1	2	4	>10	>10
Uchibo area, Chiba Prefecture	15:20	0.5	1	2	4	4
Central Hokkaido	15:30	1	2	6	8	8
Pacific coast of Aomori Prefecture	15:30	1	3	8	>10	>10
Ibaraki Prefecture	15:30	2	4	>10	>10	>10
East Pacific coast of Hokkaido	15:30	0.5	1	3	6	6
Sagami Bay and Miura Peninsula	15:30	0.5	0.5	2	3	3
West Pacific coast of Hokkaido	15:40	0.5	1	4	6	6
Ogasawara Islands	16:00	0.5	1	2	4	4

Continued

TABLE 1 Japan Meteorological Agency Tsunami Bulletins—cont'd

		JMA Bulletins				
	Earthquake Magnitude	Bull 1 14:50	Bull 6 15:14	Bull 10 15:31	Bull 16 16:09	Bull 31 18:47
	7.9	7.9	7.9	7.9	8.4	8.8
Shizuoka Prefecture	15:30	0.5	0.5	2	3	3
Tokyo Bay	15:40	–	0.5	1	2	2
Southern part of Mie Prefecture	16:00	0.5	0.5	2	2	2
Pacific coast of Aichi Prefecture	16:10	0.5	0.5	1	2	2
Japan Sea Coast of Aomori Prefecture	16:10	0.5	1	2	3	3
Wakayama Prefecture	16:10	0.5	0.5	2	3	3
Kochi Prefecture	16:30	0.5	0.5	2	2	2
Tokushima Prefecture	16:40	0.5	0.5	2	3	3
Miyazaki Prefecture	17:00	0.5	0.5	1	2	2

See real tsunami height data following URL: http://www.coastal.jp/tsunami2011/ (TTJS, 2011).
http://www.jma.go.jp/en/tsunami/info_04_20110311145026.html (Ozaki, 2011).
*1 mark: Arrival of tsunami inferred.

and is now placing greater emphasis on forecasting the height of tsunami waves. This is important because residents and local governments can compare the height of their seawall to the forecasted height of the tsunami in their town to decide whether or not to evacuate. For example, there were few people who had to be evacuated in Taro, a town which is famous for its 15 m high seawalls, on March 11. In all, however, it seems fair to say that the JMA should have placed more emphasis on the role of their warning system in tsunami disaster mitigation before 2011.

FIELD REPORTS FROM DISASTER AREAS

Higashi-Matsushima City (population around 34,000) is located in the transitional zone between much steeper terrain to the north and the broad, low-lying Sendai plain to the south (Figure 3). This city was particularly vulnerable, as tsunami surges attacked it from four different sources: the coast, the Naruse River, the Tona Canal, and Matsushima Bay. Parts of the city south of the Tona Canal had no direct access to high ground. One of the designated evacuation sites was the first story gymnasium in an elementary school (Figure 4). The elementary school was a three-story building and the upper floors were above the inundation zone; however, they were not used for vertical evacuation. An estimated 200 people gathered in the gymnasium after the earthquake, but it did not provide protection as the water level reached the base of the windows, and only a few people were able to get to safety on the ledge next to the windows. This site was located at the base of a hill where everyone could have reached high ground had they walked a few more minutes.

One family of survivors lived close to the evacuation site of the elementary school, but they had only recently moved to the area and did not know they were

FIGURE 3 Left: Tsunami hazard map for Higashi-Matsushima City (area of predicted inundation is shaded. Yellow: Under 0.5 m, Orange:0.5~ under1.0 m, Red: 1.0~ under 2.0 m, Wine:2.0~under 5.0 m). Right: actual 2011 inundation (shaded and 0.5–15 m). *Higashi-Matsushima City, 2008; Dengler and Sugimoto, 2011.*

FIGURE 4   The ground floor of the "tsunami evacuation building" in Higashi-Matsushima City, which was flooded by the 2011 tsunami.

supposed to evacuate to the designated building. Instead, they headed up the hillside behind their house after the earthquake and were able to see the waves approaching and to move further up the hill when it became clear that the tsunami was very large (Dengler and Sugimoto, 2011). This case is an example of how well intentioned but incorrect scientific information in some cases served to worsen rather than mitigate the 2011 disaster.

Tsunami evacuation buildings and areas are refuges for people trying to survive a tsunami. It is not sufficient for human security to choose evacuation places solely on the basis of numerical simulations conducted by scientists using some particular numerical model, which in most cases will not have been validated. Figure 3 shows the hazard map for Higashi-Matsushima City (left), and the actual inundation in 2011 (right), which covered a considerably larger area. Tsunami disaster planning must consider not only the area to which evacuees are directed, but also the means by which weak people such as children, the elderly, pregnant women and disabled people can be evacuated.

Tsunami scientists (seismologists, oceanographers, geologists, mathematicians, and civil engineers) separately helped to draw up the evacuation plans that were in effect in 2011. Many people needlessly died in 2011 because they acted on the basis of these plans, evacuating to sites that were in inundated areas. On the other hand, students in Kamaishi City survived by evacuating based on their own judgment. What can we learn from this?

Tsunami hazard maps were based on numerical models for an anticipated earthquake with a magnitude smaller than the actual M9.0 (or M9.1). This underscores the fact that even if the tsunami modeling method was perfect, which it is not, the reliability of the hazard model would still depend on the earthquake used as the input. How can we bridge the gap between hazard maps and future disasters? And how can we improve tsunami hazard maps? One point is that tsunami hazard maps should be revised to reflect the possibility of uplift or downdrop after earthquakes. For example, the ground sank to about 1.14 m below sea level in Ayukawa town, Miyagi Prefecture, but this was not included in estimates of inundation.

Ministry of Land, Infrastructure, Transport and Tourism's research shows that only around 10% of the people know about tsunami hazard maps in Japan. However, most people know about their designated evacuation places (buildings) through annual drills. Wider spread of information about tsunami hazard maps is needed, but it is also necessary that information about their inherent uncertainties must also be disseminated. In contrast to Japan, the California Emergency Management Agency makes tsunami hazard maps based on the largest geological recorded tsunami through checking the field situation with geologists, oceanographers, geographic information system (GIS) specialists, The National Oceanic and Atmospheric Administration (NOAA) staffs, and local police. This is still somewhat problematic, as there is no guarantee that future tsunamis will not exceed known previous cases, but still seems preferable to processes for making tsunami hazard maps in Japan.

FUKUSHIMA DAI-ICHI NUCLEAR POWER STATION ACCIDENT

The problems revealed by the Level 7 accident at TEPCO's Fukushima Dai-ichi nuclear power station can be viewed as due to a combination of risk-management, risk-communication, and geoethics issues. Insufficient earthquake and tsunami countermeasures led to the accident. All external power was lost due to the earthquake. The fact that Japan's regulator at that time, the Nuclear and Industrial Safety Agency, did not require safety upgrades to electric power transmission lines and transmission towers to strengthen earthquake resistance contributed to this failure. After losing all external power due to the earthquake, on-site emergency power was lost due to the tsunami. This led to a station blackout, core meltdown, and subsequent hydrogen explosions (e.g., Investigation Committee on the Accident at the Fukushima Nuclear Power Stations, 2011; National Diet 2012).

Matsuno (2013) said TEPCO delayed using seawater in emergency cooling efforts because that would make it impossible to use the reactors for future electric power generation. The Wall Street Journal (electronic edition) also said so on March 19, 2011 (Wall Street Journal, 2011). He interprets this as showing that a higher priority was placed on attempts to preserve the operability of the power plants than on the protection of citizens. Also, the accident showed grave deficiencies in severe accident management in Japan. Almost all experts would have recognized that loss of all external and on-site emergency power for boiling water-type reactors was grounds for immediate evacuation of nearby civilians, but such actions were not taken in 2011. Some local people individually and some groups in buses provided by local municipalities started evacuation at 20:50 on March 11, based on an order by Fukushima Prefecture, before the national government issued evacuation instructions. TEPCO reported a crisis (article 15) to the national government at 16:45 on March 11 (Ministry of Economy, Trade and Industry, 2011).

JAPANESE RESEARCH SYSTEM

Japan's government has been steadily cutting back automatic funding of universities for many years, and increasing the proportion of funds that must be obtained by application to competitive grant programs. Also, in the early 2000s, all national universities were removed from direct government control and became corporations under indirect government control, with steadily decreasing government support and encouragement to obtain external funding from the private sector. Electric power companies are now providing support to national university corporations including research funding and term-limited endowed chairs. This is both legal and encouraged by the government, but at the same time it may make it more difficult for university researchers to criticize electric power companies if they are dependent on continued funding from those companies. Although this poses obvious geoethics questions, this situation is unavoidable under current government policies. As students in the laboratories of professors funded by companies with agendas may not be fully cognizant of the ethical problems, it is desirable for their universities to provide ethical education programs for these students independently from their advisor's laboratory.

DISSEMINATION OF INFORMATION ON RADIOACTIVE PARTICLE DISPERSION

Before 2011, MEXT had developed a "System for Prediction of Environmental Emergency Dose Information" (SPEEDI) for predicting the dispersion of radioactive material from an accident at a nuclear power station, taking into account weather information. However, MEXT refused to disclose the SPEEDI prediction information from March 11 to May 31. In addition, the President of the Meteorological Society of Japan (MSJ) requested that the members of MSJ not disclose such information. As a result of this news blackout, many people evacuated from Fukushima in the same direction as the dispersing radioactive particles propagated. In contrast to MEXT, a volcanologist at Gunma University did provide his simulation results on dispersion of radioactivity to the public via the Internet. This shows the potential for individual scientists to act, even when the central government will not.

CONCLUSIONS

Students in geoscience should study geoethics as part of their education. But only when they becoming practicing professionals will they be faced with real geoethical dilemmas. And a crisis such as the 2011 earthquake, tsunami, and

Fukushima Dai-ichi nuclear accident, will force many geoscientists to suddenly confront previously unanticipated geoethics and risk-communication issues. One hopes that previous training will help them to make appropriate decisions under stress.

ACKNOWLEDGMENT

This research was mainly conducted when the author worked in the Earthquake Research Institute, the University of Tokyo. The author would like to express deep gratitude to Robert Geller. She also thanks Lori Dengler, Joanne Bourgeois, Brian Atwater, Hiroo Kanamori, Toshiyuki Hibiya, David Tappin and Max Wyss, as well as the officials and people of the areas affected by the 2011 tsunami for their spontaneous cooperation. She thanks graduate students of Graduate Education and Research Training Program in Decision Science for a Sustainable Society, JSPS. This research was supported by JSPS KAKENHI Grant number 25516024 and by the JSPS Institutional program for young researchers' overseas visits in the Department of Earth and Planetary Science, Graduate School of Science, the University of Tokyo.

REFERENCES

Atwater, B.F., 1992. Geologic evidence for earthquakes during the past 2000 years along the Copalis river, southern coastal Washington. J. Geophys. Res. 97, 1901–1919.

Cabinet Office Japan, 2011. Report (Draft) by the expert panel on countermeasures for earthquakes and tsunamis based on the lessons learned from the Great East Japan Earthquake (presented as annex No.3 to the expert panel held on 28 September 2011). Central Disaster Management Council, (in Japanese). http://www.bousai.go.jp/jishin/chubou/higashinihon/12/index.html.

Dengler, L., Sugimoto, M., 2011. The Japan Tohoku Tsunami of March 11, 2011. EERI Special Earthquake Report, November 2011, 15 pp. http://www.eqclearinghouse.org/2011-03-11-sendai/files/2011/11/Japan-eq-report-tsunami2.pdf.

Higashimatushima City Tsunami Hazard map No. 2, 2008. (in Japanese). http://www.city.higashi-matsushima.miyagi.jp/kurashi/pdf/hazardmap/tsunami2.pdf.

Investigation Committee on the Accident at the Fukushima Nuclear Power Stations, 2011. The Final Report 2012. http://www.cas.go.jp/jp/seisaku/icanps/eng/.

Japanese Meteorological Agency, 2011. Approach for Developing Tsunami Early Warning System. (in Japanese). http://www.data.jma.go.jp/svd/eqev/data/tsunami/newmethod.html#jinsokuka.

Kanamori, H., Miyazaki, M., Mori, J., 2006. Investigation of the earthquake sequence off Miyagi prefecture with historical seismograms. Earth Planets Space 58, 1533–1541.

Matsuno, G., 2013. Disaster prevention policy aids only atomic power plants, not the public. Kagaku 83 (5), 0524–0532 (in Japanese).

Ministry of Economy, Trade and Industry, April 10, 2011. News Release about Earthquake Damage. No.85. http://www.meti.go.jp/press/2011/04/20110410003/20110410003-1.pdf (in Japanese).

Minoura, K., Nakaya, S., 1990. Trances of tsunami preserved in intertidal lacustrine and marsh deposits: some examples from northwest Japan. J. Geol. 99, 265–287.

National Diet of Japan Fukushima Nuclear Accident Independent Investigation Commission, 2012. The Official Report of the Fukushima Nuclear Accident Independent Investigation Commission 2012. http://warp.da.ndl.go.jp/info:ndljp/pid/3856371/naiic.go.jp/en/.

Ozaki, T., 2011. Outline of the 2011 earthquake off the Pacific coast of Tohoku (M_w 9.0) — Tsunami warnings/advisories and observations. Earth Planets Space 63, 827–830.

Sugimoto, M., Dengler, L., 2011. Lessons on vulnerability from the 2011 Tohoku earthquake for Indonesia and the United States. In: The 2011 American Geophysical Union Fall Meeting.

The International Atomic Energy Agency (IAEA), 2011. Technical Briefing: Summary of Reactor Unit Status Total Deposition and Layer Concentration April 12th 2011. http://www.slideshare. net/iaea/summary-of-reactor-unit-status-12-april-2011-1430-utc.

The 2011 Tohoku Earthquake Tsunami Joint Survey (TTJS) Group. 2011. Tohoku Earthquake Tsunami Information. http://www.coastal.jp/tsunami2011/.

The Wall Street Journal (electric edition), April 19, 2011. http://jp.wsj.com/public/page/0_0_WJPP_7000-204149.html?mg=inert-wsj (in Japanese).

Umino, N., Kono, T., Okada, T., Nakajima, J., Matsuzawa, T., Uchida, N., Hasegawa, A., Tamura, Y., Aoki, G., 2006. Revisiting the three M similar to 7 Miyagi-oki earthquakes in the 1930s: possible seismogenic slip on asperities that were re-ruptured during the 1978 M=7.4 Miyagi-oki earthquake. Earth Planets Space 58, 1587–1592.

Chapter 27

"When Will Vesuvius Erupt?" Why Research Institutes Must Maintain a Dialogue with the Public in a High-Risk Volcanic Area: The Vesuvius Museum Observatory

Maddalena De Lucia

Istituto Nazionale di Geofisica e Vulcanologia, Osservatorio Vesuviano, Naples, Italy

Chapter Outline

Abstract

In areas of active but dormant volcanoes, people need to know some basic facts: What volcano is going to erupt? What kind of eruption will occur? When and where will it take place? Why will it occur? And, also, what should citizens do to save themselves and their families?

Communications travel in a complex network of connections among individuals and groups. If the main stakeholders involved do not work together to deliver coherent and complementary messages, the transmitted message may be contradictory, or inconsistent. Misunderstanding and confusion will take over, increasing rather than mitigating the risk. For this reason, citizens need to turn to scientists and research institutes working in hazardous areas as key interlocutors for matters concerning basic information concerning volcanic

Geoethics. http://dx.doi.org/10.1016/B978-0-12-799935-7.00027-7

hazard. At the same time, scientists in charge of volcano surveillance, together with emergency managers, media, and public officials, have the duty to answer citizens' need for information, exploring multiple channels and languages to communicate effectively.

The present chapter analyzes some systematic studies and multidisciplinary research projects on this topic that have been carried out in recent years. It also points out an analysis of the public use of the Vesuvius Observatory museum, which is a reference information point for people living in the Naples (Italy) area and exposed to volcano hazard.

Keywords: Dormant volcanoes; Education; Outreach; Public communication; Survey; Vesuvius; Visitor center; Volcanic hazard; Volcano observatory.

INTRODUCTION

Who? What? When? Where? Why?

Giving an answer to the "five Ws" is a basic requirement in news communication. But these items can also be hired to articulate a comprehensive communication of hazards and risks, as in this case, posed by volcanoes.

Risk could be defined, in a very simple way, as a function of three parameters: hazard, vulnerability, and exposed value. Hazard is the probability that a natural or man-made process occurs, while "*risk* describes the effects that those hazards may pose for people, property, or other features within a threatened area" (Lockwood and Hazlett, 2010). People, property, and other features in the threatened area constitute the exposed value, while vulnerability expresses the resistance level to the impact of the hazard, and the level of potential loss of value of exposed elements. "In addition to this definition, one has to consider coping capacities, i.e., community ability to recover after a disaster, which constitutes a major element in reducing, at least, long terms losses." (MIAVITA Team, 2012, p. 16). Therefore, mitigating the risk is not only reducing vulnerability and exposed value, but also increasing coping capacities of the population. An important part of this can be achieved through effective communication.

In areas of active but dormant volcanoes, people need to know some basic facts:

What volcano is going to erupt? What kind of eruption will occur? When and where will it take place? Why will it occur? And, also, what should citizens do to save themselves and their families?

People could get this information from different sources: public authorities, civil protection, scientists, news media, educators, volunteer associations, religious organizations, and other local groups and communities. Communication processes do not follow a linear path, but instead travel in a complex network of connections among individuals and groups. If the main stakeholders—the first four in the list—do not agree or do not work together to deliver coherent and complementary messages, the transmitted message may be contradictory, or inconsistent. Misunderstanding and confusion will take over, originating mistrust and skepticism in the population, increasing rather than mitigating the risk. This could be crucial in high-risk volcanic areas, where communities have to prepare themselves in quiet times, to minimize losses during volcanic events. For this reason, citizens need to turn to scientists and research institutes working in the area as key interlocutors for matters concerning the status, the physical behavior of a volcano, and its effects, which concern volcanic hazard. At the same time, scientists in charge of volcano monitoring, together with other actors like emergency managers, media, and public officials, have the duty to answer citizens' need of information and cannot disregard citizens' right to be informed and to be involved in risk mitigation. They (scientists and other public officials) must play an active role in volcano hazard communication, exploring multiple channels and languages and establishing interactions, to get a truly effective communication. In addition, effectiveness of the transmitted information must be evaluated by adequate assessments of knowledge and perception.

Only very recently systematic studies and multidisciplinary research projects on this topic have been carried out. In the past, many expert volcanologists have described, in scientific papers and book chapters, guidelines or personal viewpoints about how to build effective communications during both volcanic rest and unrest periods. Most of these papers are based on long field experience. A quick analysis of some of these papers, as well as of research institutes' Web pages dealing with public communication, is presented in the next paragraph.

In this framework, a permanent visitor center can be a powerful tool for effective communication when managed by a scientific research institute staff, possibly assisted by communication experts. In Italy, the museum of Vesuvius Observatory, the oldest volcano observatory in the world (founded in 1841), has become a reference information point for nearly 2 million people living in the Naples area and exposed to three active volcanoes: Somma-Vesuvius, Campi Flegrei, and Ischia. The museum, with more than 10,000 visitors per year, is also the place of preservation of a unique cultural heritage, made of old scientific instruments, rocks, and minerals, produced by past eruptions of the volcano. The evaluation/analysis of/by the public of the museum, carried out during 2005–2009, could give a base of data to plan a better organization, as well as to enhance cultural and educational enjoyment. In addition, it could allow museum staff to improve the quality of communication, becoming the base on which setting up new facilities and initiatives more appealing to new audiences.

Results show what the main streams of public are and define main categories on the basis of age and provenance, suggesting also how to attract new categories of audiences. In addition, they highlight also the need to implement evaluation studies. The second part of this chapter will focus on the results of the study.

Communication aspects concerning actions and citizens' behaviors to be implemented during volcanic unrest and eruptions are not covered by this work, as they are carried out mainly by emergency managers and public officials and not by scientists.

PREVIOUS STUDIES AND ANALYSIS ON PUBLIC COMMUNICATION OF VOLCANIC HAZARD

Until few decades ago, public communication guidelines in volcanology have been mainly based on volcanologists' experience. Many big eruptions of the past (Mount S. Helens, USA, 1980; Nevado del Ruiz, Colombia, 1985; Pinatubo, Philippines, 1991; Soufriere of Montserrat, 1995; Mt. Usu, Japan, 2000) have been the training ground for scientists and civil protection operators. Only in the last 15 years, teams of psychologists and volcanologists carried out systematic multidisciplinary research about social vulnerability of communities exposed to volcanic risk, their perception and psychological consequences of this exposition, and challenges in education.

Many volcanologists published their experiences in scientific papers or books (for example, Newhall and Punongbayan, 1996); in addition, in 1999, the international organization International Association of Volcanology and Chemistry of the Earth's Interior (IAVCEI, www.iavcei.org) published a detailed pamphlet on professional conduct of volcanologists during volcanic crises (IAVCEI Subcommittee for Crisis Protocols, 1999). The guidelines, nonprescriptive, posed some remarks to avoid or reduce frictions, misunderstandings, and problems between scientists. Some ethical items are focused on communication. The paper specifies that in a volcanic crisis, scientists in charge with the emergency have to speak with a "single voice." Failing to follow this rule could adversely affect important decisions about hazard mitigation, and lead to misunderstanding and to delay. Any public controversy among scientists causes the public and officials not to believe all scientists, so, it is preferable to avoid public discussions between scientists. This does not mean that multiple working hypotheses cannot be presented. In public talks, a neutral team leader, or spokesperson, can express different points of view, informing officials about the possibility of resolution of different issues. Uncertainties could be expressed, but in some countries they can be mistaken as scientific incompetence, while in others they are welcome. Very important public statements must first be written, so the team could check errors and correct messages that can be misunderstood.

In 1996, in a comprehensive text about monitoring and mitigating volcanic hazard (Scarpa and Tilling, 1996), D. W. Peterson, researcher of the US Geological Survey (USGS), describes more explicitly the social role of scientists in helping the population to deal with hazardous volcanoes. Communication of volcanologists with the public is necessary. It is crucial for volcanologists to care about the level of public awareness and preparedness in areas of high volcanic hazard. They are in fact those who mostly have knowledge, training, and skill to know the existence and the nature of hazards. Thus, volcanologists have to foster people's awareness of the volcano's potential hazard through education. Communication with the public has to be carried out through informal education, as it has become after the 1985 Nevado del Ruiz disaster and the

1991 Pinatubo eruption: programs in schools and other social groups, museum exhibits, and public talks, to provide awareness of the potential hazards of volcanoes among citizens. Moreover, Peterson states that, as most volcanologists are paid by public funding, they have an ethical obligation to express the results of their research to the citizens in a clear and understandable way, and must not escape this duty (Peterson, 1996)

In the same paper, the author defines what media and citizens need to know in volcanic risk areas: what may occur in the future or what is going to occur at the neighboring volcanoes, and when and how volcanoes will erupt. According to this author, citizens need to perform effective exercises to deal with volcanic emergencies. Moreover, a broad definition of the public is given. For volcano scientists the public is made up of all nonexpert persons. Some of them, as public officials, or relief organizations, have specific responsibilities in a volcanic crisis, and an active role, and need appropriate guidance from volcanologists. It is also essential for scientists to establish firm and mutual relations with local and national journalists, and with social leaders (clergy, associations) who can spread information to a wider audience.

Peterson highlights that misunderstanding by the audience could be a major problem. The basic information people have to know about a volcano is its eruptive history, its past and current behavior. A complex and detailed treatise is not necessary: few, essential facts are enough. Basic information is the location and the shape of volcanoes, the presence of vents, main eruptions and their styles, as well as erupted volumes and duration of rest periods between eruptions. This information could be popularized by means of schematic but attractive brochures to be distributed. However, people need to understand limits and uncertainties of scientific research, accepting that even unrest may not always be followed by an eruption, without being disillusioned about the quality of scientific studies. Not only the content of the message, but also the style of communication has to follow specific attributes to be effective. It has to be "specific," "consistent," "accurate," "certain," and "clear" (Peterson, 1996, p. 709), especially when it deals with warning messages. It was thought at that time that effective communication could not contain elements of uncertainty, because it could be misunderstood or worse diminish the credibility of scientists.

According to this author, public education about volcanic hazard is a long-term and continuous process, which has to be implemented during rest periods, also to minimize a reluctant or even hostile attitude toward mitigation. Educational activities aimed at students and at teachers and journalists are most easily performed where there are volcano observatories with a scientific staff that can be involved. The observatory could be opened to the public on special days, with exhibits, panels, and other outreach materials. People could meet staff members and ask for information about the volcano and the work of scientists. This outreach activity could be effective also because people could understand how scientists carry out their work, and a relationship of empathy and trust with researchers is established.

In the year 2000, the Vesuvius Observatory and the Italian Civil Defence inaugurated the permanent exhibition "Vesuvius, 2000 years of observations," in the old building of the Vesuvius Observatory, in Italy. In the introduction of the catalog of the exhibition (Carapezza et al., 2000), Franco Barberi, well-known volcanologist and chief of the Italian Civil Defence department, wrote that the exhibition has two main goals. The first is to increase the knowledge and culture of civil protection. As Mount Vesuvius is a high-risk volcano, the effectiveness of the emergency plan drawn up by Civil Protection in case of eruption depends largely on the active and rational participation of citizens, which is in turn linked to the knowledge of the behavior of the volcano. The second is to attract visitors interested in volcanoes. The exhibition was designed to be a pleasant destination for school groups, families, and groups of citizens of the Vesuvian area, becoming a great opportunity to increase people's knowledge, culture, and perception of volcanic hazard. This statement is perfectly consistent with what Peterson stated before.

In the *Encyclopaedia of Volcanoes* (Johnston and Ronan, 2000), David Johnston, volcanologist, and Kevin Ronan, psychologist, examine interactions of natural processes with personal knowledge and beliefs and community characteristics. According to the authors, quiescent times give the best chance to build up risk mitigation and get the society ready for a period of volcanic unrest. Then, in well-prepared communities, "during the crisis period, risk education and effective interventions are critical in reducing the adverse effects of volcanism in society." Moreover, the authors point out that, although citizens tend to underestimate the hazard during periods of long volcanic unrest, the population's response to a volcanic crisis will be much better if the knowledge and understanding of the hazard is greater.

A significant note about public education is included in *Volcanoes, Global Perspectives* (Lockwood and Hazlett, 2010). In the chapter "Mitigation of Volcanic Risk," the authors state that human memory is short and often the awareness of hazards and risks arising from volcanoes are not considered as a priority for exposed populations. "In the real world, the entities involved in public education are often complex, especially in rural communities, and religious beliefs and cultural traditions may be very strong" (Lockwood and Hazlett, 2010, p. 431). Moreover, often, political and economic interests may hinder the development of educational programs, because of the fear of decrease in the value of property. Sociological studies aimed at identifying the level of awareness and the readiness to take into account the directions of emergency programs could be used to evaluate the effectiveness of programs of education and communication. So, "efforts to improve communications between volcanologists, government authorities and the public remain one of our most important obligations" (Lockwood and Hazlett, 2010, p. 431).

Public education tools have to be designed "to make information available in terms that are readily understood by ordinary citizens" (Lockwood and Hazlett, 2010, p. 431), like traffic lights codes—everyday life tools—for alert levels in emergency plans. Similarly, hazard and risk maps, as well as notices for popular distribution, should be designed with expressive and easily understood graphics. According to the authors, "successful management of volcanic crises is best achieved when there is a prior awareness of potential hazard by *all* the above (scientists, emergency managers, political authorities) and the roles of these key players are well defined in advance" (Lockwood and Hazlett, 2010, p. 455).

Effective channels of communication are the most important factors in successful crisis management. Media convey the bulk of information to the public, but the communication (that is mutual comprehension), as well as citizens' responses to hazard alerts, is enhanced by public education programs about volcanic risk and hazards. Scientists have to maintain their public credibility in the communication network. However, volcanologists cannot predict exactly what could occur in volcano unrest; there will always be uncertainty in their assumptions. Moreover, volcanologists have "to reach consensus among themselves" (Lockwood and Hazlett, 2010, p. 458) and then communicate their statements to government authorities. Authorities have to face science uncertainty, but they have to decide about evacuation, to save lives. "This does not mean, however, that volcanologists can simply provide their factual observations and recommendations to government authorities and then walk away from all responsibility as to how and if their recommendations are acted on" (Lockwood and Hazlett, 2010, p. 458).

In recent years, new models of communication with the public are taking place, no more top-down, but among peers. A new reciprocal relationship between the scientific community and the public is going to be established. If citizens directly benefit from the results of scientific research, it will be more likely that they will consider it necessary, and offer support.

In the last 5 years, some European projects funded under the 7th Framework Programme, such as Vuelco (www.vuelvco.net) and MiaVita (http://miavita. brgm.fr), and important national projects, such as Streva in the United Kingdom (http://streva.ac.uk), have taken into account, in a multidisciplinary framework, the application of scientific results to decision making and management in unrest situations; the development of tools and methods to mitigate risks on active volcanoes; the resilience of exposed populations, and therefore, how to increase it even considering the communicative aspects. In a recent paper about scientific management of volcanic crises (Marzocchi et al., 2012), the authors argue that even a good warning system and an efficient structure of decision making may not be successful unless accompanied by effective public education and communication.

The author invites to refer to specific project Web sites for further details: Vuelco (www.vuelvco.net); MiaVita (http://miavita.brgm.fr); Streva (http:// streva.ac.uk).

Nowadays, in many countries with volcanoes at rest, communication about volcanoes is mainly Web communication, carried out by research institutes and civil defense departments. The case of the USGS Volcano Hazard Program is reported in the following (http://volcanoes.usgs.gov/).

In the United States, since 1974, the year of the emanation of the Disaster Relief Act, amended in 1988 in the Stafford Act, a public law about the disaster relief and emergency assistance that "constitutes the statutory authority for most Federal disaster response activities" (www.fema.gov), the USGS has to provide communication about the status of monitored volcanoes. Community emergency response plans are based on assessments of scientific results and are the basis for community preparedness actions. Information about volcano status are made available both to officials and the public by means of the Volcano Alert Level System (standardized code based on four different alert levels), via Volcano Activity Alerts (periodic bulletins), regular updates (Volcano Notification System) and volcano observatories Web sites, as, for example, Cascades Volcano Observatory (http://volcanoes.usgs.gov/observatories/cvo/).

The USGS volcano observatories support community resilience by specific outreach and education programs, such as Living with a Volcano in Your Backyard, an Educator's Guide with Emphasis on Mount Rainier (Driedger et al., 2005). This program has been specifically designed to reach not only information distributors, such as teachers and other educators, but also public officials, media, park guides, and voluntary associations. The USGS strongly supports teacher's and educator's activity to improve communities' preparedness, providing educational resources, workshops, and training courses. Moreover, in USGS volcano observatories' (such as Cascades Volcano Observatory) Web sites, there are special pages dedicated to media, in which journalists can find links to events calendar, social networks, and USGS newsroom, as well as special resources (the downloadable *Media Guidebook for Natural Hazards in Washington*, for example), information for documentary filmmakers, and about media training. Many of these activities are carried out in collaboration with other public agencies (National Disaster Preparedness Training Center and Federal Emergency Management Agency).

THE CASE STUDY OF THE MUSEUM OF VESUVIUS OBSERVATORY

The Vesuvius Observatory (Osservatorio Vesuviano, OV, Italy) was inaugurated in 1845, when Vesuvius volcano was in a state of semipersistent activity. It is the first volcanological observatory worldwide. Now it is a department of the Istituto Nazionale di Geofisica e Vulcanologia (INGV). The historical old building has become a visitor center (Figure 1) while the scientific headquarter and monitoring center is in Naples at a distance of some tens of kilometers. The INGV–OV carried out many public outreach activities that include visitor center activities, temporary exhibitions, school projects, training courses, public lectures, and participation in science fairs and festivals (www.ov.ingv.it).

In 2000, a new exhibition, "Vesuvius: 2000 years of observations" has been set up inside the visitor center (museum), with two main goals: (1) dissemination of scientific knowledge about Neapolitan volcanoes (Vesuvius, Campi Flegrei caldera, and Ischia island), their status, and their dynamics, and about

FIGURE 1   The historical old building of Vesuvius Observatory. *Photo by G. Vilardo.*

research, eruption forecasting, and monitoring activities; (2) increase in volcanic hazard awareness in people living in Neapolitan high-risk volcanic areas such as Vesuvius, Campi Flegrei, and Ischia (Carapezza et al., 2000).

The OV visitor center (museum) preserves valuable collections of rocks and minerals and historical scientific instruments, such as the first electromagnetic seismograph, constructed by Luigi Palmieri in 1856 (Borgstrom et al., 1999), that was long used by OV researchers in the past.

The visitor center provides guided tours led by museum trained educators (geologists, physicians). At the moment, the visitor center is in restoration. Reopening is scheduled for the end of 2014.

The visitors of the museum are quite variable: adults, young people, children, seniors, scientists, teachers, journalists, professionals, clerks, housewives, families, tourists, disabled, and immigrants.

There is a multiplicity of needs, motivations, and expectations, and then the educational approach varies according to different targets (category of visitors). The visiting path for students coming from Naples or from other Italian regions could be quite different because in the first case hazard-related topics are emphasized.

The educational actions currently consist of guided tours, but over time, the educational offer will expand. In addition, evaluation and qualitative studies must be implemented.

Basic information about the public's perception of the museum has been collected through questionnaires distributed to visitors and compiled on site.

ANALYSIS BY/OF THE PUBLIC OF THE MUSEUM OF VESUVIUS OBSERVATORY

This section presents the results of the survey carried out between the years 2005 and 2009 among the visitors of the museum. Numerical data regarding people influxes, the origin, and kind of visitors are showed (De Lucia et al., 2010).

Categories

In the whole period (2005–2009), the main category of visitors consisted of school groups (76% of the total number of visitors), followed by groups of adults (13%), university groups (8%), and others (individuals, families, and small groups) (Figure 2).

Among school groups, the most important categories are represented by intermediate schools (49% of school group students) and high schools (45% of school group students) (Figure 3). Primary schools (6%) are poorly represented because in the years 2005–2007 they were not accepted in the museum—during that period there was a selection in school admittance on the basis of the age of students because of limits of capacity and availability of the museum.

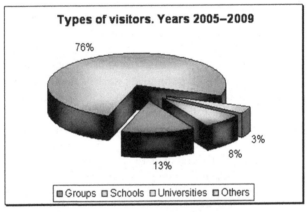

FIGURE 2   Categories of visitors to the museum during 2005–2009.

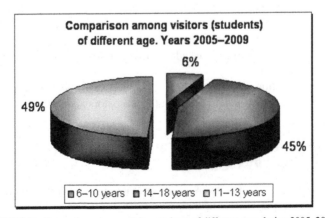

FIGURE 3   Comparison among visitor students of different ages during 2005–2009.

Temporal Distribution

In the first 4 years of the survey, the annual number of museum visitors was about 10,000 (Figure 4). In 2009, the annual visitors number chart shows a strong increase (+50%), because of more promotional activities (concerts, evening and night openings, cultural events) implemented during the year.

The distribution of guests throughout the year is strongly affected by the location of the center, 600 m above sea level on the Vesuvius volcano, as well as by the prevalence of scheduled school groups, which preferably come in the spring months (Figure 5) because of unfavorable weather conditions in winter months. The strong presence of school groups explains the incomprehensible lack of people during the months of June, July, September, and October, and highlights a critical point that will be discussed in the conclusions. However, Figure 5 shows an increasing trend over the years in particular for the spring months. In August the visitor center is closed.

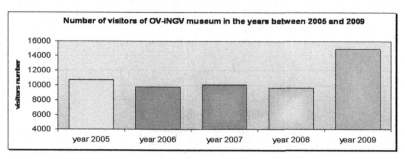

FIGURE 4   Number of visitors to the museum during 2005–2009.

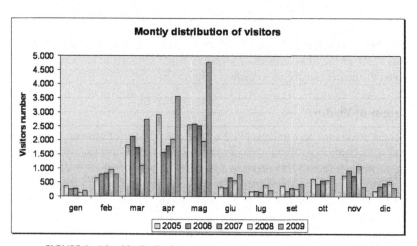

FIGURE 5   Monthly distribution of visitors to the museum during 2005–2009.

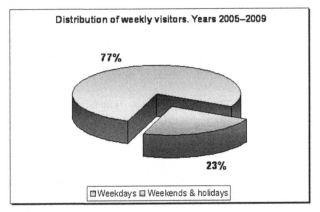

FIGURE 6   Week day frequencies of visitors to the museum during 2005–2009.

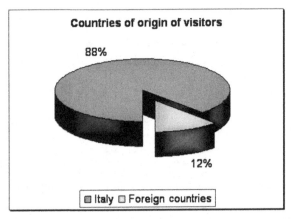

FIGURE 7   Italian versus foreign visitors to the museum during 2005–2009.

Adult groups, families, and individuals visit the museum mostly on week-ends (23% of the total number of visitors), while school and university groups attend the center mostly on weekdays (Figure 6).

Origin of Visitors

Visitors are prevalently Italians (88% of the total number of visitors) (Figure 7). Only 12% come from abroad, mainly from France (about 2800 guests in 5 years) and Germany (about 1300 guests in 5 years) (Figure 8).

A large part of Italian visitors (72%; 63% of total number of visitors) comes from Campania, which is the region surrounding Naples, in which the three volcanoes of Vesuvius, Campi Flegrei caldera, and Ischia island are located. The remaining 28% (25% of total number of visitors) comes from other regions of Italy (Figure 9).

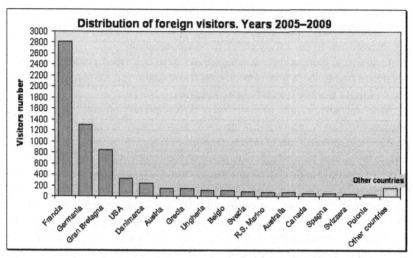

FIGURE 8   Distribution of foreign visitors to the museum during 2005–2009.

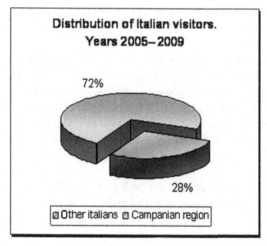

FIGURE 9   Campanian versus other Italian region visitors to the museum during 2005–2009.

Visitors come mainly from the province of Naples, which is the most populated and also the most exposed to volcano hazard.

CONCLUSIONS

Scientists have the ethical duty to answer citizens' need of information about the
status and the physical behavior of a volcano and its possible effects, together with
other actors such as emergency managers, media, and public officials. They cannot disregard citizens' right to be informed and to be involved in risk mitigation.

Scientists and other public officials must play an active role in volcano hazard communication, exploring multiple channels and languages and establishing interactions, to get a truly effective communication.

In the past, many expert volcanologists have described guidelines or personal viewpoints about how to build effective volcanic hazard communications. Most of these were based on field experience.

More recently, a more pronounced multidisciplinary approach is emerging, as demonstrated in recent EU projects. Research based on integrated methods produce in-depth analysis of the interactions of natural processes with sociological, cultural, and psychological features of the communities, and in the developing of multiple and tailored mitigation strategies, including communication aspects.

New models of communication with the public are being developed, no more top-down, but among peers. All the stakeholders of the community are involved, with different but active roles. As stated on the USGS Web site, "We are all in this together. Everyone has a role in preparedness" (http://volcanoes.usgs.gov/observatories/cvo/prepare.html).

A new reciprocal relationship between the scientific community and public is going to be established. If citizens directly benefit from the results of scientific research, it will be more likely that they will consider it necessary, and offer support.

In this framework, a permanent visitor center, managed by a public scientific research institute, is a powerful tool for effective communication, better if assisted by communication experts. In Italy, the museum of Vesuvius Observatory is a reference information point for nearly 2 million people living in the Naples area and exposed to three active volcanoes. The analysis of the visitors of the museum gives a base of data to plan a better organization, as well as to enhance cultural and educational enjoyment and to improve the quality of communication.

The numerical survey carried out on the visitors of the museum has shown that the main category consists of school groups. These (students and teachers) are specific targets on which to focus exhibition features and paths, as well as educational workshops.

Low numbers of visitors are recorded from June to October: there is a need to arrange tailored activities to attract nonschool groups and individuals.

The age of school visitors is mainly 12–18 years: much effort must be made to catch the attention of visitors between 6 and 11 years, through exhibits suitable for children of that age.

The survey could give useful information regarding organizing promotional campaigns in less informed communities in volcano hazard areas, as inferred from the ratio between the number of visitors and resident population. Moreover, qualitative evaluation studies must be implemented.

REFERENCES

Borgstrom, S., De Lucia, M., Nave, R., 1999. Luigi Palmieri: first scientific bases for geophysical surveillance in Mt. Vesuvius area. Ann. Geophys. 42 (3), 587–590.

Carapezza, M.L., Lucia, C., Orsi, G., Quareni, F., Galanti, E., Macedonio, G., Nazzaro, A., Cuna, L., Sardella, F., Di Vito, M.A., 2000. The Vesuvius: 2000 years of Observations. Presidenza del Consiglio dei Ministri Dipartimento della Protezione Civile (DPC) – Osservatorio Vesuviano. Supplement a DPC informa, n.23.

De Lucia, M., Ottaiano, M., Limoncelli, B., Parlato, L., Scala, O., 2010. Il Museo dell'Osservatorio Vesuviano e il suo pubblico. Anni 2005 – 2009. . Rapporti Tecnici Istituto Nazionale di Geofisica e Vulcanologia, n. 150, 31 pp.

Driedger, C., Doherty, A., Dixon, C., 2005. Living with a Volcano in Your Backyard – an Educator's Guide with Emphasis on Mount Rainier. U.S. Geological Survey and National Park Service. General Information Product 19 http://vulcan.wr.usgs.gov/Outreach/Publications/GIP19/.

IAVCEI Subcommittee for Crisis Protocols, 1999. Professional conduct of scientists during volcanic crises. Bull. Volcanol. 60, 324–334.

Johnston, D.M., Ronan, K.R., 2000. Risk education and intervention. In: Sigurdsson, H. (Ed.), Encyclopaedia of Volcanoes. Academic Press, New York, p. 1456.

Lockwood, J.P., Hazlett, R.W., 2010. Volcanoes: Global Perspectives. Wiley – Blackwell. 541 pp.

Marzocchi, W., Newhall, C.G., Woo, G., 2012. The scientific management of volcanic crises. J. Volcanol. Geotherm. Res. 247–248, 181–189.

MIAVITA Team, 2012. Handbook for Volcanic Risk Management: Prevention, Crisis Management, Resilience. Orleans, 198 pp http://miavita.brgm.fr/.

Newhall, C.G., Punongbayan, R.S., 1996. Fire and Mud: Eruptions and Lahars of Mount Pinatubo. University of Washington Press, Philippines. 151 pp.

Peterson, D.W., 1996. Mitigation measures and preparedness plans for volcanic emergencies. In: Scarpa, R., Tilling, R.I. (Eds.), Monitoring and Mitigation of Volcano Hazards. Springer-Verlag, pp. 701–718.

Scarpa, R., Tilling, R.I. (Eds.), 1996. Monitoring and Mitigation of Volcano Hazards. Springer-Verlag, p. 841.

Chapter 28

Voluntarism, Public Engagement, and the Role of Geosciences in Radioactive Waste Management Policy-Making

Nic Bilham

The Geological Society of London, Burlington House, Piccadilly, London, UK

Chapter Outline

Abstract

In the UK, as elsewhere in Europe, earlier "technocratic" approaches to radioactive waste management (RWM) have been abandoned. Policy-makers have recognized that for any RWM program to succeed, sustained engagement with stakeholders and the public is necessary, and any geological disposal facility must be constructed and operated with the willing support of the community which hosts it. This has opened up RWM policy-making and implementation to a wider range of (sometimes contested) expert inputs, ranging across natural and social sciences, engineering, and ethics. Geoscientists and other technical specialists have found themselves drawn into debates about how various types of expertise should be prioritized, and integrated with diverse public and stakeholder perspectives. Successful implementation of an RWM program will depend on effectively communicating geoscientific concepts and information to nongeologists. Adherence to ethical and professional standards will be vital in building and sustaining public trust in the geoscientists involved, and confidence in the information and expert judgment they provide.

Keywords: Geological disposal facility; Policy-making; Radioactive waste management; Public engagement; Geoscience communication.

Geoethics. http://dx.doi.org/10.1016/B978-0-12-799935-7.00028-9
    351

FAILURE OF THE TECHNOCRATIC APPROACH
TO RADIOACTIVE WASTE MANAGEMENT

In 1976, the Royal Commission on Environmental Pollution report "Nuclear Power and the Environment" (known as the "Flowers Report") for the first time identified long-term management of radioactive waste as a significant policy problem for the UK. It recommended geological disposal as the long-term solution to this problem, and that a geological disposal facility (GDF) be constructed as soon as reasonably possible.

Throughout the 1980s and 1990s, decision-making about radioactive waste management (RWM), including where to locate a GDF, was portrayed as a wholly technical matter by government and the bodies charged with implementing policy—initially the UK Atomic Energy Authority and then, following its establishment in 1982, the Nuclear Industry Radioactive Waste Management Executive (NIREX). Opposition from local communities and green NGOs, on the other hand, was founded principally on social and ethical concerns, and on challenges to the assumptions framing NIREX's scientific model. Several attempts were made to identify geologically suitable sites for the construction of a GDF, including exploratory drilling, but limited progress was made. Public distrust in this technocratic approach was exacerbated by secrecy—only in 2005, when ownership of NIREX passed to government, did it finally publish the list of potential sites it had considered during the 1980s and early 1990s (CoRWM, 2006; Simmons and Bickerstaff, 2006).

In 1994, NIREX applied for planning permission for a Rock Characterization Facility (RCF) at Longlands Farm, Sellafield, to test the geology at the site along with other scientific and engineering issues. This was refused by Cumbria County Council—a decision upheld at a public enquiry. With the final refusal of the planning application by the Secretary of State for the Environment shortly before the general election in 1997, the UK's RWM program was effectively suspended. A GDF siting process based on technical criteria, with identification of a suitable geology in the forefront and public acceptance something of an afterthought, had failed.

Such technocratic approaches to policy problems considered to be scientific or technical in nature, and the model of scientific advice to policy-makers which underpinned them, were widespread. RWM was not the only such policy area to be in crisis during the final months of the Conservative government.

Bovine spongiform encephalopathy (BSE) had first come to public attention in the UK in the 1980s, when it was dubbed "mad cow disease," and graphic footage of cattle suffering from this debilitating disease filled television news broadcasts. It was thought to be transgenic in origin, and there were concerns that it might be transmissible to humans. These claims were categorically denied by government, citing evidence from its expert scientific advisors. Yet shortly before the final refusal of planning permission for the Sellafield RCF, government was forced to admit that its reassurances had been false. By that time, it was clear that variant Creutzfeldt–Jakob disease, a transmuted form of BSE affecting humans, was being contracted by some people who had eaten BSE-infected beef. Some tens of people have died since as a result, though there were fears at the time that the number of deaths might be much higher.

The BSE crisis has been described by the science policy scholar Erik Millstone as "the most serious failure of UK public policy since the Suez invasion of 1956" (Millstone and van Zwanenberg, 2005). Certainly it is hard to overstate the extent to which it shook established structures of scientific advice to policy-makers, and its communication to the public. Crucially, uncertainty had been covered up over a period of years. The Phillips inquiry (1998–2000) was a powerful critique of these structures, and was highly influential in shaping extensive reform of scientific advice for policy-making, in the UK and internationally. Stilgoe and Burall have paraphrased its main conclusions as follows:

- "Trust can only be generated by openness."
- "Openness requires recognition of uncertainty, where it exists."
- "The public should be trusted to respond rationally to openness."
- "Scientific investigation of risk should be open and transparent."
- "The advice and reasoning of advisory committees should be made public."

(Stilgoe and Burall, 2013)

Several other European countries had already come to the conclusion that a purely technocratic approach to RWM would not succeed, and had embarked on processes in which engagement of the public in decision-making and informed voluntarism on the part of the host community for any GDF played a central role. Some of these programs were making real progress, while managing to maintain a much higher level of consensus than had been the case in the UK. In light of this, and the wider rethink of scientific policy-making structures precipitated by the Phillips inquiry, it was time for a fresh start to RWM in the UK.

DEVELOPING A NEW RWM FRAMEWORK FOR THE UK

The Committee on Radioactive Waste Management (CoRWM), established by the UK government in 2003, was far from being a traditional government scientific advisory committee. Its membership included natural scientists, engineers, social scientists, and those with experience of local government, with a wide variety of perspectives on RWM. CoRWM was asked to recommend a long-term solution for management of higher level wastes, looking at all viable options from scratch—rather than simply assume geological disposal should be the solution, as had previously been the case—and to make recommendations about implementation (including how to engender public trust). Also, crucially, CoRWM's review process should itself be carried out in such a way as to start to engage the public and build confidence in the outcome.

CoRWM engaged extensively with public audiences and a diversity of stakeholders, as well as drawing on a wide range of specialist expert inputs. It took an innovative approach to its deliberations, almost all of which were carried out in public, which differed sharply from the largely closed and technically led approach which had previously dominated. Some in the scientific community who adhered to a more traditional view of scientific input to policy-making (notably the House of Lords Science and Technology Committee (2004)) were critical of what they saw as a lack of scientific expertise and rigor on CoRWM's part. On the other hand, some social scientists who advocated reform of these advice structures, while generally supportive of CoRWM's work, felt that technical factors continued to be too dominant (e.g., Chilvers and Burgess, 2008). Nonetheless, by the time it reported in 2006, CoRWM was widely praised for the approach it had taken, and its recommendations commanded a high level of support. The key ones were the following:

- geological disposal of higher level wastes as soon as practicable (several decades at least);
- a robust program of surface storage in the interim;
- the principle of voluntarism on the part of any potential host community for a GDF;
- partnership of the host community with the implementing body; and
- provision of community benefits for participating in the process and for eventually hosting a GDF.

(CoRWM, 2006)

Government accepted these recommendations, and following further consultation announced a staged process for siting and implementation of a GDF in 2008. Whereas geological factors had previously been paramount in siting of a GDF, the first step in the new framework was an open invitation for any local community in England and Wales to express an interest in entering the process. There was no initial nationwide screening for geological suitability—a step which was included at the start of the process in some European nations, but which it was suggested would be difficult to achieve in the UK given its heterogeneous geology and the variety of different geological settings in which a GDF could be hosted, which include hard rocks (such as granite), clay, and salt. Nor was any public information provided on relevant aspects of geology to guide the volunteering process, as had been done in some other European countries. Only once a community had expressed an interest in entering the process would there be initial screening to rule out areas which were clearly geologically unsuitable. It was envisaged that as the staged process moved ahead, and as potential sites within a volunteer community's area were narrowed down, more and more detailed geological investigations would be carried out. Geological considerations would continue to be fundamental to the development of the safety case for construction and operation of a GDF, in conjunction with the engineering of a repository design suited to the geological setting—but geological factors would now be in dialog with societal concerns throughout the process.

Notably, a number of those who had worked on the previous RWM program under NIREX, including geoscientists, remained involved in the newly launched program, some of them working for the new implementing body—the Nuclear Decommissioning Authority (NDA). They generally embraced the change in approach to siting of a GDF, recognizing the fundamental importance of community voluntarism if such a facility is to be acceptable to its host community in the long term. Their willingness to accept such change is indicative of the cultural change brought about by CoRWM, as well as the depth of the crisis in the program which preceded it.

Only one local community, in West Cumbria (represented by two borough councils, and Cumbria County Council above them), formally entered the process, although there were preliminary informal discussions with one or two other communities. In January 2013, the County Council decided not to proceed to Stage 4 of the process, which would have initiated the identification of potential sites within the area, and desk-based geological investigations (although the more local borough councils were in favor of proceeding). Several reasons for this decision were cited, including uncertainty about the community's future right of withdrawal from the process, security and timing of community benefits, and also doubts about the likelihood of finding a geologically suitable location within the area. There was no clear NDA strategy for communicating the geological requirements of a GDF, how the repository design and engineering would interact with the geology to provide protection, what was already known about the relevant geology of West Cumbria, and how remaining geological uncertainty would be addressed later in the process. There was no effective response to two geologists who opposed locating a GDF in West Cumbria, and who cited geological arguments in arguing their case publicly, strengthening local opposition—although there was no suggestion that their views were widely supported by the geological community. This was exacerbated by the lack of clarity about what geoscientific work would be done in Stage 4 of the process, a concern which had been raised by others in the geological community. A 2011 government consultation document referred to this work only in general terms, and did not mention hydrogeology (the very matter on which the two geologists campaigning locally were focusing). In fact, NDA already had detailed plans for consideration of a wide variety of geological factors, but these remained under internal review at the point that West Cumbria withdrew from the process. It seems that while NDA had learnt some of the lessons spelled out by the Phillips inquiry, lack of transparency continued to dog the process.

With no local communities remaining in the process, government launched a review of the GDF siting process in 2013. It is expected that they will set out a revised process in the summer of 2014. As consultation documents published during the review make clear, this will not revisit the fundamental conclusions of CoRWM (geological disposal, underpinned by public engagement and community voluntarism). But it will reexamine what geoscience information might be provided in the early stages of a volunteer-led process, how plans for later geoscientific work can most effectively be communicated, and how geoscience communication can best support public engagement with the important societal challenge of responsibly managing our radioactive waste.

DISCUSSION AND CONCLUSIONS

Although it did not lead to successful identification of a location for a GDF, the siting process which was attempted in the UK from 2008 to 2013 represented a significant step forward in comparison with the previous discredited process. There are grounds for optimism that with relatively minor modifications it can now evolve into an effective participatory process, underpinned by effective public communication of the geoscience and other technical inputs, and sustained by public engagement with these technical aspects of the program as it moves forward. In particular, the review provides an opportunity to renegotiate the roles of geoscience and geoscience communication (and of their practitioners) in the process, and how these factors interrelate with other technical, social, and ethical factors.

Successful siting and implementation of a GDF will depend on effective integration of voluntarism and public engagement on the one hand, and geoscience on the other. In retrospect, however, the 2008–2013 siting process in the UK seems inadvertently to have introduced something of a disconnect between these aspects. Why did this happen?

It may be explained in part by overcompensation for the failure of the previous technocratic approach. Although the importance of public engagement and voluntarism is recognized in all nations where RWM programs are progressing well, what this means in practice varies considerably from country to country. In some cases, a host rock type has been specified before communities are invited to participate. In others, communities may not be asked to respond to an open invitation to participate, but may be involved in a later stage of the selection process. The UK government and NDA were so keen to demonstrate that the siting process was not technically led that they were reluctant to say anything much about the geological requirements of a GDF and what geologies might meet those requirements in the early stages, for fear of being thought to be imposing a technically led choice by stealth.

There has also been a tendency to view the integration of geological and other technical considerations, societal concerns and public engagement as a "zero-sum game"—as factors which can and must be "traded off" against each other, as though greater emphasis on geology would mean less importance being accorded to public engagement and public confidence, for instance. This way of thinking may have been encouraged by the use of techniques such as Multi-Criteria Decision Analysis in RWM decision-making in the UK. While such techniques can be useful tools for testing different stakeholder perspectives on the weighting of competing factors, they cannot and should not be relied upon to neutralize ethical and political judgments about how competing specialist and public inputs should be used and integrated. Nor should their use be taken to imply that these inputs can simply be traded off.

Far from constituting a zero-sum game, these factors are mutually dependent. High-quality geoscience, effectively communicated, will be essential to building and maintaining public confidence throughout the siting, construction, and use of a geological repository. Whatever the reason for this disconnect, insufficient attention has hitherto been paid to developing strategies to communicate scientific concepts and information to nonspecialists in a way which assists their consideration of the issues which matter to them.

Geoscientists have been in the vanguard of those arguing not for a technically led process, but one led by public engagement and voluntarism, informed by better communication of the relevant geology. Such a process will confront them with novel challenges. Their role is evolving from an essentially technical and scientific one, into one which will also test their abilities to communicate their knowledge and understanding to nongeologists, putting them at the heart of efforts to engage the public, and build the trust of potential volunteer communities in the process. Geoscientists are used to working with uncertainty, with incomplete data and timescales which, to most people, are unimaginably long. To those not used to thinking and working in this way, uncertainty can be mistaken for simple ignorance, which may critically undermine confidence in the safety of a GDF. So it will be vital for geoscientists working in the program to communicate how they characterize and manage uncertainty, and how this can be constrained and reduced by further data gathering, for example.

In participating in public deliberations about RWM, which have social and ethical as well as technical dimensions, it is also important that geoscientists recognize that their own statements (including those which may seem to be entirely technical) are value-laden, and are framed by their own political, personal, institutional, and ethical assumptions. This raises the need to maintain professional and ethical standards within recognized frameworks which can inspire the confidence of members of the public with whom they come in contact. There is a vital role to be played here by learned and professional societies. As in the case of the Geological Society of London in the UK, their members are typically bound by a code of conduct, which links professional formation and development to a framework for ethical professional behavior. These organizations can also help to provide access to impartial technical advice, identify authoritative individuals to peer review the work of others, and comment on the use of geoscience within RWM policy-making and implementation on behalf of their professional community.

This paper has focused on RWM in the UK, but the challenges facing geoscientists as they take on novel roles in this process are shared by those in other nations. Nor will these challenges be limited to RWM. They will increasingly come to the fore in other areas of public interest, such as shale gas and water usage, where we need to manage our use of resources and the attendant environmental impacts sustainably, under conditions of uncertainty and competing public and stakeholder values. Geoscientists can make an invaluable contribution to enabling informed public debate and decision-making about these fundamental twenty-first century challenges.

REFERENCES

Chilvers, J., Burgess, J., 2008. Power relations: the politics of risk and procedure in nuclear waste governance. Environ. Plann. A 40, 1881–1900.

CoRWM, 2006. Managing Our Radioactive Waste Safely. CoRWM's Recommendations to Government. Committee on Radioactive Waste Management, July 2006. CoRWM, London.

House of Lords Science and Technology Committee, 2004. Radioactive Waste Management. 5th Report of Session 2003–2004. House of Lords, London.

Millstone, E., van Zwanenberg, P., 2005. Mad Cows and Englishmen: BSE, Risk, Science and Governance. Oxford University Press, Oxford.

Simmons, P., Bickerstaff, K., 2006. CARL Country Report. United Kingdom: Summary. CARL, Antwerp.

Stilgoe, J., Burall, S., 2013. Windows or doors? experts, publics and open policy. In: Doubleday, R., Wilsdon, J. (Eds.), Future Directions for Scientific Advice in Whitehall. Alliance for Useful Evidence, London.

Chapter 29

Nuclear Waste Repositories and Ethical Challenges

Peter Hocke
Institute for Technology Assessment and Systems Analysis, Karlsruhe Institute of Technology, Karlsruhe, Germany

Chapter Outline

Abstract

The analysis of the concrete management of radioactive waste and the ethical discussion on this topic show specifically that the reflection on side effects of human action in most cases is not as clear in its differentiations as it would be possible and necessary. Collective actors calculating the "costs" and "risks" of side effects in relation to future generations misunderstand that the testing of "claims"—isolated from contexts—does not lead to processes of societal deliberation. A discussion of central positions in German philosophical ethics in the field of radioactive waste management, which are connected with the research team of Chr. Streffer, C.F. Gethmann and colleagues, and with R. Spaemann, shows fundamental differences in their lines of argumentation. As contexts, especially expressed in technological numbers, also play an important role in the debate, the limited amount of waste in countries with phase-out decisions concerning nuclear power makes it easier for all stakeholders to develop modes of civil governance and societal deliberation. These modes of governance and deliberation are important for open assessments of ethical and nontechnical knowledge in the same way as knowledge from natural science and engineering, in cases where a decision for underground repositories for nuclear waste has been made.

Keywords: Ethics; Nuclear debate; Philosophy; Radioactive waste management; Stakeholder.

Geoethics. http://dx.doi.org/10.1016/B978-0-12-799935-7.00029-0
359

INTRODUCTION

The optimistic position supporting the idea that advanced technologies are neutral and do not cause unintended side effects has become fragile, especially in the last decade. The struggle in most Western societies about the use of civil nuclear power as one of the catalysts has shown that a robust integration of ethical concepts of long-term responsibility for developments in research and economy is rather complicated. Technologies for solving political or socioeconomic problems (such as high-tech weapons for peacekeeping or mining activities to gain cheap resources from everywhere in the world) have provoked debate about the relevant criteria for taking decisions on technological developments and assessing their side effects.

The idea that ethical reflection itself designs universal and in some way abstract general tools to react on problems, which are framed as "risks" or "hazards," has predominated in science for decades. Some researchers in applied sciences argue that concepts integrating ethical reflection can influence social processes by assisting responsible collective actors like governments or industry to solve sociotechnical problems like air pollution or unintended impacts of waste disposal in a direct way. By them qualified scientific risk assessment and classical representative forms of government were seen as instruments with a specific catalytic effect. This expected effect is to avoid environmental or geological problems such as the destruction of landscapes. Meanwhile, this argument is under debate. The ethics of nuclear waste management is an instructive example of this.

In the next section, I will sketch the social challenge of radioactive waste management as a "double-complex" problem and section 3 describes German positions in the philosophy of ethics as brought up by Streffer et al. (2011) and by Spaemann (2003, 2009). Following their definition of the problem, I will discuss the limitations and challenges of the concepts of radioactive waste management in countries committed to Western patterns of argumentative science and democratic politics. This discussion is based on research and experience in technology assessment and social sciences. In an empirical sense, this essay will reflect on the gradual opening of nuclear waste management for nontechnological perspectives, particularly in Germany and Switzerland, as the author is more familiar with than with the developments in other countries.

THE SPECIFIC COMPLEXITY

In the beginning, the management of nuclear waste resulting from the use of nuclear power for electricity generation, which in many industrialized countries started in the 1960s (see Högselius, 2009), was considered a problem to be solved by science in a linear way. In the course of time, however, it became evident that the challenge of managing nuclear waste under high-quality standards is much more difficult than expected for several reasons. With the ban on nuclear waste dumping at sea, which was done by several Western countries up to the 1970s, and after several conceptual failures in reprocessing of high-level nuclear waste, a new concept became dominant: The current generation will care for its own waste from nuclear power plants within its lifespan on its own territory (see Grunwald, 2010, p. 74).

The technological concept of translating this decision into collective action in countries like Sweden, Germany, or Switzerland was to construct an underground repository, which would be closed after an operation phase of some decades and would be maintenance-free from then on.[1] However, it had not been expected that the long-term management of some types of nuclear waste is much more difficult than that of others. While low- and medium-level radioactive waste can be isolated from humans and the environment when some time has passed and the challenges for isolation are limited, high-level and heat-generating waste (mostly spent fuel from nuclear power plants) have very different temporal requirements: Safe isolation must be ensured for a time span ranging from some ten thousand to a million years.[2] Given these assessed time spans, every approach to constructing underground repositories for such types of waste is really ambitious. Consequently, the main focus of attention involved geology in a special way. The time spans analyzed for the past are extrapolated into the future and the intended isolation period. By the proponents of maintenance-free underground repositories, the criteria for assessing the suitability of potential host rocks were based on the assumption that certain geological structures such as salt formations in the deep underground are inherently stable and self-repairing. It was argued that the salt would have already disappeared in the past if this would not work. In this line of argumentation the crucial point is that the site selection process is now formed by collective human action, including a specific form of responsibility for choosing a suitable site.

Under these conditions, ethical reflection, social sciences, and technology assessment gain in interdisciplinary importance. On the one hand, concrete science-based decisions related to the identification and selection of suitable siting areas are technologically and conceptually challenged by experts, since knowledge about the conditions in deep geological formations in specific areas is limited. On the other hand, in every potential siting area the population affected has concerns regarding their own safety, their health, and the level of hazards they are willing to accept. Natural-science-based siting decisions do, in the most cases, not take account of this form of social complexity and the expectable social cleavages stimulated by most siting projects nowadays. In some countries, radioactive waste management activities in this field led to intense national conflicts (like in Switzerland and Germany in the 1980s and 1990s). In other countries (such as Sweden) the conflict was less intense and

---

1. AkEnd, 2002, p. 24.—AkEnd was an independent expert committee on a site selection procedure for repository sites in Germany with high reputation. It existed from 1999 to 2002 and consisted of members who represented different views on the use of nuclear energy.
2. For the classification of waste see Streffer et al., 2011, pp. 112–116.

a national consensus about the siting procedure could be reached. The national debates on this topic can also be differentiated by the way ethical considerations were addressed. The thesis here is: If the issue of fair and adequate procedures for site selection is not addressed, conflicts are more likely than consensus. To discuss this thesis, it makes sense to look at the philosophical discussion of the issue.

NORMATIVE ETHICS AND SPAEMANN'S CONTRADICTION

A screening of literature to identify the state of research on this ethical issue in academic philosophy in the German (and also in the international) context showed that the corpus of literature is very limited (see Hocke, 2013). A debate in the classical sense cannot be identified. However, the positions of Streffer et al. and Spaemann are very instructive as they represent different schools of ethical argumentation. The small ethical debate on fair and adequate procedures to reach robust and informed decisions in the siting, closure, and postclosure process for extraordinary timescales is fought with arguments and concepts of linear reduction of problems (Boetsch, 2003; Birnbacher et al., 2006). From my point of view, philosophical ethics cannot establish normative criteria, which—as in numerical safety analysis—can be used to generate self-steering mechanisms of adequate decision-making by following ethical standards.

The differences in ethics arise from the "qualification of claims" required for a reasoned postulation. In this context, the team Streffer et al., including the German philosopher Carl Friedrich Gethmann, stresses the importance of claims of action with respect to radioactive waste management, characterizing them as universal (2011, pp. 230–240). These universal claims are reconciled with Kant's categorical imperative, stating that "wherever our actions are meant to be guided by ethical ('sittliche') considerations, no one is available as a 'mere means' for other people's ends, but each person must always be considered 'an end in himself' at the same time" (Kant cited in Streffer et al., 2011, p. 237). Following from the claim of a universal law, the respective universal norm involves a moral "obligation" or claim, and these claims can serve as criteria.

The concept of Streffer et al. is based on two assumptions: on the one hand, ethical reflection must be independent from social and political concerns (Streffer et al., 2011, p. 234). On the other hand, the fundamental positions in the conflict are explained by the need to develop instruments which allow not only a safe, but also a nonviolent solution of the nuclear waste disposal problem. So not every moral concept (individual and collective) is relevant for qualified decisions, but the qualification of claims justifies actions and decisions.

In the context of these ethical and applied aspects of nuclear waste management, Streffer et al. mention three systematic misunderstandings. The first is the requirement for "immediate response" to the uncomfortable situation with problematic nuclear waste on the surface by the responsible actor now. Their argument is that immediate response is only necessary in specific emergencies. According to them this does not apply to the disposal of high-active waste (Streffer et al., 2011, p. 243).

"Fairness" is the second misunderstanding, which needs to be put into perspective in conflicts like this one. In these authors' view, fairness does not always provide a sustainable foundation for successful conflict resolution. Instead, they suggest using certain forms of bartering and, in some sense, of trading. The "bartering of burdens and benefits" is acceptable if it leads to an "optimized overall distribution" (Streffer et al., 2011, p. 244).

Third, Streffer et al. argue that the "polluter pays" principle is sometimes overstressed. According to the authors, temporal and spatial limits of responsibility by generations are historically contingent. However, "mechanisms of exchange across these boundaries could be established" (Streffer et al., 2011, pp. 245–246). In a certain way, their argumentation is based on social analogies, which cannot be discussed in detail here. However, it should be noted that they differentiate between the competence of science on the cognitive and of lay people on the evaluative level (Streffer et al., 2011, p. 258).

The approach of the German philosopher Robert Spaemann differs from that of Streffer et al. in a number of ways. Two of them are particularly worth mentioning. The first is Spaemann's essayistic style of argumentation, which differs to the claimed systematic approach of Streffer et al.; in this context, Spaemann underlines the importance to clarify now what constitutes "good" life.

The second is his analytical angle: His subject is the human being with its problems in decision-making in the context of underground repositories. Here, his perception and conceptualization of the "following generations" is interesting. Beside his general positions his key concepts are the "idea of man" and the "power of judgment". The power of judgment depends on whether the person is a scientist, an expert, or a citizen. Spaemann stresses the importance of "respect" and the "dignity of man" in an argumentative way (Spaemann, 2003, p. 29). The central characteristic of men to be highlighted in the context of nuclear waste management is human competence, or the ability to resolve conflicting interests by "sympathy" for the other person with his or her legitimate interests (Spaemann, 2003, 2009).

In his particular way of discussing nuclear waste problems and conflicts, he argues that people should be drawn away from the comparative perspective between our generation and the following ones. The ethical problem in nuclear waste management is for Spaemann that the hazard lies in the "interest of current generations". This hazard will be discharged into deep underground and then it will "threaten the life of generations in the far future" (Spaemann, 2003, p. 29; translated by the author). Generations in the far future are in this sense only "virtual partners in the discourse"; reciprocal treatment and interactive communication with them are missing (as they are not possible), and in this way communication is structured by the dominating power of the present generation (Spaemann, 2003, p. 31).

Following this concept, he differentiates between "generations in the far future" and "generations close-by". As the intrusion by man into geological spheres and by this the intrusion into the physical "infrastructure of life" is very intensive, the long-term impacts resulting from this action are serious ones in comparison with classical long-term impacts of human action. His diagnosis is that the present generation is in a "perplex situation," like it was described in the scholastic school of philosophy. It is not possible to make a correct decision because former wrong decisions force us to manage the problem. However, today mankind has the choice to decide between "better" and "worse" alternatives. The crucial question for him is that of a "good live" and, in this sense, an established one in the philosophy of ethics (Spaemann, 2003, p. 36).

RESPONSIBILITY IN LATE-MODERN SOCIETIES

On the argumentative level, Spaemann's approach focuses on mankind in a humanistic way, which is oriented toward the dignity of life and prioritizes for the next generation over generations in the far future. From a theoretical perspective, the concept of normative ethics used by Streffer et al. underlines abstract obligations and claims and can therefore contribute to the debate about radioactive waste management in a different way; and by this it fits into a technological concept of waste management where geology (and not collective action of concrete stakeholders and individuals) provides the suitable safety in host rocks. But these discussions leave a mixed impression. It is not obvious whether they make sense for experts working in interdisciplinary teams on research issues of nuclear waste management or for experts at the front end of decision-making. Complex social processes and the managing of arguments from influential and very different stakeholders structure their business.

Spaemann's strength is his interpretation of long-term responsibility of current generations and their possibilities in line with the classical humanistic tradition. Why not? Life and societal self-organization serves the people and not the well-being of subsystems (like engineering communities or economy) and their systemic attempts to solve every messy problem. Streffer et al., on the other hand, argue that conflict resolution has to be developed not by structures of formal power and authorities, but on the base of convincing arguments. And they miss the concrete interests of social actors as they stress the universal law by Kant and the argument of a moral obligation.

The bridging idea which has not yet been explained is the "civility" in conflict resolution and interest aggregation. Civility is substantial in the evaluation of different alternatives of radioactive waste management and final disposal in repositories. It is not about choosing the best technological solution, but citizens, stakeholders, politics, and industry should be involved to a considerable extent in order to solve this problem of long-term chemotoxicity and hazardous radiological processes with convincing arguments, public deliberation, and civil standards for balancing different interests.

It remains to be seen whether late-modern societies are prepared to solve such "wicked problems" of planning (Rittel and Webber, 1984), where no simple and really satisfactory solutions for the complex long-term problems can be expected. In this context, the responsibility of involved scientists and experts, the interested public, and stakeholders as well as processes to integrate these actors gain in importance. They have to face the fact that they are confronted with a "transgenerational project" with extreme challenges of long-term stewardship and "middle-term" technological and social monitoring on the one hand. On the other hand, the question is how to develop professional solutions, where regional interests (nice landscape, no waste), political interest groups (for and against nuclear energy or the siting of repositories) and sometimes also the industry are engaged in this waste management, and where the integration of civil society and classical governmental organizations is constitutive for debate and planning. Do these governmental organizations and actors from industry and civil society have the relevant knowledge from all disciplines? Are they willing to cooperate in a next step and appreciate the advantages of compromise?

Too many failures in the history of technology governance, too dramatic accidents, and too unprepared authorities belong to our experience. They all challenge decisions about responsible innovations, for extremely long timescales as well as the confidence in the "safety" of the involved subsystems. In the same context, well-known stakeholders argue for sustainability as a supranational strategy with more or (often) less success; and the same collective actors would have to agree on a sociotechnical long-term "experiment" for highly active nuclear waste, which is known as a serious problem with unclear impacts—an experiment, which has not been finished somewhere, up to now.

Land-grabbing and mining activities in the field of uranium supply, as well as political instrumentalization of nuclear interests—also in the civil use of nuclear energy—show the challenges, which have to be addressed from a social and technological perspective. Interdisciplinary governance research with its fondness for the need of early participation of the affected and interested public e.g., tries to address these challenges (Kuppler, 2012). However, as every concept from social sciences has its shortcomings, no one knows whether constructive collective action in "baby-steps" of reforming technological politics will be effective enough or whether the highly dynamic processes of economics, the North–South conflict or other global developments, will overwhelm these small fields of collective action as the one in nuclear waste management.

CONCLUSIONS

Does the context matter? Especially in the case of this type of waste management the context plays an important role. From an ethical perspective on the one hand it does not make a difference, which amount of nuclear waste canisters with which type of waste (high- or low- and medium-level radioactive nuclear waste) are in the "game". One central question is whether there will be a robust decision to site a repository for this type of dangerous waste on the own territory or not. On the other hand the context does matter, as the general problem of long-term effects has something to do with quantities. Are negative effects of the stored waste to be expected only in a limited geological area or in a larger one, in one or in many regions, nowadays or in the far future—all these questions have to do with quantities. The space potentially affected by the migration of stored nuclear waste to the environment differs depending on whether the amount is limited by a phase-out planned now for a certain number of reactors on a limited national territory or not. If the responsible government has decided to increase the number of nuclear power plants within the near future, the question is open, which amount of this waste has to be managed. In this case the number of tons of burnt fuel will be limited. In the other case it can be extrapolated by factors such as the type of nuclear power plants and the type of radioactive waste management. But in every case the interpretation of those numbers happens in a broader context. A central part of this context is the role that renewable energies will have in the close future. In addition, the results can differ very much from nation to nation.

So the ethical question of "obligations" and the concrete arguments by Streffer et al. underline another type of reflection than Spaemann's orientation on the responsibility of the next generations and the impossibility to reflect their expectations within our generation, as dialog with them is not possible. At a societal level, ethical reflection is of particular importance if there are serious tensions between the positions of the "collective actors" interested in this topic and all other stakeholders (see Chapter 28). If there are tensions and societal conflicts, the sociopolitical actors have to decide on the mode of conflict resolution. Are proactive civil procedures the favored path for decision-making? Or is reliance on the established (power) structures of late-modern societies dominant? Every decision in this context is related to the concrete relevance, which the current society gives to civility. A limited amount of waste makes the problem of radioactive waste management not really easier. But it gives stakeholders, industry, citizens, and governmental organizations the chance to decide now and to limit the "evil" of struggle about the site and risk of contamination of surface and underground in a serious way. In countries like Germany and Switzerland, decision-makers in radioactive waste management and society are in a more or less comfortable situation. The amount of nuclear waste (especially in its high-active form) is limited, and the transformation of the energy system to a modern ecological mode of electric power production offers a chance for politics, industry, and society as unintended side effects are reduced. Still, the task remains challenging also in this situation. The following generations in the sense of Spaemann will be the evaluators of their success—on the infrastructural level and on the level of political culture and civility.

REFERENCES

AkEnd, 2002. Site Selection Procedure for Repository Sites. Recommendations of the AkEnd (Committee on a Site Selection Procedure for Repository Sites), 248 pp.

Birnbacher, D., Gelfort, E., Schwager, J., Schwarz, D., Tietze, A., 2006. Ethik und Kernenergie. VDI Energietechnik, Düsseldorf.

Boetsch, W., 2003. Ethische Aspekte bei der Endlagerung radioaktiver Abfälle. Abschlussbericht (Schriftenreihe Reaktorsicherheit und Strahlenschutz BMU-2003-619). BMU, Bonn.

Grunwald, A., 2010. Ethische Anforderungen an nukleare Endlager. Der ethische Diskurs und seine Voraussetzungen. In: Hocke, P., Arens, G. (Eds.), Die Endlagerung hochradioaktiver Abfälle. Documentation of the Conference "Internationales Endlagersymposium" Berlin, 30.10. to 1.11.08 (pp. 73–84). Karlsruhe. download 28.2.14 http://www.itas.fzk.de/deu/lit/2010/hoar10a.pdf.

Hocke, P., 2013. Endlagerung hochradioaktiver Abfälle. In: Grunwald, A. (Ed.), Handbuch Technikethik. Metzler, Stuttgart, pp. 263–268.

Högselius, P., 2009. Spent Nuclear Fuel Policies in Historical Perspective: An International Comparison. Energy Policy 37, pp. 254–263.

Kuppler, S., 2012. From government to governance? (non-)Effects of deliberation on decision-making structures for nuclear waste management in Germany and Switzerland. J. Integr. Environ. Sci. 9, pp.103–122.

Rittel, H.W.J., Webber, M.M., 1984. Planning problems are wicked problems. In: Cross, N. (Ed.), Developments in Design Methodology. John Wiley & Sons, Chichester, pp. 135–144.

Spaemann, R., 2003. Ethische Aspekte der Endlagerung. In: Baltes, B. (Ed.), Ethische Aspekte der Endlagerung (Schriftenreihe Reaktorsicherheit und Strahlenschutz, BMU—2003-620). BMU, Bonn, pp. 25–36.

Spaemann, R., 2009. Moralische Grundbegriffe. Beck'sche Reihe, München.

Streffer, C., Gethmann, C.F., Kamp, G., Kröger, W., Rehbinder, E., Renn, O., Röhlig, K.-J. (Eds.), 2011. Radioactive Waste. Technical and Normative Aspects of its Disposal. Springer, Berlin.

Section VI

LOW INCOME AND INDIGENOUS COMMUNITIES

Chapter 30

Mining in Indigenous Regions: The Case of Tampakan, Philippines

Daniel Hostettler
Fastenopfer, Lucerne, Switzerland

Chapter Outline

Abstract

The neoliberal globalization of trade and investment allows capital and products to flow with almost no restrictions between countries and markets. At the same time, global demand for natural resources soared. The high turnover in the raw material sector makes it profitable to invest in remote and inaccessible regions. Often, the people affected by business intervention in these areas are indigenous communities. The international human rights framework recognizes a series of specific rights for indigenous people. Indigenous people are entitled to set their own priorities for the development of their communities. If measures coming from outside affect their specific values and way of life, indigenous peoples must be consulted in order to obtain their free, prior, and informed consent. Thus, the appropriate way of trying to reconcile the often diverging interests of business and indigenous peoples would be to engage in such a consultation process, allowing indigenous peoples to know the risks and opportunities of a project from the very beginning. Recent case studies on conflicts concerning raw material projects in indigenous territories show that due consultation processes with indigenous peoples did not take place or were not in compliance with international human rights standards. A Human Rights Impact Assessment carried out in 2013 suggests that this may also be the case for the Tampakan

Geoethics. http://dx.doi.org/10.1016/B978-0-12-799935-7.00030-7

Copper–Gold Project of Swiss-based Glencore Xstrata, in Mindanao, Philippines (INEF, 2013). The initial consultation processes with the local indigenous people may not have met international standards. Today the context of the mining project is so violent that the conditions for a correctly carried out consultation process are not given, according to the above-mentioned assessment. It follows that without an adequate de-escalation strategy and an implementation of due consultation processes with the local indigenous people, the violence threatens to grow.

Keywords: Free prior and informed consent; Glencore Xstrata; Indigenous people; Mining; Raw material; Tampakan.

INCREASING DEMAND FOR NATURAL RESOURCES

Global economic development in the last two decades has created a huge demand for natural resources. In emerging economies such as China, India, and Brazil, the need for raw materials increased sharply. According to the WTO, worldwide imports of fuels and mining products rose from around USD 500 billion (1990) to more than USD 4 trillion (2012)[1], which represents an increase by a factor of 8. At the same time, neoliberal globalization, with its deregulating trends, led to the gradual removal of national legal barriers, such as import duties, opening up markets to unprotected competition. Today, the flow of both capital and goods between countries and markets is largely unrestricted. The consistently strong demand and unimpeded access have, in recent years, encouraged a price trend that made raw material extraction profitable even in the most remote and inaccessible terrains. The consequences are investment in ever more fragile environments and socially sensitive regions. Many of these regions, which are mostly in developing countries, are inhabited by indigenous communities.

For many countries in the Global South, natural resource extraction is a key element in the State's economic policy. The sale of natural resources and the accompanying investment of foreign capital lead to what is often considerable economic growth. However, paradoxically, a large proportion of the population continues to live in poverty in many of these countries. There is no correlation between increasing raw material exports and the fight against poverty. On the contrary, there seems to be a link between an abundance of natural resources and underdevelopment given that, according to Revenue Watch, two-thirds of the world's poorest people live in resource-rich countries (Revenue Watch Institute, 2010)—and indigenous peoples

---

1. Imports by China of fuels and mining products rose during the same period from USD 2.8 billion to more than USD 500 billion, those of India from USD 8.4 billion to more than USD 200 billion: http://stat.wto.org/Home/WSDBHome.aspx?Language=E.

worldwide are among the poorest of the poor (World Bank, 2010)[2] (e.g., Zúñiga et al., 2015).

Many countries in the Global South lack the means and the capacity, but also often the will, to effectively regulate and monitor the activities of transnational mining corporations. Weak State structures and deficiency of the rule of law, inadequate enforcement bodies, corruption (e.g., Bilham, 2014, 2015), as well as frequently a deep-seated cultural racism toward indigenous peoples are reasons why the interests of those directly affected are scarcely taken into account, despite relevant national legislation.

According to a survey carried out by the Special Rapporteur on the rights of indigenous peoples, James Anaya, the majority of consulted indigenous communities do not see any positive impact of mining operations within their territories whereas these operations are among the most significant sources of abuse of the rights of indigenous peoples worldwide (Anaya, 2011). There is indeed growing awareness among the public and politicians of the need for special protection of the rights of indigenous peoples, which has been enshrined in international agreements. However, numerous examples in recent years show that, in practice, these are being implemented rather tentatively (e.g., Indigenous Peoples Links, 2013).

INDIGENOUS PEOPLES' RIGHT TO DETERMINE THEIR OWN DEVELOPMENT

Because of their history, values and way of life, the UN human rights framework recognizes specific rights for indigenous peoples. Two agreements are fundamental in this.

ILO (1989), the International Labour Organisation (ILO) adopted Convention 169 Concerning Indigenous and Tribal Peoples in Independent Countries. The convention states that indigenous peoples shall have the right to decide their own priorities for the process of development and to exercise control over their own economic, social, and cultural development. From this self-determination of their own development, the convention derives that indigenous peoples must, in case of measures which may affect them directly, be consulted through appropriate procedures. Such consultations shall be undertaken by the State with the objective of achieving agreement or consent to the proposed measures. The ILO Convention 169 states expressly that the rights of the people concerned to the natural resources pertaining to their lands shall be specially safeguarded. The right of indigenous peoples to be consulted is also referred to in issues relating to land and resources. In article 16 on the issue of relocations, the term of *free and informed consent* is introduced.

---

2. "Indigenous Peoples worldwide continue to be among the poorest of the poor and continue to suffer from higher poverty, lower education, and a greater incidence of disease and discrimination than other groups."

The right to free, prior, and informed consent (FPIC) is incorporated in the United Nations Declaration on the Rights of Indigenous Peoples, adopted in 2007, as a general duty which States must comply with before adopting and implementing legislative or administrative measures that may affect these peoples. This includes any measure that negatively affects the specific values and way of life of the indigenous peoples. A significant and direct impact on indigenous peoples' lives or territories is not permitted without FPIC by those affected (Anaya, 2009).

The right to FPIC is not an abstract right, but rather the precondition for respecting a large number of other rights, such as the right to participation, the right to information, and the right to self-determination. As a result of its incorporation and elaboration by other UN Treaty Bodies and Special Procedures, the right to consultation of indigenous peoples is now widely recognized, even though the 2007 UN Declaration is not binding. However, there are no consolidated standards as to what exactly FPIC requires. Correspondingly, different interpretations exist. According to First People Worldwide,[3] the following definition may be possible:

Definition of Free, Prior, and Informed Consent

Free
Communities must be free to participate in negotiations that affect them without force, intimidation, manipulation, coercion, or pressure by the government, company, or organization seeking consent.

Prior
The community must be given a sufficient amount of time to review and consider all necessary information and to reach a decision before the implementation of the project begins. Because every community is different and has different decision-making processes, the community and only the community must decide how much time it needs.

Informed
The interested parties must provide adequate, complete, relevant information to the community so that it can assess the potential pros and cons of a particular action. Information must be provided in a form that is easily accessible to the community, including translated documents and media and descriptions of proposed actions that can be understood by a layperson. [...] It is also crucial that the community has access to independent, neutral counseling and the necessary legal and/or technical expertise to understand all of the potential results of the proposed action.

Consent
The community must have the option of saying "yes" or "no" to the project before planning begins, along with a detailed explanation of the conditions under which consent will be given. This decision must be respected absolutely

---

3. http://firstpeoples.org/wp/fpic-101-an-introduction-to-free-prior-and-informed-consent/.

by all interested parties. The community must also be given the opportunity to provide feedback at every stage of project development and execution to ensure that the conditions of consent are met. If the conditions of initial consent are not met, the community must have the option of withdrawing its consent and all interested parties must immediately cease any part of the project to which the community had not agreed.

The right to free, prior and informed consent gives indigenous peoples—at least in theory—a powerful tool that is intended to help redress the imbalance of power between indigenous communities and the much stronger State structures and economic forces. Ultimately, what is at stake is the protection of their interests in an environment that is unfavorable for them.

DIVERGING INTERESTS BETWEEN INVESTORS AND AFFECTED COMMUNITIES IN MINING PROJECTS

Large-scale projects in indigenous areas repeatedly lead to serious environmental destruction and . dominates (Ruggie, 2006).[4] As numerous studies (e.g., First Peoples, 2013; Wetzlmaier, 2012) show, mining projects can lead to a wide range of negative impacts on indigenous peoples, which all ultimately mean the violation of the human rights of the affected population. James Anaya meticulously lists the results of a survey relating to this issue (Anaya, 2011). The negative effects include, for instance,

- indigenous peoples' loss of control over land and natural resources,
- devastating impacts on indigenous peoples' subsistence economy,
- environmental degradation,
- water contamination, which has harmful effects on the health of people and cattle,
- adverse impact on indigenous peoples' social structures and cultures, particularly as a result of relocations or forced emigration,
- negative restrictions resulting from nonindigenous migration into indigenous lands, owing to the construction of roads and other infrastructure in previously isolated areas.

Added to this is frequently the militarization of the region by regular troops or private security forces in response to resistance, which, in addition to repression, leads to many other harmful side effects (Anaya, 2011). From this analysis, Anaya draws the conclusion that in its prevailing form, the model for advancing natural resource extraction within the territories of indigenous peoples appears to run counter to the rights of indigenous peoples.

---

4. "The extractive sector—oil, gas, and mining—utterly dominates this sample of reported abuses with two-thirds of the total."

In the same report, however, he also comes to the conclusion that among indigenous peoples and governments, as well as the corporations involved, which are often transnational, awareness is growing with regard to indigenous rights. Above all, there is growing recognition of the need for appropriate consultation processes. If no such consultation processes take place, mutual agreement is difficult to achieve. This leads to conflict situations, which—as experience shows—can easily result in violence on both sides.

Responsibility for the correct implementation of consultation processes that comply with the conditions of FPIC rests first and foremost with the State. However, the companies also have an obligation to act with due diligence to ensure that the human rights of those affected are respected—independently of States' abilities and/or willingness to fulfill their own human rights obligations (Ruggie, 2011). A company's obligation to exercise due diligence also includes verifying whether the State concerned has fulfilled its obligation to consult, whether the consultation process meets international standards, and whether the indigenous population has given its consent. Anaya asserts that companies should refrain from making investments if the State has not fulfilled these obligations (Anaya, 2010).

THE CASE OF THE TAMPAKAN COPPER–GOLD PROJECT IN THE PHILIPPINES

The following example of the Tampakan Copper–Gold Project demonstrates the complexity regarding FPIC processes in mining projects in indigenous regions. A Human Rights Impact Assessment (HRIA) published in 2013 (INEF, 2013) comes to the conclusion that the consultation processes with the indigenous communities does not meet international standards, and that consequently, the conditions for consent by those affected according to FPIC are not given.

The Tampakan project is located in the south of Mindanao and is managed by Sagittarius Mines Inc. (SMI), in which the Swiss-based company Glencore Xstrata owns the majority part of the controlling equity interest. Tampakan is set to become the largest copper mine in Asia, with an annual output of 375,000 tonnes of copper and 360,000 ounces of gold, over a period of 17 years. The entire project would occupy an area of roughly 10,000 ha, 11 indigenous communities would be directly or indirectly affected and 5000 people would have to be resettled. The project has completed the exploration phase, but the construction of the infrastructure for the open-pit mine is currently blocked due to an open-pit ban by the district government. At present, it is not possible to predict when the actual mining phase can start.

Tension has been triggered in the region as a result of the project. The indigenous peoples' attitude to the project is divided. Following attacks by the communist New People's Army and armed indigenous groups on the company infrastructure, the region is now heavily militarized. Army, police, and paramilitary units, partly funded by the company, have been moved to the region to protect the company.

Armed clashes since autumn 2012 alone have resulted in eight deaths among indigenous peoples and security forces. In 2013, the population of the most affected community found themselves constrained to live during several months as refugees in neighboring villages, for fear of attacks by the security forces.

In response to requests from concerned partner organizations to get actively involved, three aid organizations (Misereor, Fastenopfer, and Bread for all) decided in 2012 to commission an HRIA of the project. The UN Guiding Principles on Business and Human Rights stipulate that companies have a responsibility to respect human rights, and that consequently they must act with due diligence. A company can only claim not to be violating human rights, if it carries out a proper assessment. The Guiding Principles formulate concrete instruments of how a company should exercise due diligence. The key instrument for this purpose is the HRIA (Ruggie, 2011). SMI itself has not carried out such an assessment (INEF, 2013).

As described above, it is first and foremost the State's responsibility to guarantee the enforcement of indigenous rights and to ensure companies act with due diligence. There is no evidence that the Philippine State has so far adequately fulfilled this obligation in the Tampakan case. It should have made sure that the FPIC processes comply with international standards as well as Philippine legislation. This would have required an active role by the State: the processes should have been adapted to the specific case, analyzed and identified the indigenous social structures and decision-making mechanisms, as well as developed adequate methods of involving the indigenous population. The State should further have ensured that the imbalance of power between the company and the communities is redressed, it should have guaranteed that information is imparted in a balanced way, and that the potential risks of the project are sufficiently taken into account. Furthermore, the State should have guaranteed that the ongoing consultations take place in a peaceful context, something that has not be given after the different arm clashes in recent years and the following militarization of the region.

The Philippine State has taken on few of these tasks. A large proportion of the shortcomings listed below in the FPIC process of the Tampakan project is due to failures by the State. However, these shortcomings do not release the company from its responsibility. Rather, the State's shortcomings oblige the company to take account of these in its assessments and to adapt its conclusions and strategies correspondingly—always with the aim of respecting human rights, and specifically indigenous rights.

With regard to FPIC, the HRIA commissioned by the aid agencies arrived at the following conclusions (INEF, 2013):

Free

An FPIC process requires that the affected indigenous peoples are able to accept or reject the project without any external pressure or manipulation. In the Tampakan mining project, this independence of the population as a precondition for making a free decision is not given according to the HRIA report.

State provision of services is largely lacking in the indigenous communities of the region. Since the company was required to contribute to local development, it engaged in social projects in the health and education sector, which largely responded to the needs of the population. In interviews with those affected, the importance of this infrastructure was repeatedly stressed—as well as the fear of losing it again. By operating the social projects, SMI has temporarily taken over the provision of State services that were lacking, and also made it clear that the continuation of the social projects is tied to the continuation of the mining project (INEF, 2013). This means that the population's attitude to the project must be strongly influenced, and the possibility of an independent and free decision is very limited.

SMI could have realized, when carrying out the initial analyses of the local context which formed the basis for identifying social projects, that the company was acting as a substitute for the State and thereby creating a high dependency situation for the population. Other forms of financing the social projects, which would not have created a direct dependency of the population on the company, would have been possible, for example, through the creation of long-term funds.

Prior

The interviews in the communities carried out for the HRIA showed that, despite the many consultation meetings that were held before and during the exploration phase, a large proportion of the population has no clear picture of what an open-pit mine on this scale means for the region, in practice. The HRIA report concludes that while the population was given prior information, the fact that the company was the decisive provider of information meant that not sufficient weight was given to the risks of the project. Some of these were only raised in later studies. Thus, a study carried out in 2013 by Ateneo University of Davao came to the conclusion that the original calculations regarding the impact of the project on the region's water resources do not meet international standards. The impacts are potentially worse both in qualitative and quantitative terms than the studies carried out by SMI suggested (Vidal et al., 2013).

Finally, "Prior" is an ongoing process. When new information or findings emerge, those affected must each time be provided with adequate information without delay, so that they can gain a continually updated picture of the situation and form their opinion about it.

Informed

In compliance with Philippine law, the SMI carried out a wide-ranging Stakeholder Engagement Programme, which, in addition to many consultation and scoping meetings, also included an education programme and visits to key facilities and other mine sites. However, the interviews with the affected indigenous communities present a mixed picture with regard to the effectiveness of these consultations.

The main criticism expressed by the interviewed indigenous people was that, while SMI collected a lot of information during the consultations, it did not give adequate feedback. Thus, they felt there was an insufficient response to questions and fears. In the interviews, those affected repeatedly expressed their fears regarding environmental degradation, food and water supplies, the destruction of cultural heritage and disruptions of the social fabric, which the company's consultation processes were seemingly unable to allay.

In view of the many uncertainties about the meaning and scope of the project, which emerged in the interviews, it is clear that the condition of "informed" was not adequately met. The interviewees were not aware of the possibility of alternatives to open-pit mining. They stated that no information was provided about other options, only about the variant of open-pit mining that is most favorable for the company. But the interviewees also had little knowledge about the meaning of open-pit mining and its impact on the environment.

Criticism was expressed in the interviews that the company's main aim seemed to be to gain legitimacy for the project through the consultation processes, rather than provide transparency about the impacts of the project on the affected population.

Consent

Not all affected communities are opposed to the project. However, in view of the shortcomings cited above regarding the requirements of an FPIC, the consent of the pro-mining communities is, from a human rights point of view, at least questionable. The HRIA concluded that the consultation processes exhibit so many deficits that it is not possible to talk of consent in accordance with FPIC in the Tampakan mining project.

CONCLUSIONS

The HRIA concluded, "The conditions with regard to communities and State authorities required for handling a transnational enterprise and a project of this magnitude in a responsible and sustainable way are currently not fulfilled in the Tampakan area" (INEF, 2013). There must be a fundamental rethinking of the Tampakan mining project, taking account of the latest studies and findings. It goes without saying that this also involves implementing FPIC processes with the indigenous population, which allow them to be owners of their own development and be able to decide themselves what position they wish to take with regard to the mining project.

The Philippine State, the company, and, if necessary, international bodies should guarantee that these processes are carried out correctly. Not least, the State in which the parent company has its head office, namely Switzerland, has a duty to ensure that the company meets its obligation to respect human rights, according to the UN Guiding Principles (Ruggie, 2011). A reconsideration of the Tampakan Mine project could help to find a way out of this deadlocked situation and create the space to redefine roles and responsibilities. This could allow establishing a more favorable basis than the current confrontation by arms for the correct implementation of future FPIC processes in Tampakan.

REFERENCES

Anaya, J., 2009. Special Rapporteur on the Situation of Human Rights and Fundamental Freedoms of Indigenous Peoples. Report submitted to the Human Rights Council (A/HRC/12/34).

Anaya, J., 2010. Special Rapporteur on the Situation of Human Rights and Fundamental Freedoms of Indigenous Peoples. Report submitted to the Human Rights Council (A/HRC/15/37).

Anaya, J., 2011. Special Rapporteur on the Rights of Indigenous Peoples, Extractive Industries Operating within or Near Indigenous Territories. Report submitted to the Human Rights Council (A/HCR/18/35).

Bilham, R., 2014. Aggravated earthquake risk in South Asia: Engineering vs. human nature. In: Wyss, M., (Eds.), Earthquake hazard, risk and disasters. Elsevier, Waltham, Massachusetts, pp. 103–141.

Bilham, R., 2015. Mmax: ethics of the maximum credible earthquake. In: Wyss, M., Peppoloni, S. (Eds.), Geoethics: Ethical Challenges and Case Studies in Earth Science. Elsevier, Waltham, Massachusetts, pp. 119–140.

First Peoples Worldwide, 2013. Indigenous Risk Report for the Extractive Industries.

ILO, 1989. Convention Concerning Indigenous and Tribal Peoples in Independent Countries.

Indigenous Peoples Links, 2013. The Ecumenical Council for Corporate Responsibility: Making Free Prior & Informed Consent a Reality, Indigenous Peoples and the Extractive Sector. Middlesex University School of Law.

INEF, 2013. Institute for Development and Peace: Human Rights Impact Assessment of the Tampakan Copper-gold Project.

Revenue Watch Institute, 2010. Transforming Resource Wealth into Well-being.

Ruggie, J., 2006. Special Representative of the Secretary-general on the Issues of Human Rights and Transnational Corporations and Other Business Enterprises. Report submitted to the Human Rights Council (E/CN.4/2006/97).

Ruggie, J., 2011. Special Representative of the Secretary-general on the Issues of Human Rights and Transnational Corporations and Other Business Enterprises. UN Guiding Principles on Business and Human Rights (A/HRC/17/31).

United Nation, 2007. Declaration on the Rights of Indigenous Peoples.

Vidal, L., Hilario-Patiño, M., Simpol, L., 2013. Mining and Water Use in South Cotabato and Davao del Sur.

Wetzlmaier, M., 2012. Cultural impacts of mining in indigenous peoples' Ancestral Domains in the Philippines. ASEAS – Aust. J. South-East Asian Stud. 5 (2), 335–344.

World Bank, 2010. Indigenous Peoples, Poverty, and Development.

Zúñiga, F.R., Merlo, J. Wyss, M., 2015. On the vulnerability of the indigenous and low income population of Mexico to natural hazards. A case study: State of Guerrero. In: Wyss, M., Peppoloni, S. (Eds.), Geoethics: Ethical Challenges and Case Studies in Earth Science. Elsevier, Waltham, Massachusetts, pp. 381–391.

Chapter 31

On the Vulnerability of the Indigenous and Low-Income Population of Mexico to Natural Hazards. A Case Study: The Guerrero Region

F. Ramón Zúñiga,[1] Janette Merlo[2] and Max Wyss[3]

[1]Centro de Geociencias, National Autonomous University of Mexico, Mexico D.F., Mexico; [2]Department of Environmental Sciences and Hazard Management, School of Sciences, University of Colima, Colima, Mexico; [3]International Centre for Earth Simulation, Geneva, Switzerland

Chapter Outline

Abstract

This chapter deals with some of the issues behind the vulnerability of that part of the Mexican population who are most exposed to natural hazards due to the site and conditions of their dwellings. In the case of Mexico, as is the case in many other countries, the people most vulnerable to natural hazards, in general, are the ones with the lowest income. This fact, in turn, translates in countries such as Mexico, in a large percentage also being indigenous. The issue of ethnicity plays various roles dealing with ethics when facing measures to issue warnings or carry out contingency procedures. Although it would be intuitive to assume that low-income people are most vulnerable to natural hazards, it is necessary to quantify the degree of vulnerability compared to the rest of the population, so measures can be taken to reduce the unfair difference. In this chapter, we want to draw attention to the different causes for the vulnerability of this segment of the population in Mexico, and some of the reasons why the ways to help them improve their living conditions continue to be neglected. We also address the question of whether

Geoethics. http://dx.doi.org/10.1016/B978-0-12-799935-7.00031-9

there is a difference in vulnerability, which corresponds directly to the ethnicity status. We estimate the degree of vulnerability by means of an index derived from figures on demography (Social Vulnerability Index). Overall, we can see a clear coincidence between the level of social marginalization in the indigenous regions with vulnerability, as expected. Our results also show that the vulnerability of the central and coastal regions is lower than that of the Montaña (mountain) region and the recovery time is an important factor to be considered as a source of the difference between the three areas due to the isolation and difficulty of access to the mountain region. We estimate that the fatality rate in the rural population would be 20% larger than in the urban population in case of a hypothetical earthquake with magnitude M8.6 off the coast of Guerrero. We discuss some of the repercussions of a lack of planning strategies to mitigate damage and/or lack of enforcement of planning and regulations in cases where they exist. The recent catastrophic results of the "Manuel" storm of September 2013 are examples of poor preventing strategies and lack of enforcement of hazard mitigation practice.

Keywords: Earthquakes; Emergency warning; Hazards; Low income; Urban planning; Vulnerability.

INTRODUCTION

Recently, important advances have been made in the evaluation of what is now known as social vulnerability, which deals with aspects other than the biophysical or built environment. One of the main reasons why this aspect had been neglected in the past is the difficulty in quantifying socially created vulnerabilities (Cutter, 2003; Schmidtlein et al., 2008). This effort has lead to research, which has provided means for comparing the differences in vulnerability between different localities in a particular region or country, and has even lead to the creation of institutions fully devoted to this type of studies such as the Hazards & Vulnerability Research Institute at the University of South Carolina, USA.

Vulnerability varies over time and space, as Cutter et al. (2003) have pointed out, so it is necessary to have a method that can be used to derive inferences about the differences in vulnerability as they pertain to time and space.

We also agree with Cutter et al. (2003) in that "Social vulnerability is partially the product of social inequalities—those social factors that influence or shape the susceptibility of various groups to harm and that also govern their ability to respond. However, it also includes place inequalities—those characteristics of communities and the built environment, such as the level of urbanization, growth rates, and economic vitality that contribute to the social vulnerability of places. To date, there has been little research effort focused on comparing the social vulnerability of one place to another."

Mexico comprises a large population of indigenous people, who in general suffer a lack of social services and infrastructure, as compared to the rest of the population. In many instances, indigenous communities are isolated and far from the emergency services, so response in an emergency phase is usually too slow. However, the condition of income or ethnicity not always bears negatively on the vulnerability of a community to any type of hazard. For example, some indigenous style of dwelling structures are more resistant to the impact of earthquakes due to the natural flexibility of their construction materials (Figure 1), but can offer little protection from storms and hurricanes.

In this work, we present figures on demography for one of the states of Mexico (Guerrero) with the largest proportion of indigenous communities as a particular example of the conditions that render indigenous people highly vulnerable to natural hazards. Then we make use of the Social Vulnerability Index (SOVI) method, proposed by Cutter and colleagues, in order to quantify a measure of the vulnerability of the localities of Guerrero state. We then attempt to correlate the index of vulnerability with the social status (i.e., income and ethnicity) of the localities identified as being most vulnerable to natural hazards. We discuss some of the possibilities of using this method for evaluating a particular characteristic of hazard, such as earthquakes or flooding, and point

FIGURE 1 Typical dwellings of many indigenous communities in Mexico. (a) Example of "palapa" with wood planks walls. (b) Palapa with walls of palm leaves.

out questions, which, from our perspective, should be addressed in order to minimize the risk of communities that are susceptible of suffering serious harm. Finally, we conduct an exercise estimating the casualties that are expected to result from an earthquake with magnitude M8.6 along the coast of Guerrero. This last section compares the relative impact of earthquakes on large and small communities.

SOME BASIC DEMOGRAPHICS OF THE STATE OF GUERRERO, MEXICO

The states of Mexico hardest hit by large earthquakes (mostly related to the subduction process of the Cocos and Caribbean plates under the North-American plate taking place along the Pacific Margin) and climate phenomena are the regions with the largest proportion of indigenous inhabitants, namely the coastal sates of Guerrero, Oaxaca, and Chiapas. Among these, the state of Guerrero has the lowest Human Development Index (HDI) in the country. This index was developed by the United Nations Development Programme to have an objective means to quantify poverty. The HDI combines indicators of life expectancy, educational attainment, and income into a composite index. The HDI sets a minimum and a maximum for each dimension, called goalposts, and then shows where each country stands in relation to these goalposts, expressed as a value between 0 and 1. Guerrero has an HDI of 0.6151 compared to the highest value of 0.7683 for Jalisco, considering indigenous population only (Informe PNUD, 2010).

Guerrero has 3,388,768 inhabitants who live in 81 municipalities distributed as 58% in urban communities and 42% in rural communities (following the classification of the National Institute of Statistics and Geography, a rural community is that which comprises less than 2500 people, with the opposite being an urban community). The state is known for its ethnic diversity and its large indigenous population, with an estimated half million people who speak a vernacular or indigenous language and live in one of the 76 municipalities of this state.

THE SOVI FROM ETHNIC FACTORS

The method for constructing the SOVI for the regions of Guerrero based on ethnicity, is based on the steps proposed by Cutter et al. (2000). In general, the SOVI is constructed by the normalization of each variable with respect to its maximum and minimum value for each municipality, also considering whether its contribution is positive or negative.

In our case, we selected four variables from available data, which we think represent the specific vulnerability of indigenous communities (after data from the National Committee for the Development of Indigenous Communities Report, CDI, 2008). The variables selected represent the levels of population dispersal and the economic and social characteristics of the regions for the indigenous population. They also provide dimensions to some educational and cultural issues. The selected variables are percentage of indigenous population, illiteracy percentage, distribution of the population in households with indigenous population by urban area, and percentage of population with high levels of social marginalization. These criteria have been considered due to the characteristics of the region and the data available.

DISCUSSION

The pattern of regional vulnerability as shown by the SOVI results (Figure 2(a)) closely follows that of the regional distribution of indigenous population (Figure 2(b)). We emphasize that the variables employed in the SOVI calculations have been normalized with respect to the total population, but in order to show that the total population is not related to the vulnerability, we also show the total population regional distribution in the histogram of Figure 2(b). Thus, this can be seen as an indication of how indigenous people are shaping the distribution of vulnerability in the state of Guerrero due to the conditions under which they live.

The above findings are also evident when considering the distribution of the social marginalization index (hereafter SMI) for all the population of the regions (after data from the Comisión Nacional de Pueblos Indígenas, 2008). The SMI is based on the following variables: illiterate population of 15 years of age or more; population with 15 years of age and over, who did not completed primary education; occupants in dwellings without drainage or toilet; occupants in dwellings without electricity; occupants in households without tap water; homes with some level of overcrowding; occupancy in houses with dirt floor; population in towns with less than 5000 inhabitants; people employed with income of up to two minimum wages. As illustrated in Figure 2(c), this index also matches the most vulnerable region as identified by their SOVI. The Montaña region, shown as the most vulnerable area in our results, is the region with the largest indigenous percentage of the population and with the highest SMI and poverty. The region with the lowest SMI (i.e., the Norte region), on the other hand, also matches the region with the lowest SOVI and has the lowest indigenous percentage except for the regions marked as "other," which basically have no significant indigenous population (and therefore was excluded from the SOVI calculation for indigenous regions). The indexes of the two intermediate regions do not match as closely, possibly due to the limitations in the variables employed for the analysis.

These results are reflected in the recent experiences of the vulnerable regions following the disasters triggered by climate phenomena. A total of 1.2 million people were affected when Hurricane Manuel made landfall on September 15, 2013 on Mexico's southwestern coastline, mostly in the state of Guerrero. The Subsecretary of Civil Protection of Guerrero reported (Periodical 24 hours, 2013) that in the Centro region, there were 1000 houses affected, 981 houses were

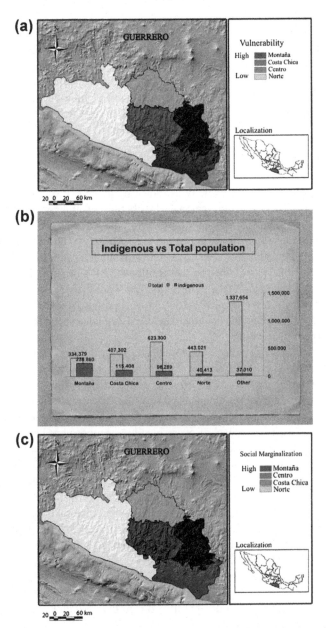

FIGURE 2 (a) Map of the vulnerability index (SOVI) based on ethnic factors at the regional level (see text). (b) Regional distribution of indigenous versus total population in Guerrero for 2010 (indigenous population data from National Institute of Statistics and Geography and total data from Sistema Estatal de Información Estadística y Geográfica (INEGI, 2010). Note that the index of vulnerability (Figure 2(a)) correlates well with the indigenous population but not so with total population. (c) Map of the social marginalization index (SMI) at the regional level. The regions with the highest and lowest SMIs coincide with the highest and lowest indexes of vulnerability (Figure 2(a)). The regions Centro and Costa Chica are not coincident, although this could be due to both having small variations in the indigenous population (Figure 2(b)).

damaged in the Costa Chica region, 2000 houses in the Costa Grande region, and 1400 houses in the Montaña region. However the emergency situation in the Montaña region was exacerbated by the late arrival of government assistance and the slow recovery time due to the isolation of the communities in that region.

In the Central and Costa Chica region, urban development for touristic purposes has taken an important role in the vulnerability issue because even though the levels of marginalization are often lower due to access to better infrastructure, low-income people favor the election of unsafe but cheap places to live on unstable ground or on the slope of rivers. Thus, in these regions there is a trade-off in vulnerabilities because of the context in which populations are developed although poverty is still the main factor that drives these communities as the most vulnerable. As for the region of Tierra Caliente and Costa Grande, the demographic data indicate no indigenous communities that inhabit the area. The results, which show these regions as least vulnerable, come as no surprise since they include the most developed towns due to the recent tourism boom in this area. The Tierra Caliente region comprises the state's municipalities with the largest population. Therefore, some of them have high rates of vulnerability, which would be necessary to analyze as a particular case.

ESTIMATING THE RELATIVE IMPACT OF EARTHQUAKES ON LARGE AND SMALL COMMUNITIES

While sophisticated models are constructed for large cities to estimate losses in future earthquakes, this is not so for equally large urban populations. Data on vulnerability, such as those compiled in this work, are currently not being used in estimating future earthquake losses and in designing strategies to protect the population. The difference in impact of earthquake disaster between large and small communities has not been quantitatively estimated, as far as we know. In this section, we will theoretically estimate the casualties that would result from an earthquake with magnitude M8.6 in Guerrero.

The hypothetical earthquake model assumes a rupture of $L = 440$ km along the Pacific subduction zone section off Guerrero. This length equals that of the 1787 San Sixto earthquake, which ruptured the plate boundary to the east of Guerrero.

Our results are only order-of-magnitude estimates. It is the sort of first-order review of a problem that is called "back of the envelope calculation." We submit that our estimate has the essential quality of the back-of-the-envelope approach. Although approximate, it cuts straight to the heart of the problem.

The computer program QLARM (Trendafiloski et al., 2009) has been used for 11 years to estimate damage, fatalities, and injured in real time after earthquakes worldwide (Wyss and Wu, 2014) and for scenario calculations like the one here (Wyss, 2005, 2006). The population data for Mexico are based on the 2010 census, in which small and large settlements are considered. The data on building stock come from the Web site of the World Housing Encyclopedia (www.world-housing.net). The distribution of dwellings in QLARM is modeled separately in three size classes of settlements: population smaller than 2000, between 2000 and 20,000, and larger than 20,000 because villages do not have skyscrapers. The aforementioned limits have been proposed by Satterthwaite (2006) for developing countries. They may not be the best choice for Mexico, but we will use them in the absence of specific classification for the area studied.

Before venturing to estimate losses in hypothetical future earthquakes in Mexico, we have verified that QLARM correctly calculates intensity of shaking for 10 large earthquakes during the last decades, and that fatalities are also estimated correctly within the margins of error of about a factor of two (Wyss, personal comm.). The mean damage calculated in the case of a hypothetical M8.6 earthquake offshore Guerrero is mapped in Figure 3. The mean damage grade in each settlement is color coded on a scale from 0 to 5 (total destruction). Major damage is expected in the settlements shown by red dots along the coast. Based on the damage caused by the strong ground shaking, the percentages of fatalities and injured are calculated, resulting in casualty counts for each settlement. We have had good success in estimating the sum total of casualties in large earthquakes worldwide (Wyss and Wu, 2014), but the numbers for single settlements can be off substantially, depending on the circumstances (Wyss et al., 2011).

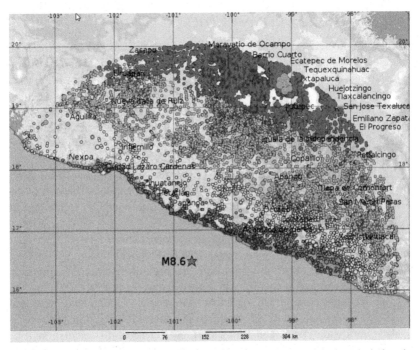

FIGURE 3   Map of mean damage (on a scale of 0–5) by settlement due to a hypothetical earthquake of M8.6, the rupture of which runs along the entire coast of Guerrero. Red is grade 4 (ruinous), blue signifies minor damage such as cracks in walls. The calculation is restricted to a radius of 400km from the epicenter.

In this thought experiment, we have summed the casualties (fatalities and injured) separately for the affected settlements in each of the three size classes defined above. In these overall sums, we do not consider distance as a parameter, assuming the large and small settlements are mixed approximately equally at all distances. The contrast in losses expected between small and large settlements is quite distinct. The numbers of settlements within 240 km from the epicenter and the population in them are listed for the three size categories in column 1 and 2. The mean numbers of injured and fatalities expected are shown in the columns so labeled. The key column is the one showing the percentage killed (the last column). These results show that there are 20% more people killed in small settlements than in large ones. On the contrary, there are about 40% fewer people expected to be injured in small settlements than in large ones. This is so because in large cities more people survive, but with injuries, due to their more resistant homes.

The estimate of the difference in the suffering of the population in small and large settlements is a function of the model for the building stock in the computer tool we use. Our model is primitive and in its reliability far from what a researcher would wish. Nevertheless, first of all we have tested the performance of QLARM in historic earthquakes, so our estimates are relatively reliable. Secondly, it is a generally observed fact that the built environment in cities is more resistant to strong shaking than in small settlements.

CONCLUSIONS

With respect to estimating the relative impact of earthquakes on large and small communities, we reach two conclusions. (1) The sum of the numbers of fatalities in small settlements equals approximately that of the large cities. This means that the preoccupation of disaster mangers with large cities is partially misplaced. In many earthquake disasters, there are as many suffering people in the countryside as in the cities. Nevertheless, it is true that the largest concentration of victims is found in the cities. (2) The percentage of fatalities is 20% more in small settlements compared to large ones. This is a result that has been shown quantitatively here for the first time, as far as we know.

Regarding the vulnerability of indigenous regions, overall, we can see a clear coincidence between the level of social marginalization in the indigenous regions with vulnerability, as expected, and these results are confirmed after the experience of Hurricane Manuel, and with the damage inflicted by earthquakes that have occurred in the past. Our results also show that the vulnerability of the central and coastal regions is lower than that of the Montaña region and the recovery time is an important factor to be considered as a source of the difference between the three areas due to the isolation and difficulty of access to the mountain region.

When dealing with the vulnerability of indigenous communities, we are faced with several questions such as: Are we really considering the differences in ethnicity that affect vulnerability, should government agencies treat indigenous communities different than the rest of the population, are indigenous and low-income people liable for their living conditions with respect to the hazard of their environment. Even though these are questions that cannot be answered directly by our results, the SOVI method provides an objective way of quantifying differences, which can be used as base for further analysis. These results can also be used toward the planning of strategies for disaster prevention or emergency contingencies. The results of the vulnerability index, quantified by the SOVI, are verified to some extent by the extent and area of the damage inflicted by recent meteorological events, but more detailed analysis is needed so we can establish a firm basis for calculating vulnerability to particular hazards. To date, Mexican municipalities are working toward the generation of a Hazard Atlas, which delineates the principal hazards to which the communities are exposed. However, in the case of Guerrero, only 10% of the municipalities have produced such a document (Urgen a Municipios, 2012), even though it is a requisite to be able to have access to federal disaster prevention funds. The methods used here are an example of how to advance in not only cataloging the current hazards, as is done up to now in the case of Hazard Atlases, but quantifying the vulnerability of the communities.

ACKNOWLEDGMENTS

The authors wish to thank the editors for their careful revision and suggestions that greatly improved the presentation and technical details of our work. This study was partially funded by the National Autonomous University of Mexico (UNAM) under the Program for Research and Technological Innovation (PAPIIT) Grant IN112110 and by the National Council of Science and Technology–Mexico (CONACyT) Grant CB-2009-01-129010. J.M. also benefitted from an award by the Mexican Academy of Sciences for a summer internship.

REFERENCES

Cutter, S.L., 2003. The vulnerability of science and the science of vulnerability. Ann. Assoc. Am. Geogr. 93, 1–12.

Cutter, S.L., Boruff, B.J., Shirley, W.L., 2003. Social vulnerability to environmental hazards. Soc. Sci. Q., 242–261.

Cutter, Susan L., Mitchell, Jerry T., Scott, Michael S., 2000. Revealing the vulnerability of people and places: a case study of Georgetown County, South Carolina. Ann. Assoc. Am. Geogr. 90 (4), 713–737. http://dx.doi.org/10.1111/0004-5608.00219.

Comisión Nacional para el Desarrollo de los Pueblos Indígenas Región Sur. Tomo 2. Chiapas, Guerrero y Morelos [texto]: Condiciones Socioeconómicas y Demográficas de la Población Indígena / Comisión Nacional para el Desarrollo de los Pueblos Indígenas. – México: CDI: PNUD, 2008. p. 176 ISBN 978-970-753-138-3 (Región Sur, tomo 2) ISBN 978-970-753-139-0 (Obra completa).

Informe sobre Desarrollo Humano de los Pueblos Indígenas en México, El reto de la desigualdad de oportunidades, 2010. 978-92-1-326051-7. Programa de las Naciones Unidas para el Desarrollo PNUD, UN Development Programme.

INEGI, National Institute of Geography and Statistics, 2010. Online data base. http://www.inegi. org.mx/.

Satterthwaite, D., 2006. Outside the Large Cities; The Demographic Importance of Small Urban Centres and Large Villages in Africa, Asia and Latin America. Human Settlements, Working Paper Series Urban Change No. 3. IIED, London.

Schmidtlein, M.C., Deutsch, R.C., Piegorsch, W.W., Cutter, S.L., 2008. A sensitivity analysis of the social vulnerability index. Risk Anal., 1099–1114.

Suman 11,500 casas dañadas y 25 ríos desbordados en Guerrero, Periodical 24 hours, September 17, 2013, http://www.24-horas.mx/suman-11-mil-500-casas-danadas-y-25-rios-desbordados-en-guerrero/ (last accessed 06.05.14.).

Trendafiloski, G., Wyss, M., Rosset, P., Marmureanu, G., August, 2009. Constructing city models to estimate losses due to earthquakes worldwide: application to Bucharest, Romania. Earthquake. Spectra 25 (3), 665–685.

Urgen a municipios crear Atlas de Riesgo en Guerrero, June 24, 2012. Novedades de Acapulco.

Wyss, M., 2005. Human losses expected in Himalayan earthquakes. Nat. Hazards 34, 305–314.

Wyss, M., 2006. The Kashmir M7.6 Shock of 8 October 2005 Calibrates Estimates of Losses in Future Himalayan Earthquakes Proceedings of the 3rd International ISCRAM Conference Van de Walle, B., Turoff M. (Eds.), Newark, NJ, USA.

Wyss, M., Elashvili, M., Jorjiashvili, N., Javakhishvili, Z., 2011. Uncertainties in teleseismic epicenter estimates: implications for real-time loss estimates. Bull. Seismol. Soc. Am. 101 (3), 1152–1161.

Wyss, M., Wu, Z., 2014. How many lives were saved by the evacuation before the M7.3 Haicheng earthquake of 1975? Seismol. Res. Lett. 85 (1), 126–129.

Chapter 32

Ethics and Mining—Moving beyond the Evident: A Case Study of Manganese Mining from Keonjhar District, India

Madhumita Das
Department of Geology, Utkal University, Bhubaneswar, Odisha, India

Chapter Outline

Abstract

The mining and metallurgical sector is essential for the development and economic growth of a developing country like India. It is evident that social responsibility has moved up the agendas of mining companies, with increased pressure on industries to be more responsive to meeting the needs of local people. It has become a common practice to scrutinize the mining projects for assessing the social and environmental risks, prior to decisions being made regarding project finance, and even for setting up pilot projects. Dubna manganese mines of Keonjhar district of Odisha is one of the important manganese ore-producing regions of India, which forms a part of Precambrian sedimentary formation known as the Iron Ore Series developed in Singhbhum–Keonjhar–Bonai area. The present study is an attempt to evaluate the impact of mining on environment with special reference to impact of manganese on the health of mine workers. The study indicates that the existing legislations have not provided justice to communities suffering from health problems due to mining, there are no laws that specifically protect the rights of women's health in mining, either as communities or as workers.

Geoethics. http://dx.doi.org/10.1016/B978-0-12-799935-7.00032-0

Keywords: Dubna manganese mines; Exposure; Manganism; Mine worker

INTRODUCTION

Over a long period of time, mines in many parts of the world are considered to be an important source of revenue and livelihood generation. The mining and metallurgical sector is essential for the development and economic growth of a developing country like India. However, mineral resources have intrinsic characteristics that make them not only useful but also problematic. Use of mineral resources is fundamental to human well-being as they are essential to virtually every sector of the economy, are the basis for human-built environment, and provide desired services (Shields, 1998). The products of the industrial sector, including metallic and nonmetallic minerals, construction and building materials, and fertilizers, not only are crucial for developmental activities but also regulate foreign exchange earnings.

Mineral deposits are unique in their occurrence and nature, having different qualities in terms of grade or volume. As a result, the sitting of mineral operation depends upon the location and character of deposits not wholly on the preferences of the general public. Mining is the extraction of valuable minerals or other geological materials from the earth from an ore body, lode, vein, seam, or reef, which forms the mineralized package of economic interest to the miner. Mines may be proposed in areas where extraction is not a preferred land use such as adjacent to or beneath an existing community or in an environmentally sensitive area. Because these practical aspects have not always been well managed, there are cases where the extraction, use, and disposal of mineral resources have negatively impacted societies and the environment (IIED, 2002). Modern mining processes involve prospecting for ore bodies, analysis of the profit potential of a proposed mine, extraction of the desired materials, and final reclamation of the land after the mine is closed. But in addition to generating wealth, mining can also be a major source of degradation to the physical and social environment unless it is properly managed. These guidelines are designed to assist regulators, particularly in developing countries, to encourage sustainable mining while at the same time protecting the environment. This study is an effort to illustrate the plight of men and women working in mines, analyze the issues, and suggest the way forward.

THE MISSING ETHICS OF MINING

Global population is estimated to reach 9 billion by 2050 and this population growth alone will exert substantial pressure on the supply and availability of resources such as food, fresh water, energy, and minerals. The demand for mineral resources particularly for iron, manganese, chromium, aluminum, copper, lead, zinc, nickel, tin, and many others has been exponentially increased to cater to the mineral-based industries. At one end, we need intensive investigation, exploration, and mining, and on the other, there is antipathy of an exploited and broken environment. Hence, it is necessary to balance the linkage in earth–ocean–climate relationship and bring a harmony between mineral development and environment.

Minerals in general and mining in particular have played the fundamental role in the evolution of civilization. Major divisions of history are named after the dominant mineral products, the Copper, Bronze, and Iron Ages. More than ever, human civilization depends on minerals to sustain its existence. The unusual growth of population, speed of transportation, and sudden increase in electronic gadgets depend on the development of mining (Siegel, 2013). But ethics of mining has never been taken as a serious thought by planners, administrators, or mine owners. Some view the concept as a contradiction in terms, others are distressed that mining continues to exist whatsoever.

MINING AND HEALTH

Mining is a hazardous task, particularly underground mining. The physical risks on mine workers are tremendous, and are a challenge to manage. Any equipment or method that is applied to loosen and remove rock can do a lot of damage to a person, even if it does not kill the person outright. Mining also causes widespread environmental damage, including pollution of the ecosystem, changing the landscape, and contamination of air, water, and soil. Overall, it affects the physical, mental, and social health of the local mining community. The nature of mining processes creates a potential negative impact on the environment both during the mining operations and for years after the mine is closed. This impact has led to most of the world's nations adopting regulations to moderate the negative effects of mining operations. Safety has long been a concern as well, and modern practices have improved safety in mines significantly. Environmental issues can include erosion, formation of sinkholes, loss of biodiversity, and contamination of soil, groundwater, and surface water by chemicals from mining processes. In some cases, additional forest logging is done in the vicinity of mines to increase the available room for the storage of the created debris and soil. Contamination resulting from leakage of chemicals can also affect the health of the local population if not properly controlled. Extreme examples of pollution from mining activities include coal fires, which can last for years or even decades, producing massive amounts of environmental damage.

The mining sector is also accountable for some of the largest releases of heavy metals into the environment of any industry. It also releases other air pollutants including sulfur dioxide and nitrogen oxides in addition to leaving behind tons of overburden, hazardous waste tailings, slag, and acid drainage. Around the world, unsafe mining and smelting practices have been responsible for a continuing series of environmental and human health disasters, which cause great human tragedy and undermine social stability, economic development, and sustainability goals.

TABLE 1 Global Lead and Mercury Emissions from Mining

Release of Metal	Amount/Year
Lead from smelting and refining nonferrous metals (e.g., gold, lead, zinc, copper)	28,000 metric tons/year (Jimenez-Soto and Flegal, 2011)
Mercury from smelting and refining nonferrous metals	710 metric tons/year (Pirrone et al., 2010)
Mercury into the environment from artisanal/small-scale gold mining	Ranges from 400 to 1102 metric tons/year (Pirrone et al., 2010, Telmer and Veiga, 2009)
Lead from mining and smelting	980 metric tons/year (U.S. Geological Survey report, 2009)
Mercury from mining and smelting	9 metric tons/year (U.S. Geological Survey Report, 2010)

Occupational and environmental exposure to heavy metals, silica, and asbestos can occur during mining and milling operations. The smelting operation is associated with the highest exposures and environmental releases. The human health hazards caused by exposure to heavy metals, including lead, cadmium and mercury, have been reported globally. These metals are linked with a range of neurological deficits in both children and adults in addition to a range of other systemic effects. Exposure to airborne silica and asbestos can cause lung cancer, pneumoconiosis, and numerous other health effects. Airborne emissions from metal mining and smelting in Australia, Canada, and the United States, countries with some of the world's best environmental controls, amount to more than 80% of lead production in these countries (U.S. Geological Survey Report, 2009, 2010). The examples of global emissions are shown in Table 1.

LIVING IN THE MIDST OF MANGANESE DUST AND CONTAMINATED SEWAGE

Constant contact with dust and pollution and indirectly through contamination of water and air cause severe health hazards to the mine workers. As majority of the workers are contract laborers, and paid on a daily wage basis, there is no economic security or compensation paid due to loss of workdays on account of health problems. Meager or no compensation is given during pregnancy period to women workers. Even during pregnancy, women have to work in hazardous conditions amidst noise and air pollution, which have adverse affects on their offspring.

The majority of the health problems in mining regions are caused by unimpeded pollution and high levels of toxicity, mine tailings, and mine accidents. The health and safety problems of manganese mining varies from other mining, from the technology used, type of mining cast to the size of operations. The lands, water bodies, air, and environment are polluted due to constant release of chemical wastes, dust generated by blasting and excavation, and dumping of mine wastes and overburden in the surrounding lands, springs, and rivers.

MINING IN ODISHA

The mineral resources of Odisha form a very significant constituent of India's mineral wealth. Many of the minerals are known to be in plentiful supply. Odisha's resources of bauxite, chromite, coal, iron ore, manganese, and nickel ore are formidable constituting approximately 50%, 98%, 25%, 35%, 27%, and 91%, respectively, of the total resources of the country. Besides metallic and fuel minerals, Odisha has enormous industrial mineral potential, viz., china clay, fire clay, graphite, limestone, dolomite, pyrophyllite, kyanite, sillimanite, andalusite, soapstone, steatite, talc, quartz, quartzite, and some dimension and decorative stones such as granite, sandstone, laterite, and khondalite (Das and Goswami, 2009).

Mineral resources lead to increased exploitations and development of large mines to meet the demand of a number of existing and proposed industries in Odisha. More than 1200 sq km area of the state is under mining leases, which accounts for more than 0.7% of the total geographical area of the state. Of this, 400 are operated mines with 828 sq km lease area. The total forest area of the state is 58,135 sq km, i.e., 37.3% of the geographical area of the state. Open forest spreads over 20, 866 sq km. Substantial zones of mining activities come under these degraded and open forest areas of the state. Though afforestation has been taken up by the mines, in last 50 years, more than 10% of open forest area has been lost due to mining activities.

A CASE STUDY OF MANGANESE MINES

The present study is an attempt to evaluate the effect of manganese on the health of mine workers of Dubna manganese mines of Bonai–Keonjhar mineral belt of Odisha.

This is the most important manganese ore-producing region of the country. India is well known for its bountiful manganese ore reserves of the order 167 MT. Odisha has a share of 49 MT, representing 29.34% of India's total reserve. The manganese deposits of Odisha are concentrated around this particular Bonai–Keonjhar belt. It is associated with banded iron formation (BIF) of North Odisha. It belongs to the Iron Ore Super Group. This belt forms a part of synclinorium (60 × 25 km) with a low plunge toward north northeast. The BIF

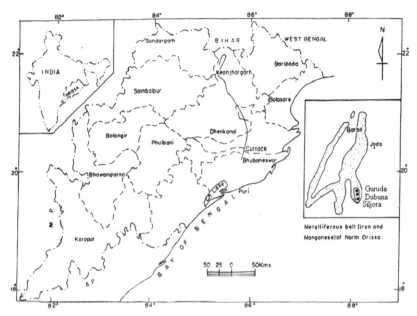

FIGURE 1 Location of investigated Dubna manganese mines, Keonjhar district, Odisha, India.

broadly defines the outline of the synclinorium. Manganese ore is confined to the top lateritised shale and lower lithomarge horizons in the core zone of the synclinorium. The major lithologic units of the area include BIF, banded shales and mixed facies formation of the Iron Ore Group (Mohapatra et al., 2006; Goswami et al., 2008).

The working manganese mine, Dubna, is owned by Odisha Mining Corporation. The mine is bounded by latitude 21°54′04″ and 21°51′47″ North and longitude 85°21′37″ and 85°24′33″ East (Figure 1) of Survey of India Toposheet No. 73 G/5 (R.F.C. 1: 50000). Manganese mineralization in the upper shale formation above the BIF is syngenetic—remobilized in origin (Goswami et al., 2008; Mohapatra et al., 2006; Das, 2001). The upper shale is ferruginous with pinkish color, whereas the lower shale is variegated, tuffaceous, and banded. The mining in both the cases is opencast semimechanized type. Excavation and removal of overburden rocks are done by dozers, pay loader, hydraulic excavators, and Down-The-Hole (DTH) drills.

The manganese ore deposits occur mostly in lateritised shales and phyllites as boulders, pockets, lenses, and discontinuous patches. In the lease hold area, the manganese ore occurs in three types, viz., float ore, laterite ore, and shaly ore. The iron ore deposits are on the top level and overlain by manganese ore.

The general lithological succession may be described as under:

Laterite/morrum

Floats (may be manganese ore/iron ore)

Ferruginous laterite/banded hematite jasper)

Manganiferous laterite

Manganiferous shale

Barren shale

MANGANESE TOXICITY

Manganese is a mineral element that is both nutritionally essential and potentially toxic. It functions as an antioxidant. It helps in metabolism, bone development, and wound healing. Low levels of manganese in blood are associated with several diseases such as osteoporosis, diabetes mellitus, and epilepsy. The human body (adult) contains approximately 10–11 mg of manganese, most of which is found in the liver, bones, and kidneys. This trace element is a cofactor for a number of important enzymes and is actually required for various enzymatic reactions. In other words, manganese plays an important role in a number of physiologic processes as a constituent of some enzymes and an activator of other enzymes (Goswami et al., 2008, 2009). On the contrary, the excess manganese interferes with the absorption of dietary iron. Long-term exposure to excess levels may result in iron-deficiency anemia, chronic liver disease, metal fume fever, chemical pneumonia, and manganism (Bencko and Cikrt, 1984). Due to the similarities in symptoms, researchers classify manganism as a Parkinson's syndrome.

In the study area, there was a reduction in the number of children fathered by male workers during the time when they were exposed to manganese and high infant mortality rate in case of female manganese mine workers. In this particular mine, few pregnant tribal women are also working for their livelihoods (Das, 2001). Hence, there is a significant risk that their newborns will be seriously affected by neurological problems. It is evident that the newborn brain may be more susceptible to manganese toxicity due to a greater expression of receptors for the manganese transport protein (transferrin) in developing nerve cells and the immaturity of the liver's bile elimination system (Rodríguez-Agudelo et al., 2006; Inoue, 2007; Roth, 2006; Fitsanakis et al., 2006; Saarkoppel et al., 2005; Bowler et al., 2007; Schaumburg et al.,

TABLE 2 Manganese Level Reported in the Patients of Manganism

Laboratory Test	Manganese Level Reported in the Patients of Manganism
Urinary levels	>10 mg/l (normal reference range, 0.5–9.8 mg/l; up to 50 mg/l for occupational exposure)
Whole blood levels	>19 mg/l (normal reference range, 8.0–18.7 mg/l)
Serum levels	>1.3 mg/l (normal reference range, 0.3–1.3 mg/l)
Neuropsychiatry testing	Decreased memory, decreased reaction time, decreased motor coordination
Computed tomography scan	Radiodense accumulations of manganese in affected areas, abnormalities in globus pallidus, caudate nucleus, putamen

Donaldson (2001)

2006; Dobson et al., 2004). Tolerable upper intake level of manganese in case of children of 1–3 years is 2 mg, whereas in case of children of 4–8 years, it is 3 mg. It is 6 mg in case of children of 9–13 years and in case of adolescents of 14–18 years, it is 9 mg (Myers et al., 2003; Levy and Nassetta, 2003; Iashchenko, 1998; Koudogbo et al., 1991). So the tolerable upper intake level of manganese of children is much less than that of an adult (10–11 mg). The children population of the villages at and around Dubna mine is 600. Hence, they are at significant risk of being affected by several serious diseases.

Epidemiological studies in the Western Pacific region have revealed clusters of neurological disorders like Parkinsonism, Alzheimer's disease, and motor neuron disease in diverse geographic areas. Their location in Guam, Kii Peninsula, Japan, and Western New Guinea suggests that an environmental agent is a likely contributor to their etiology (Inoue, 2007). Moreover, research work at California University on "Manganese Toxicity" has inferred that higher level of manganese in the blood is a marker for violence (Table 2, Donaldson, 2001).

AIR AND NOISE QUALITY

Constant exposure to the suspended particle in the air due to various mining activities has a detrimental effect on the general health of the mine workers. The constant blasting and drilling operation, loading and unloading of ores and overburden, and operation of petrol- and diesel-driven vehicles in opencast mines generate a huge amount of dust. The common pollutants are suspended particulate matter (SPM) and manganese metal dust. Blasting and diesel equipments are major sources, which generate harmful gases like CO_2, NO_x, Soot, Pb, SO_2, hydrocarbon (HC), and polycyclic aromatic hydrocarbons (PAH) (Das et al., 2012). Manganese metal dust readily becomes airborne as a cloud. Chronic inhalation of dust can cause manganese poisoning (manganism), a serious cumulative disease. Such type of dust, when deposited on land, harms the soil quality and its agricultural production and the health of grazing animals is also affected.

High-capacity pneumatic drills and blasting of tons of explosives are acute noise-prone activities. Hence, it causes hearing loss and other health effects (Prashant et al., 2000). In addition, it influences work performance and makes communication more difficult.

CHEMICAL QUALITY OF WATER

Water samples from shallow aquifers (mining ponds) and deeper aquifers (tap water and tube wells) in and around Dubna manganese mines are collected and analyzed. For the heavy metal analysis, the samples are preserved by adding 5 ml concentrated nitric acid in 1 l of sample to maintain the pH below 4.0, following the procedure suggested by Agemian and Chau (1975). The samples are then filtered through Whatmann filter paper No. 40 and the filtrate is directly used for analysis. Various physicochemical parameters are analyzed by following standard analytical procedures (APHA, 2005; Gray, 2006; Gupta, 2004; Jaiswal, 2004) and compared with World Health Organization (WHO) standard.

RESULTS AND DISCUSSION

Water discharged on the surface during mining ultimately takes root through groundwater and surface water and contaminates both. Sometimes, this water may be highly acidic and on discharge to agriculture lands, it affects badly. Manganese mine water may also contain high level of suspended solid particles especially manganese dust, which harms man, animal, and agriculture. Runoff water from overburden dump around Dubna and Siljora mines and mine runoff water also affect the receiving water bodies. Within the Dubna mining premises, there is one mining pond, namely, Betjhar.

The elements present in the water in Dubna mining area are calcium, manganese, chloride, fluoride, sulfate, phosphate, nitrate, iron, zinc, sodium, potassium, and manganese. This analysis indicates that the water contains impurities, but within the permissible limit prescribed by WHO standard (Tables 3 and 4) for drinking water purpose.

Staining is more severe in case of excess of manganese than iron. For that reason, a much stricter drinking water limit has been imposed at 0.05 mg/l in Europe as prescribed by European Union standards, the concentration at which laundry and sanitary ware become discolored (Gray, 2006). Manganese levels in almost all the water samples collected in the present study are higher than this value (0.05 mg/l). In most of the cases, it is more than 0.2 mg/l and approaches 0.5 mg/l. Although WHO (2004) has a health-related guide value of 0.5 mg/l, a much lower value must be set in practice. As manganese is soluble in water, it is observed that during rainy season the water is more turbid and has more

TABLE 3 Range of Chemical Constituents in Different Aquifers in Dubna

Physicochemical Parameters	Shallow Aquifer	Deeper Aquifer
pH	6.2–6.8	7.0–7.6
Specific conductance μs/cm at 25 °C	75–140	76–190
TDS (total dissolved solid)	45–116	58–124
Calcium (mg/l)	3–10	11–36
Magnesium (mg/l)	4–11	6–20
Chloride (mg/l)	6–11	5–17
Sulfate (mg/l)	1.79–2.32	1.58–3.91
Phosphate (mg/l)	1.04–1.35	1.14–1.25
Nitrate (mg/l)	1.24–1.74	0.494–0.57
Iron (mg/l)	0.005–0.016	0.018–0.029
Zinc (mg/l)	0.023–2.06	0.002–0.003
Sodium (mg/l)	2–5	1–3
Potassium (mg/l)	0.5–1.0	0.5–1.0
Manganese (mg/l)	0.15–0.35	0.25–0.36

associate salts, giving a high total dissolved solids as well as total suspended solids, causing health hazards (Das, 2001).

Mine workers of these mines and people of nearby surroundings are suffering from a variety of early symptoms of manganese-related Parkinson's, such as weak appetite, nervousness, irritability, hallucinations, tremors, lethargy, impaired concentration, speech difficulties, short-term memory failure, loss of coordination, muscle weakness, muscle pain, trembling fingers, stiffness of limbs, difficulty with fine movements, stuttering, hoarse voice, urinary problems, and respiratory problems (Das, 2001).

MINING LAWS AND REGULATIONS: ARE THEY IMPLEMENTED?

The Mines Act, 1952, has laid out guidelines for safety of workers employed in mines and regulates the working conditions and amenities for them. To ensure the implementation of the Mines Act, 1952, the Union Legislature has framed the Mines rules, 1955; Metalliferous Mines Regulations, 1961; and the Maternity Benefit (Mines) Rules, 1963. The Mines Act, 1952, prescribes the duty of the owner to manage mines and mining operation and the health and safety in mines. It also prescribes the number of working hours in mines, the minimum wage

TABLE 4 WHO and EU Standards for Investigated Parameters

Parameters	WHO Standards	EU Standards
pH	No guideline	No guideline
Specific conductance μs/cm at 25°	250	250
TDS (total dissolved solid)	No guideline	No guideline
Calcium (mg/l)[a]	70	70
Magnesium (mg/l)[a]	35	35
Chloride (mg/l)	250	250
Sulfate (mg/l)	500	250
Phosphate (mg/l)	No guideline	No guideline
Nitrate (mg/l)	50	50
Iron (mg/l)	No guideline	0.2
Zinc (mg/l)	3	Not mentioned
Sodium (mg/l)	200	200
Potassium (mg/l)	No guideline	No guideline
Manganese (mg/l)	0.5	0.05

[a]*Highest desirable limits according to Bureau of Indian standard standards.*

rates, and other related issues. The Mines Act was amended in 1983 to cover all personnel solely engaged on work relating to mines within the scope of the Mines Act. This suggests that safety of workers engaged in overground activities and related activities was not considered by the state till 1983 (MMP, 2003).

The Air (Prevention and Control of Pollution) Act, 1981 (14 of 1981), and the Environment (Protection) Act, 1986 (29 of 1986), state that air pollution due to fines, dust, smoke, or gaseous emissions during prospecting, mining, beneficiation, or metallurgical operations and related activities shall be controlled and kept within "permissible limits." While the existing legislations in India have not provided justice to communities suffering from health problems due to mining, there are no laws that specifically protect the rights of women's health in mining, either as communities or as workers. Occupational health issues of women mine workers (either during their work on during accidents or disasters) also need to be addressed urgently. In many cases, no proper medical records are maintained and no health checkups are conducted either by companies or governments. Moreover, occupational illnesses are often suppressed (Figure 2). First, illness or accidents may not be reported. A miner might be aware of the disease or illness, but may be afraid of reporting the disease because of fear of

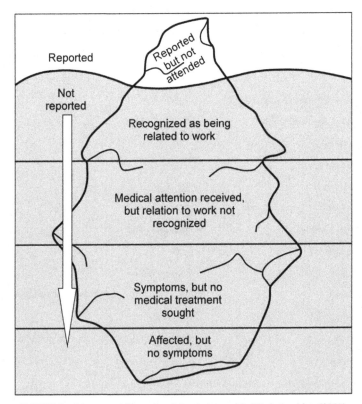

FIGURE 2 Disease and illness in mining industry. *Modified after Metz (2002).*

losing her job, or other job-related benefits. Therefore, the disease or illness is not reported. Second, medical attention might be received, but neither the attending physician nor the miner associates the disease with the work environment. Third, the miner may have symptoms of a disease, but no medical attention is sought, and the disease or illness is not reported (Metz, 2002).

CONCLUSIONS

There should be epidemiologic studies regarding adverse effects on health in this manganese mining belt of Odisha. Occupational medicine physicians can play critical roles in preventing the adverse health effects of manganese. The study regarding the health status of the people at and around these mines indicate that in spite of the satisfactory health and sanitation facilities provided by the mining authorities, malnutrition persists to the agony of the local tribal mine workers. A number of practical measures should be undertaken in a serious manner. Manganese from water should be removed executing treatments such as phosphate compound treatment, oxidizing filters, pressure-type aeration followed by filtration, and chemical aeration followed by filtration. In order to avoid scarcity of water, rainwater harvesting in mine pits and digging more tube wells would be desirable. More precautions have to be taken to suppress the dust by sprinkling water at regular intervals, undertaking plantation (of especially dust-absorbent plants) in and around the mine area. At a legislative level, a legal rule mentioning permissible limit of manganese content (0.05 mg/l) in water should be included in Indian environmental acts such as The Water Act, 1974, and The Metalliferous Mines Regulations, 1966. Similarly, legal rule mentioning airborne exposure permissible limit of manganese (5 mg/m^3) should be included in The Metalliferous Mines Regulations, 1966. Pollution Control Boards should be vigilant that the recommended airborne exposure limit of manganese dust should not be exceeded at any time. The Maternity Benefit (Mines) Rules, 1963, should be facilitated to all pregnant women workers; otherwise, newborns will be seriously affected by neurological problems. Mine authorities should be careful that the respective prescribed sound level should not be exceeded at any time in core zones, residential areas, and sensitive areas.

As the manganese mining is confined to a tribal belt, a regular social impact analysis is required all through the life of the mine, particularly those who are displaced by mine management. Environmental action plan (EAP) such as domestic effluent treatment plant, workshop effluent treatment plant, and mine water discharge sedimentation plant should be set up and run. Moreover, other EAP, viz., dust suppression majors, tree plantation, overburden dump reclamation, topsoil storage and spreading for bioreclamation, and environmental monitoring should be undertaken. Similarly, rehabilitation action plan should be carried out successfully by the mining authorities. Hence, the government, mining authority, and public should take remedial measures for reduction of the adverse impact of manganese mining on women workers, rehabilitation of mined land, reuse of mine wastes and mine water, etc. (Aswathanarayana, 2003). Hence, a perceptive strategy should be implemented in mining industries of India to promote job-led economic growth through the adoption of employment generating, economically viable, and ecofriendly technologies.

REFERENCES

Agemian, H., Chau, A.S.Y., 1975. An atomic absorption method for determination of 20 elements in the lake sediments after acid digestion. Anal. Chem. Acta 80, 61–66.

APHA, AWWA, WPCF, 2005. Standard Methods for the Examination of Water and Wastewater, 21st ed. American Public Health Association, Washington DC, USA.

Aswathanarayana, U., 2003. Natural resources and environment. Geol. Soc. India, 58–59 (Bangalore, India).

Bencko, V., Cikrt, M., 1984. Manganese: a review of occupational and environmental toxicology. J. Hyg. Epidemiol. Microbiol. Immunol. 28 (2), 139–148.

Bowler, R.M., Roels, H.A., Nakagawa, S., Drezgic, M., Diamond, E., Park, R., Koller, W., Bowler, R.P., Mergler, D., Bouchard, M., Smith, D., Gwiazda, R., Doty, R.L., 2007. Dose-effect relationships between manganese exposure and neurological, neuropsychological and pulmonary function in confined space bridge welders. Occup. Environ. Med. 64 (3), 167–177.

Das, M., 2001. Impact of Mining on Health of Women – A Case Study of Dubna Guruda Manganese Mines, dist., Keonjhar, unpublished project report submitted to NCW, New Delhi, pp. 1–31.

Das, M., Goswami, S., 2009. A profile of the industrial mineral resource potential of Odisha. In: Srivastava, K.L. (Ed.), Economic Mineralization. Scientific Publishers, pp. 369–382.

Das, M., Goswami, S., Pradhan, A.A., 2012. Impact of mining on environment: a case study of manganese mining in Keonjhar district, Odisha. In: Proc. Minetech Vol. 12: Golden Jubilee Workshop on Mining Technology, pp. 11–20.

Dobson, A.W., Erikson, K.M., Aschner, M., 2004. Manganese neurotoxicity. Ann. N.Y. Acad. Sci. 1012, 115–128.

Donaldson, J., 2001. Manganese madness clues to the etiology of human brain disease emerges from a geological anomaly. Med. Geol. Newsl. 4, 8–10.

Fitsanakis, V.A., Zhang, N., Avison, M.J., Gore, J.C., Aschner, J.L., Aschner, M., 2006. The use of magnetic resonance imaging (MRI) in the study of manganese neurotoxicity. Neurotoxicology 27 (5), 798–806.

Goswami, S., Das, M., Guru, B.C., 2008. Environmental impact of siljora opencast manganese mining, keonjhar: an overview. Vistas Geol. Res. 7, 121–131.

Goswami, S., Mishra, J.S., Das, M., 2009. Environmental impact of Manganese mining: a case study of Dubna opencast mine, Keonjhar district, Odisha, India. J. Ecophysiol. Occup. Hlth. 9, 189–197.

Gray, N.F., 2006. Water Technology: An Introduction for Environmental Scientists and Engineers. Elsevier, Amsterdam, Netherlands. pp. 1–645.

Gupta, P.K., 2004. Methods in Environmental Analysis: Water, Soil and Air. Agrobios, Jodhpur, India. pp. 1–408.

Iashchenko, A.B., 1998. The diagnosis of the initial subclinical stage of manganese toxicosis. Lik Sprava 5, 65–68.

IIED, 2002. Breaking New Ground – The Report of the Mining, Minerals and Sustainable Development Project, 1st ed. Earthscan Publication Ltd., London, U.K., Sterling VA, USA.

Inoue, N., 2007. Occupational neurotoxicology due to heavy metals - especially manganese poisoning. Brain Nerve 59 (6), 581–589 (in Japanese).

Jaiswal, P.C., 2004. Soil, Plant and Water Analysis. Kalyani Publishers, Ludhiana, India. pp. 1–201.

Jiménez-Soto, M.F., Flegal, A.R., 2011. Childhood poisoning from a smelter in Torreon, Mexico. Environ. Res. 111, 590–596.

Koudogbo, B., Mavoungou, D., Khouilla, J.D., ObameNguema, P., ObiangOssoubita, B., 1991. Manganese impregnation in miners of the Ogooue mining company at Moanda (Gabon). J. Toxicol. Clin. Exp. 11 (3–4), 175–181.

Levy, B.S., Nassetta, W.J., 2003. Neurologic effects of manganese in humans: a review. Int. J. Occup. Environ. Health 9 (2), 153–163.

Metz, E.A., 2002. Occupational Ergonomics: Ergonomics Issues in the Mining Industry. SME http://www.smenet.org/gpac.

Mines, Minerals and People, 2003. Background Paper by Mines, Minerals and People (MMP) for the Indian Women and Mining Seminar, Delhi April 2003. Published by MAC on 2003-0-15.

Mohapatra, B.K., Sarangi, S.K., Mohanty, B.K., 2006. Manganese ore. In: Mahalik, N.K., Sahoo, H.K., Hota, R.N., Mishra, B.P., Nanda, J.K., Panigrahi, A.B. (Eds.), Geology and Mineral Resources of Odisha. Society of Geoscientists and Allied Technologist, Bhubaneswar, India, pp. 334–358.

Myers, J.E., teWaterNaude, J., Fourie, M., Zogoe, H.B., Naik, I., Theodorou, P., Tassel, H., Daya, A., Thompson, M.L., 2003. Nervous system effects of occupational manganese exposure on South African manganese mineworkers. Neurotoxicology 24, 649–656.

Pirrone, N., et al., 2010. Global mercury emissions to the atmosphere from anthropogenic and natural sources. Atmos. Chem. Phys. 10, 5951–5964.

Prashant, S., Arora, H.L., Verma, A.K., 2000. Occupational noise exposure in iron ore mines. Indian Jr. Occ. Env. Med. 4 (2), 1–7.

Rodríguez-Agudelo, Y., Riojas-Rodríguez, H., Ríos, C., Rosas, I., Sabido Pedraza, E., Miranda, J., Siebe, C., Texcalac, J.L., Santos-Burgoa, C., 2006. Motor alterations associated with exposure to manganese in the environment in Mexico. Sci. Total Environ. 368 (2–3), 542–556.

Roth, J.A., 2006. Homeostatic and toxic mechanisms regulating manganese uptake, retention, and elimination. Biol. Res. 39 (1), 45–57.

Saarkoppel, L.M., Rushkevich, O.P., Kir'iakov, V.A., Sineva, E.L., Kazantsev, D.P., 2005. Occupational hazards in metal mining industry. Vestn. Ross. Akad. Med. Nauk 3, 39–42 (in Russian).

Schaumburg, H.H., Herskovitz, S., Cassano, V.A., 2006. Occupational manganese neurotoxicity provoked by hepatitis C. Neurology 67 (2), 322–323.

Shields, D.J., 1998. Nonrenewable resources in economic, social and environmental sustainability. Nonrenewable Resour. 7, 251–261. spl. issue: sustainability for non-renewable resources – Alternative perspective.

Siegel, Shefa, 2013. The missing ethics of Mining. J. Ethics Int. Aff. 27 (1), 3–17.

Telmer, K.H., Veiga, M.M., 2009. World emissions of mercury from artisanal and small scale gold mining. In: Pirrone, N., Manson, R. (Ed.), Mercury Fate and Transport in the Global Atmosphere. Emissions, Measurements and Models, NY:Springer, New York, pp. 131–172.

U.S. Geological Survey, January 2009. Mineral Commodity Summaries. http://www.minerals.usgs.gov/minerals/pubs/mcs.

U.S. Geological Survey, January 2012. Mineral Commodity Summaries. http://www.minerals.usgs.gov/minerals/pubs/mcs.

WHO, 2004. Guidelines for Drinking-water Quality. Vol. 1, Recommendation, third ed. Geneva: World Health Organization.

Chapter 33

Geoethics and Geohazards: A Perspective from Low-Income Countries, an Indian Experience

Shrikant Daji Limaye

Ground Water Institute (NGO), Pune, India

Chapter Outline

Abstract

Geoethics is relatively a new branch in geosciences while geohazards are being faced since the dawn of civilization. Due to advancements in the knowledge about Earth-related phenomena, it has now been possible to forecast some disasters with fair accuracy, while others still remain mostly unpredictable. Considering principles of ethics in mitigating activities is possible before and after the occurrence of a disaster. Considering principles of ethics during a disaster is especially important, as pointed out by Skandari in this book. The aim of mitigation is to minimize the loss of lives and damage to property. The long-term objectives are reduction, if possible, of future losses in similar events and education of the population concerning preparedness and environmental restoration. In low-income countries, the main hurdles toward achieving these objectives are financial constraints, lack of infrastructure and technical manpower, lack of coordination between agencies like government departments, NGOs, and social workers, and the overriding factor of corruption due to which the aid meant for the affected people is pocketed by the relief providers. This paper discusses the Indian experience, which is similar to that in other low-income countries.

Keywords: Geoethics; Geohazards; Low-income countries; Mitigation; Principles; Practice.

Geoethics. http://dx.doi.org/10.1016/B978-0-12-799935-7.00033-2

INTRODUCTION

Geoethics may be defined as application of ethics in our approach to and our actions for (1) protecting the Earth-environment; (2) conserving the Earth-resources and geoheritages for future generations; and (3) forecasting and mitigating geohazards. While dealing with geohazards, the principles of geoethics provide a framework for a meaningful cooperation between government (as the main caretaker), industry, and the affected population (including its sympathizers), the NGOs (as facilitators or liaisons in mitigation efforts) and the geoscientists (as advisors and technical experts).

The practice of geoethics means creating preparedness in the society for facing geohazards, issuing warnings of an impending calamity, and mitigating their after effects when they occur. While the principles of geoethics are based on considerations touching on geosciences, sociology, and philosophy, their application involves additional controlling parameters such as ecology, cultural practices, infrastructure, legal aspects, supervision, governance, accountability, transparency in operation and finally the technical feasibility, economic viability, and social acceptability of the application. These are the parameters due to which the practice of geoethics while dealing with geohazards in a low-income country like India differs from that in a developed country.

Ideally, if geoethical practices are observed while mitigating geohazards the relief providers and the affected population should be on the same side, which means that, while working together, they both win or lose the battle against the natural calamity. But in some low-income countries, due to corrupt practices the relief providers often stand to gain from a calamity by not providing proper help to the affected population, but siphoning off some of the funds intended for aid. On their part, the affected people may also give inflated statements of the losses incurred, knowing that full compensation would never be granted. The local political leaders take the calamity and its remediation as an opportunity to amass a good amount of relief funds and win more votes. Some amount from these funds may not reach the intended beneficiaries or the affected people.

Unfortunately, such opportunities come often in the form of various types of geohazards in India and other South and South-east Asian countries. In India, 59% of its territory is prone to earthquakes, 10% is prone to floods and river erosion, and 68% to droughts. Along the Indian coast, 76% of the coastline of 7516 km is vulnerable to cyclones and tsunamis (Issues of India – Climatic Vulnerability of India's Coastal Region, 2014).

HISTORICAL BACKGROUND OF GEOETHICS

Geohazards have been plaguing civilization since its early days. Initially, the responsibility of facing such hazards and then coming back to normal life, rested basically on the individual family. Later on, when the civilization and the structure of governance gradually developed, people started facing hazards as a group and a much larger community along with the government got involved in mitigation efforts. Observing ethics or geoethics while dealing with the three phases of geohazards: (1) prediction and creating awareness or preparedness; (2) facing the calamity when it arrives; and (3) recovering from a calamity was a much later thought in geosciences, which evolved following the examples in other sciences.

In sciences such as physics, biology, and chemistry, ethical principles attracted the attention of social leaders and philosophers long ago, while using laboratory research for field trials and applications of new inventions on a large scale. Using the research in nuclear physics for obtaining atomic energy and not for making atomic bombs and using chemical explosives for mining and not for making bombs was the priority of philosophers and social leaders. In Italy in late 1960s, Felice Ippolito, who was a geologist and engineer with philosophical inclination, began to reflect on the philosophical, sociological, and historical aspects of geosciences, considering the ethical value of the geological knowledge while using nature or the natural resources to meet the needs of the society (Ippolito, 1968). This could be interpreted as the beginning of geoethics in Europe. Ippolito developed philosophical and ethical topics related to geosciences such as the relationship between man and nature; between scientific community, nature, and society; and the position of geology between exact sciences like mathematics and natural sciences like biology. He had realized that, even if scientific research is not associated with ethical values, the application of such research must necessarily ensure ethical considerations about the effects on society. His reflections on the ethical and social implications of geological activities showed for the first time the importance of the role played by geoscientists in promoting and observing geoethical principles and values. It is therefore necessary to create awareness about geoethics in young geoscientists through relevant material in university curricula.

The International Geological Congress (IGC) is held every four years and attracts over 6000 geoscientists. The Association of Geoscientists for International Development (AGID) has been holding sessions on geoethics at IGCs for the past over 12 years. In the annual Mining Conference at Pribram (Czech Republic) sessions on geoethics are being held regularly. The European Geosciences Union's (EGU) General Assembly in Vienna also holds geoethics sessions each year. Recently, during the IGC at Brisbane in August 2012, two international associations have been established for geoethics. International Association for Promoting Geoethics (IAPG) is one of them and has its head office in Rome. In many countries, geoscientists who are working primarily in the mining industry promote geoethics to ensure sociofriendly and ecofriendly behavior on the part of mining companies (Limaye, 2012). Thus, although geoethics is a relatively new branch in geosciences, its principles are slowly gaining recognition. However, much remains to be done regarding the application of geoethics especially while dealing with geohazards in low-income countries.

GEOHAZARDS AND APPLICATION OF GEOETHICS

When applying geoethics in low-income countries like India, there is the need to understand the socioeconomic and sociopolitical environment in which the government authorities and the geoscientists working in government and NGOs have to work. The controlling parameters in this milieu have already been mentioned in the introduction. A few examples would illustrate this milieu. Although these examples are from India, similar conditions do occur in other low-income countries.

Creating Preparedness or Awareness in the Society

In Japan there is a saying that "a natural disaster strikes after the society has forgotten the previous one." It is difficult to explain to the population the need for disaster preparedness and for construction of disaster prevention infrastructure, because geoscientists do not know when a natural disaster will happen. The media, which have a significant influence on the social mind-set have therefore the responsibility of not allowing society to forget past events and of promoting a defense strategy. This could be done through popular science articles in newspapers. TV channels have a better access to people who cannot read newspapers or pamphlets issued by governments or NGOs. They are able to educate the viewers in cities and villages alike, especially when they use famous actors and actresses to convey the message. This is an effective method of spreading awareness messages and alerts on a mass scale. Satellite pictures are able to give advance notice of an approaching cyclone or a tornado so that people could be warned on TV regarding the precautions to be taken. The telephone directory of the Los Angeles area carries instructions on "what to do" and "what not to do" in the event of an earthquake. This could also be done in low-income countries by the telephone departments.

Events like an earthquake, tornado, volcanic eruption, or tsunami occur without or only short notice. The awareness has therefore to be continuous. Preparedness has to be like a round the clock vigilance.

The Indonesian tsunami of 2004 devastated coastal areas in South-eastern India and Eastern Sri Lanka. Afterward, there was a proposal to install tsunami warning units in the district and county government offices, as if the next Tsunami was going to strike on week days and during office-hours only. The warning system ideally needs to give an alert to the houses of the village chiefs, in every village in the coastal area. The lessons learned and decisions taken from an earlier geodisaster also form the basis of getting prepared for the next one provided there is a mechanism for follow-up on these decisions.

Ideally, a coastal village or a group of neighboring villages should have a concrete school building—a strong two-storied structure with ground floor housing a dispensary cum hospital and the first floor providing classrooms for the local school. In the event of an earthquake, flood, or tsunami, such a structure could provide temporary housing for the affected people.

The need to relocate fishermen's villages at least 500 m away from the coast was evident after the Tsunami hazard of 2004 in India, Bangladesh, Thailand, Indonesia, and Sri Lanka. The fishermen's villages in India consist mainly of huts constructed on coastal sand dunes with coconut leaves, bricks, and mud. But the fishermen did not agree to move away arguing that a similar calamity may not strike again in the next 20 years or so. They insist on staying within a 100 m or so from the high-tide line. They however, agreed to give a piece of land to construct a strong concrete building as a shelter 500 m away, as mentioned above but funds were not readily available for constructing such shelters in coastal villages in India, except in the states of Andhra Pradesh and Odisha, where shelters have been constructed at convenient locations, with the help of the Red Cross Society of India.

These cyclone shelters, are managed by community-based Cyclone Shelter Management and Maintenance Committees (CSMMC). Community members are trained in search and rescue and first-aid and various rescue equipment is supplied to these shelters. This equipment usually includes power saw, siren, kitchen utensils, flexi water tank, solar lights, stretchers, life buoys, life jackets, and generators.

Incidentally, till date, people in coastal villages are unwilling to vacate their houses, fearing theft or loss of their belongings. Usually able men stay behind, sending women and children to the shelters.

Due to such preparedness in the society, the destruction caused by cyclone Phailin in October 2013 was much less than the earlier cyclones. In Bangladesh, the Government and NGOs are active in building concrete shelters which often double as school or community center. There are about 4000 such shelters including government office buildings and 1500 more are planned. The role of coastal mangroves as energy dissipaters has been proved in many cyclones and mangrove plantations are being protected and promoted wherever possible so as to act as energy absorbers for any Tsunami attack in future.

In Maharashtra state in India, the basaltic strata of late Cretaceous age was considered to be stable for ages. In 1993, an early morning earthquake near Latur city strongly hit 52 villages and resulted in the death of 7928 persons and 15,824 cattle (Website, 2014). The village dwellings were mainly boulder and mud-type constructions. People sleeping by the walls had boulders falling on them. As a precaution against repetition of such a tragedy in future earthquakes, people were advised to use bricks and not boulders in constructing houses. During rehabilitation, some houses for poor people were constructed in small size stones, wire mesh and a mixture of cement, ash, and clay. Geodesic domes would have offered a strong option for housing, but could not get social acceptance, as people are used to rectangular rooms. After rehabilitation an extensive "training and awareness" program was initiated and a DMIS (Disaster Management Information System) was installed.

Coastal erosion and sea-level rise are long-term geohazards and provide so much time for precautionary measures that implementing them is often neglected due to other pressing needs, which require funding. Droughts give a fairly early warning of about a year to get prepared. In India, when monsoon rains (June–September) fail after a good start, the depleted water resources recover only partially during the summer season (February–May). The warning of water scarcity till the next monsoon season is thus explicit. Planning of water conservation in agricultural or domestic use could be started accordingly.

Desertification gives warning signals over a few years and mitigation could be planned locally by the concerned society through participation in government schemes. On the contrary, flash floods due to cloud bursts are sudden events and the devastation takes place in a few hours to a few days. Major rivers inundating their flood plains are responsible for loss of life and property almost every year, but the problem arises because people have encroached upon the flood plains due to population pressure. Preparation to face the floods means packing up and getting ready to move away temporarily, after receiving flood alerts.

Watershed development through forestation and soil and water conservation practices has been a well-known recipe for long-term protection from droughts. If these measures are implemented sincerely, a degraded watershed could be rejuvenated within about five years and the villages within the watershed could be made drought-proof, at least for getting adequate domestic water supply. There are a few examples of successful drought-proofing of villages even in semiarid, hard-rock regions, but these examples are not replicated as they should be. This is partly due to the lethargy of people who expect the government to do everything for them. The farmers are the owners of ground water, which is located under their farms and through an ever increasing number of dug-wells, bored wells, and tube wells, pumping of ground water is increasing every year and the water table is depleting. Recharge augmentation through watershed development and structures like stream bunds and percolation tanks has traditionally been a government activity.

Ideally, all activities related to watershed development, forestation and recharge augmentation should have an active participation of the local population. Although this leads to drought-proofing, only a few watersheds are developed in this way with involvement of farmers and NGOs. It appears that the government machinery is not sincere about drought-proofing. Due to frequent occurrence of droughts, the relief funds come-in handy for starting several programs of drought-relief. The transparency of operations and accountability for use of funds is far from being satisfactory. Providing drinking water by tankers to drought affected villages is also a relief measure controlled by people having political connections.

In the past 40 years, several thousand drinking water bore wells have been drilled in villages in drought-prone areas in the hard-rock terrain in India, but the number of tankers running to provide drinking water to these villages in a drought year has not decreased. This could be compared to the "arsenic hazard" in drinking water supply wells of villages in many sedimentary basins. Here, in spite of availability of funds, the problem is still alive, as if getting local and international funds to do research on the problem is the main goal rather than actually implementing the solutions at field level.

Landslides often give a long-term warning although their occurrence is sudden, especially during the monsoon rains when the pore-pressure of water builds up quickly. Some precautionary measures are taken when soil creep is observed, but people often refuse to relocate themselves to a safer site.

The case of volcanic activity is similar. Mount Merapi in Indonesia is famous for its activity. When there is a preactivity build-up in tremors or smoke emission, people are warned about an impending calamity, although the exact time cannot be predicted. Farmers in the villages at the foot of Mount Merapi treat the mountain as God who has given them fertile soil and adequate water. They refuse to vacate the houses in the vulnerable area. Even if a few casualties occur in an eruption due to hot ash falling on the houses, they treat it as an offering to the God to keep him pleased!

Mitigating Geohazards

Staying prepared and being aware does not help reducing the chances of facing a geohazard, but it may reduce the damage, which the geohazard might inflict. The first thing to be done after the calamity is to assess the damage. This needs to be done by going to the affected areas and meeting people, even if the infrastructure is damaged. In the 1993 earthquake near Latur, India, the amateur radio operators (HAMsters) from Mumbai rushed to the scene, stayed there for 10 days and gave accurate information on the help needed.

All precautionary measures and precalamity activities are aimed at creating a dialog or an alliance between government agencies, NGOs, and the vulnerable community so that everyone knows what to do, when to do, where to do, and how to do, during the calamity and the postcalamity period. In this activity, care has to be exercised not to offend local people's beliefs and customs (Skandari, 2015).

Within the affected society, rich people often lose only a part of their belongings, while the poor section loses almost everything. Recovery should therefore start by helping the poor section with shelter, water, and food. This ideal situation is, however, not attained, due to lack of sincere volunteers and due to external influence by politicians. Rich people are sometimes in the need of security forces to protect their belongings from looters, although such cases are rare in India. The relief work is often hampered by poor people coming from nearby unaffected areas hoping to get free meals and temporary shelter for a few weeks.

After the Tsunami from Sumatra (Indonesia) devastated coastal villages of eastern Sri Lanka and south-east India in 2004, the initial priority was to send food supplies. But the Tsunami wave had traveled inland and had filled drinking water wells and village water tanks with saline water. The population required drinking water first. Till it was supplied, "coconut water" from the coconuts collected from collapsed coconut trees provided some relief.

In case of droughts due to monsoon failure, the villagers have no work to do on farms and the standard remedial measure is to employ them in road construction, watershed development, forestation, and construction of percolation tanks. The payment is given partly in cash and partly in food grains. As mentioned above these works are not carried out effectively. If they were done so, then most of the semiarid region would have been out of the grip of frequent droughts due to remedial measures carried out in previous drought years. This indicates that these measures taken by the government and donor agencies were aimed at providing temporary relief through peripheral activities rather than dealing with the root cause of draughts (Sainath, 1996). This also suggests that due to unethical practices some funds meant for soil and water conservation programs in the watersheds may not have been honestly utilized. Such unethical practices can spread like cancer in all the activities, for example, by falsifying the records, which show the number of persons engaged in soil and water conservation works, or the number of trips made by water-tankers for supplying drinking water to villages, or the depths to which drinking water bore wells are drilled.

After the initial emergency period is over, the affected society should also take an active part in the remedial measures which are meant for long-term protection. The tendency to sit idle and just watch the works being done by the government departments and donor agencies does not lead to sustainable solutions. NGOs have an important role in motivating the society, acting as liaison between the concerned parties and keeping a watchful eye on the benefit: cost ratio of the operations.

DISCUSSION AND CONCLUSIONS

Many times in this milieu described above, the geoscientists do not have a controlling role to fight against corrupt practices. They have to be either a partner or just tolerate what others are doing. However, they do have a useful role especially in the precalamity period as outlined below:

1. Educating society about geohazards like floods, landslides, earthquakes, droughts, volcanic activity, and tsunami through popular science articles in newspapers, through guest-lectures in high-schools/universities and through audio-visual media;
2. predicting the disasters whenever possible, and establishing social contacts in vulnerable areas to see if the solutions suggested for minimizing the possible damage are socially acceptable; and
3. conducting scientific research and giving expert opinion to the government.

Dissemination of results to society may be done, if possible, but without legal guarantee as to the accuracy. This precaution has become necessary after the result of the court case in Italy regarding communications by government agencies and lack of information by experts regarding the L'Aquila earthquake of 2009. This earthquake caused serious damage to several medieval hill towns in the Aquila region, killing over 300 residents, injuring over 1500 and leaving tens of thousands of people homeless. The court case in Italy has been an eye opener for the global fraternity of geoscientists about the legal and personal responsibility of geoscientists when asked to render advice.

In this court case, the lower court held senior scientists *legally and personally responsible* for creating a false sense of security in the population and convicted them on the count of manslaughter (Mucciarelli, 2015). No one expects a geoscientist to accurately predict a geohazard like an earthquake. A geoscientist's role is just to analyze available data and communicate his considered opinion to the government and if possible to the population. But really, the communication of any warning to the population about a geohazard should be the duty of government department and not of any technical expert or a group of experts (Dolce and Di Bucci, 2015). In any scientific study in natural sciences, the predictions may go wrong. Scientists have the duty to inform about this uncertainty of the results of their research, so that the population is able to understand the limitations of scientific observations and the conclusions based upon them.

In conclusion, the application of geoethics while dealing with geohazards should result in educating the society about geohazards through multiple approaches including audio-visual media; creating resilience in the society to face such calamities; building teams of volunteers for better preparedness and for communication of warnings; giving aid to the affected population with urgency and efficiency when the calamity strikes; and finally taking appropriate steps to ensure that any repetition of the calamity in future would result in much less toll on life and property. The exposure of frequent unethical practices mentioned in this paper, is for creating awareness in NGOs and donor agencies, so that they will insist on "transparency and accountability" of any remediation operations related to geohazards, sponsored by them.

REFERENCES

Dolce, M., Di Bucci, D., 2015. Risk management: roles and responsibilities in the decision-making process. In: Wyss, M., Peppoloni, S. (Eds.), Geoethics: Ethical Challenges and Case Studies in Earth Science. Elsevier, Waltham, Massachusetts, pp. 211–221.

Ippolito, F., 1968. La natura e la storia. Vanni Scheiwiller, Milano (in Italian).

Issues of India – Climatic Vulnerability of India's Coastal Region Website: http://socialissuesindia. wordpress.com/2014/01/06/climatic-vulnerable-of-indias-coastal-regions.

Limaye, S.D., 2012. Observing geoethics in mining and in ground-water development: an Indian experience. Ann. Geophys. 55 (3). http://dx.doi.org/10.4401/ag-5573.

Mucciarelli, M., 2015. Some comments on the first degree sentence of the "L'Aquila trial". In: Wyss, M., Peppoloni, S. (Eds.), Geoethics: Ethical Challenges and Case Studies in Earth Science. Elsevier, Waltham, Massachusetts, pp. 205–210.

Sainath, P., 1996. Everybody Loves a Good Drought: Stories from India's Poorest Districts. Penguin, New Delhi. ISBN: 0-1402-5984-8, 470 pages.

Skandari, M.A., 2015. What foreign rescue teams should do and must not do in Muslim countries. In: Wyss, M., Peppoloni, S. (Eds.), Geoethics: Ethical Challenges and Case Studies in Earth Science. Elsevier, Waltham, Massachusetts, pp. 193–201.

Website: http://www.latur.nic.in/html/earthquake.htm (Visited April 2014).

Index

Note: Page numbers followed by "b", "f" and "t" indicate boxes, figures and tables respectively.

Printed in the United States
By Bookmasters